# Napatree Point

## A DECADE OF STEWARDSHIP

# Napatree Point

## A DECADE OF STEWARDSHIP

COMPILED BY

PETER AUGUST, JANICE SASSI, AND DANIEL T. COLE

BRG

SCIENTIFIC

*Napatree Point: A Decade of Stewardship*



Compiled by Peter August, Janice Sassi, and Daniel T. Cole



For rights and permissions, please contact: BRG.Scientific@gmail.com or

BRG Scientific LLC

3025 Ontario RD NW #407

Washington, District of Columbia 20009

www.BRGScientific.com

First edition

ISBN: 979-8-9890369-0-5 (Hard Cover)

ISBN: 979-8-9890369-1-2 (Paperback)

ISBN: 979-8-9890369-2-9 (eBook)

Library of Congress Control Number: 2023924247

Editor: Melissa Stevens, Purple Ninja Editorial

Cover Design and Interior Formatting: Becky's Graphic Design® LLC

This volume is dedicated to the memory of Chaplin B. Barnes, one of the founders and the first executive director of The Watch Hill Conservancy. The work reported here has its origins in Chap's vision for the protection and stewardship of the Napatree Point Conservation Area.

# Contents

## Section 5 | Plants



## Section 6 | News

# PREFACE | Science, Monitoring, and Stewardship on Napatree: A Short History

Peter August & Janice Sassi
The Watch Hill Conservancy

*Chap and Grant knew Napatree was special. The habitats and wildlife must be protected, and the public should be encouraged to visit and enjoy the pristine beauty of the site. But first, Napatree needed to be understood—you can't steward what you don't understand.*

In 2005, The Watch Hill Conservancy and the Watch Hill Fire District commissioned the Rhode Island Natural History Survey (RINHS) to conduct a comprehensive assessment of the stewardship priorities in the Napatree Point Conservation Area. In 2007, Julia Royster was hired as the Napatree manager and began preparation of a stewardship plan for Napatree. The RINHS repeated the assessment in 2010. These studies established an ecological baseline for Napatree. URI Professor Peter August and Napatree Manager Janice Sassi were on the 2010 RINHS study team. Following the assessment, Chaplin B. Barnes, The Watch Hill Conservancy executive director, and Grant G. Simmons III, Watch Hill Fire District park commissioner, charged Pete and Janice to implement one of the 2010 report recommendations: to convene a science advisory team who would be available to counsel the Conservancy on emerging stewardship needs and establish monitoring programs to continually assess the health of the Napatree Point ecosystem.

The science advisors and Napatree Naturalist staff convened their first annual meeting in October 2011 to review the RINHS report and establish conservation management priorities. They discussed areas of Napatree that should be targeted for deeper assessment. Many important collaborative monitoring programs were underway, and had been for many years. The data collected were shared with any interested organization; examples include water quality data, horseshoe crab surveys, annual nesting success of Piping Plovers, and shorebird monitoring reports. When visiting scientists came to Napatree to conduct studies, the data collected usually travelled with the investigator back to the home institution, making it impossible to examine the whole Napatree ecosystem and all of its parts together. It was clear that a single, comprehensive collection of data gathered, emerging trends the monitoring showed, and the stewardship implications of these data was needed. An annual *State of Napatree Report* was the solution.[1]

The report's format was designed to make it interesting to read as well as informative. Color photos and graphic illustrations were encouraged. In 2013, the first *State of Napatree Report* was produced and

1 A catalogue of reports from each year are available at https://thewatchhillconservancy.org/napatree/napatree-resources/.

contained ten chapters spanning sixty pages. The *2022 State of Napatree Report* contains twenty-eight different chapters spanning 323 pages written by forty-six different authors. Some chapters are updated annually; for example, surveys of Piping Plovers, horseshoe crabs, bats, mammals, dune elevation, and visitor use. Other studies have a long shelf life and do not require annual update, such as the tidal regime and bathymetry of the lagoon, plant inventories, etc. This volume, *Napatree Point: A Decade of Stewardship*, is based on the *2022 State of Napatree Report*, the tenth anniversary edition.

The annual compilation of monitoring and stewardship data for a nature reserve like Napatree is unique among conservation areas. Because of Napatree's value as a migratory stopover site, its importance as a horseshoe crab breeding area, and its significance as a nesting and foraging habitat for endangered wildlife (e.g., Piping Plovers, Least Terns, American Oystercatchers), reporting the ecological condition of the conservation area has resulted in a valuable overview of the natural history of a pristine barrier spit ecosystem. Furthermore, stewardship of Napatree has created a living laboratory for training undergraduate and graduate students. Two-thirds of all the chapters have one or more student authors. Six of the chapters are the result of graduate or undergraduate research projects. Student training begins early: the Napatree Investigators summer program introduces children between the ages of seven and fourteen to coastal science and the need to conserve natural areas.

Our stewardship of Napatree has been well-received by the Rhode Island conservation community:

- Dr. David Gregg, executive director of the RINHS, comments: "The Watch Hill Conservancy has done a magnificent job documenting the fauna, flora, and geological processes of, and developing education programs about, this important natural area. There are few, if any, conservation sites in the region that have such a rich and current inventory of patterns and trends in plants, animals, and visitors. The Natural History Survey is very proud to have played a role in the formation of the Science Advisors and the production of the annual *State of Napatree Report*."
- Kate Sayles, executive director of the Land Trust Council of Rhode Island, notes: "The Watch Hill Conservancy's stewardship of Napatree Point is an excellent example of conserving the fauna and flora of a fragile ecosystem while ensuring visitors have a safe and informative experience on this pristine barrier spit! It really showcases the need to balance educating our citizens and protecting nature."
- Judith Swift, director emeritus of the URI Coastal Institute, said it well: "This compilation of studies is an excellent example of scientific transparency and permanently recording the ecological condition of the site and changes that are occurring. Napatree is also a superb training venue for future land managers. Over 60 percent of the chapters in the *State of Napatree Report* include student authors. With the excellent youth education program—the Napatree Investigators—(see chapter 3) run by The Watch Hill Conservancy, and the practical, boots-on-the-ground training Napatree Naturalists and students receive, the future of conservation land management in Rhode Island is in good hands with a serious stake in the future. This volume also serves as a valuable model for stewardship of additional natural areas in Rhode Island and along the rapidly changing Atlantic coastline, an undertaking critical to a Goldilocks level of observation and intervention.

# Introduction and Acknowledgments

Deborah Lamm
Chairman
The Watch Hill Conservancy

Grant G. Simmons III
Vice President
The Watch Hill Conservancy

*Napatree Point: A Decade of Stewardship* is a summary of the environmental and educational programs at the Napatree Point Conservation Area (NTPCA) in Westerly, Rhode Island. Scientists and naturalists working on Napatree have contributed short descriptions of their projects, the results they obtained, and the implications of their studies. This volume is based on the tenth edition of the *State of Napatree Report* (all years are available as a free download on The Watch Hill Conservancy website) and is a permanent record of the valuable science, education, and stewardship underway at Napatree. We feel it is important to have a single compendium describing Napatree natural history and programs.

Compiling this book was a team effort. Janice Sassi, Daniel Cole, and Peter August coordinated writing, review, and final production of the volume. We are especially grateful to the authors of the chapters; their findings are important to our stewardship of this conservation area. The following people contributed significantly to the work reported here: Jocelyn Lahey, Laurie Andrews, Tom Pappadia, Barbara Gearhart, Judith Swift, Amber Neville, Jaclyn Witterschein, Elin Torell, Nate Vinhateiro, and the URI Watershed Watch program. We are grateful for the guidance from the current and past Napatree Point Conservation Area science advisors: Peter August, Keith Killingbeck, Peter Paton, Hope Leeson, Bryan Oakley, Alan Desbonnet, Janice Sassi, Howard Ginsberg, Nicole Rohr, and the late Jon Boothroyd. Most importantly, we appreciate the dedication and hard work of the front line of our stewardship programs—the Napatree Naturalists. Special thanks to Ted Grand, Becky Bayne, Kayla Snyder, Melissa Stevens, Kira Stillwell, David Gregg, and BRG Scientific for making the production of this book possible.

The work described here was made possible by the support we receive from The Watch Hill Conservancy and the Watch Hill Fire District, and grants from the Alfred M. Roberts, Jr. Charitable Foundation, University of Rhode Island (URI) Coastal Institute, the RI Coastal Resources Management Council, Sounds Conservancy, Mystic Garden Club, RI Sea Grant, Eastern Connecticut State University, URI Cooperative Extension, the Betram S. and Ruth M. Grand Charitable Foundation, and the URI Environmental Data Center.

## SPECIAL THANKS TO:

**The Alfred M. Roberts, Jr. Charitable Foundation**

*The Roberts Foundation supports projects that maintain the character and sense of place of Watch Hill. It has generously provided funding for the annual operations of the Napatree Point Conservation Area. The work described here would not be possible were it not for the continued generosity of the Roberts Foundation.*

**URI Coastal Institute**

*The URI Coastal Institute has designated the Napatree Point Conservation Area as one of three Climate Response Demonstration Sites in Rhode Island. The Demonstrations Sites (Barrington/ Warren Mixed Use, Urban Watersheds, and Napatree) were chosen to showcase best management practices to enhance climate change resilience. Through the Demonstration Site project the Coastal Institute has been a generous supporter of Napatree stewardship for many years.*

**Our Volunteers**

*Volunteers make a difference on Napatree. In 2022 alone, 1,936 hours of volunteer time were provided by our interns, visiting scientists, science advisors, and citizen volunteers. This is equivalent to a full-time additional employee working to steward Napatree!*

**Our Partners**

*The work that is done on Napatree is a powerful collaboration of many organizations. The logo cloud showing our partners is an impressive demonstration of the importance of Napatree and the commitment made by many to steward the conservation area and deliver effective education programs. To our many partners, we are grateful!*

## Our Partners

# Napatree Point Conservation Area Mission and Vision

## DESCRIPTION

The Napatree Point Conservation Area (NTPCA) is collaboratively managed by the Watch Hill Fire District and The Watch Hill Conservancy. The NTPCA stewards the Napatree Point ecosystem, monitors its environmental condition, and delivers environmental education programs to children, students, and adult learners.

## MISSION STATEMENT

We protect and enhance the ecological condition and ecosystem resilience of Napatree Point to make it a safe, enjoyable, and informative destination for all visitors.

## OUR VISION

The NTPCA is recognized as a national model for natural area stewardship and is regarded as a premier destination for visitors to enjoy its dramatic natural beauty and spectacular wildlife.

## OUR CORE VALUES

- We base management and stewardship decisions of the NTPCA on the best available science and data.
- We strive to enhance the resilience to human and natural disturbances of the Napatree Point ecosystem.
- We monitor the condition of the fauna and flora and the ecosystem and geological processes of Napatree Point.
- We share data and information with the local community, students, scientists, and decision-makers and make lessons learned and best management practices available to other natural areas across the country.
- We engage and educate school-aged children in the ecology of Napatree Point and instill in them a curiosity and respect for the natural world.
- Napatree Point is a community resource and is open to all visitors.

## HOW WE ACHIEVE OUR MISSION

- Through our Investigators program, we engage and educate school-aged children of the ecology of Napatree Point, instill in them a curiosity for nature, and establish a respect for the environment.
- Through our Naturalist staff and programs, we educate visitors to Napatree Point of the natural history of the site and encourage respectful behaviors that do not disturb wildlife or damage the ecosystem.

- Through our scientific monitoring of the ecological and geological condition of Napatree, we know the driving processes and degrees of variation in the system and can assess the impact of future natural and human-caused disturbances.
- Through our restoration programs, we enhance the biological diversity and resilience of the Napatree Point ecosystem.
- Through *The State of Napatree Report*, lectures, hosted visits, and multimedia outreach materials, we share the results of our monitoring and education programs with the public, scientists, and decision-makers.

## Napatree Point Conservation Area Staff and Scientists: 2013–2023

**Leadership**

Chaplin B. Barnes
Grant G. Simmons III
Deborah Lamm

**Senior Staff**

Grant G. Simmons III
Stephen Brown
Tom Pappadia
Julia Royster
Hugh Markey
Janice Sassi
Daniel Cole

**Science Advisors**

Dr. Peter August, Landscape Ecology, GIS
Dr. Jon Boothroyd, Geology
Alan Desbonnet, Marine Science
Dr. Howard Ginsberg, Entomology
Dr. Keith Killingbeck, Plant Ecology
Hope Leeson, Plant Ecology
Dr. Bryan Oakley, Geology
Dr. Peter Paton, Avian Ecology
Dr. Nicole Rohr, Marine Science
Janice Sassi, Policy

**Napatree Naturalists**

Seaver Anderson
Noah Barreto
Juliana Berry
Joshua Beuth
Emily Bodell
Katherine Bodell
Marco Greco-Byrne
Jackie Claver
Lucianna Farone Coccia
Melissa Cote
Jennifer Croteau
Luiz Cruz
Morgan Devlin
Alex Duryea
Braden Fleming
Christian Fox
Azure Giroux
Jessica Cressman Greene
Finn Harty
Sam Holton
Alex Jakubczak
Connor Jones
John Killian
Ryan Kleinert
Grace Lentini
Paige Letzelter
James Mistrik
Christina Montello
Cy Murphy
Alyssa Peterson
Scott Rasmussen
Kevin Rogers
Laura Craver Rogers
Scott Ruhren
Sarah Schechter
Maddie Smith
Emily Watling
Max Weinstein
Jake Wilson
Matt Zucconi

*Photo Credit: Philo Willetts and Bryan Oakley, The Watch Hill Conservancy*

SECTION ONE

# OVERVIEW

*Kevin Rogers and Rey Larsen conducting a winter bird survey.*
*Photo Credit: The Watch Hill Conservancy*

CHAPTER 1

# Monitoring and Stewardship in the Napatree Point Conservation Area: Where Do the Data Go?*

*Peter August,[1] Grant G. Simmons III[2] & Janice Sassi[2]*
[1] Department of Natural Resources Science, University of Rhode Island
[2] Napatree Point Conservation Area, The Watch Hill Conservancy

## INTRODUCTION

As evidenced by the chapters in the *State of Napatree Report*, large volumes of many kinds of data are collected every year in the Napatree Point Conservation Area (NTPCA). In this chapter, we describe the principles we follow when collecting and managing data, our database management system, and the consumers of the data collected in the NTPCA. When we use the term *data*, we include measurements of things (e.g., ocean temperature, number of nests); written descriptions of events or conditions; geospatial data, such as aerial photos, GIS data, GPS coordinates; and photographs and videos. The following principles and best practices guide our data collection and management activities:

- We strive to use the most current scientific procedures and protocols to collect data.
- Data are stored in a managed DropBox cloud database, which is backed up on a regular basis. The current size of the Napatree data volume is 267 GB (gigabytes).
- We share data as much as possible, except when they must be embargoed for student theses, publications, or to ensure the safety of species and habitats on Napatree.
- We communicate the results of data collection and interpret these results in the annual *State of Napatree Report*.
- Data provide an objective, accurate, numeric foundation to support management and stewardship of the NTPCA.

What motivates Napatree stewardship and monitoring? There are many answers to this question. First

* Originally published in 2018

and foremost, The Watch Hill Conservancy owns a conservation easement over Napatree, and the easement clearly stipulates that the ecological values of the property must be protected and that the Conservancy may use scientific monitoring to support this goal. From the easement:

> WHEREAS, the Grantor (Watch Hill Fire District) and the Grantee (The Watch Hill Conservancy) have the common purpose of conserving the above-described conservation values of the Protected Property (Napatree Point Conservation Area) in perpetuity . . .
>
> It is the purpose of this Conservation Easement to assure that the Protected Property will be retained forever in its current natural and scenic condition; to protect any rare plants, animals, or plant communities on the Protected Property; and to prevent any use of the Protected Property that will significantly impair or interfere with the conservation values of the Protected Property.
>
> The right, but not the obligation, to monitor the condition of the rare plant and animal populations, plant communities, and natural habitats on the Protected Property, to manage them, if necessary, to ensure their continued presence and viability on the Protected Property and to cooperate and share management with other conservation entities such as the U.S. Fish & Wildlife Service, the Rhode Island Department of Environmental Management, the Rhode Island Coastal Resources Management Council, the Audubon Society of Rhode Island or The Nature Conservancy. Such activities shall be in accordance with management practices of Grantee.

The NTPCA is unique in its biodiversity and pristine condition. It should come as no surprise that scientists and the scientific community find Napatree to be an interesting site to understand how barrier beach ecosystems work. Recognition of the conservation importance of Napatree and the stewardship activities underway come from many sources. The NTPCA has been designated a Globally Important Bird Area by the National Audubon Society. The United States Fish and Wildlife Service has officially designated Napatree to be part of the national John H. Chafee Coastal Barrier Resources System (CBRS). The Watch Hill Conservancy and the Watch Hill Fire District's management of Napatree has been designated by the National Land Trust Alliance as one of a small number of national case studies of effective land stewardship. The University of Rhode Island Coastal Institute has designated the NTPCA as a Climate Response Demonstration Site. The Institute uses the stewardship underway at Napatree to educate land managers on how to manage conservation lands in a manner that enhances resilience to climate change impacts. There are only three such sites in Rhode Island.

## DATA TO SUPPORT STEWARDSHIP

We collaborate with many partners in our stewardship and monitoring programs. In Table 1, we review some of the ongoing data collection projects and how the information is used.

| Data Category | Purpose | Partners and Disposition |
|---|---|---|
| Water Quality | Monitor water quality in Little Narragansett Bay and the oceanside of Napatree. It is important to know water quality trends for wildlife condition and safety of swimmers. | Data are collected for the URI Watershed Watch program. They are part of a statewide water quality database and are used to assess Rhode Island's fresh and coastal water quality. The Wood-Pawcatuck Watershed Association distributes our water quality data on their web mapping application. |

| | | |
|---|---|---|
| Horseshoe Crab Surveys | Monitor the population size of horseshoe crabs. Eggs produced by spawning horseshoe crabs are a critical food resource for migratory shorebirds. | Horseshoe crab surveys follow detailed protocols established by Project *Limulus* at Sacred Heart University (SHU). SHU maintains a regional database of horseshoe crab monitoring data. Our survey data are also provided to the Rhode Island Department of Environmental Management (RI DEM). |
| Piping Plover Surveys | Monitor the breeding success of Piping Plovers, a federally listed threatened species. | The US Fish and Wildlife Service (USFWS) is the lead on monitoring Piping Plovers and tracking breeding success throughout the region. Staff from NTPCA assist USFWS in their monitoring activities. |
| Bird Surveys | Monitor seasonal and annual patterns of birds on Napatree. | Reynold Larsen has been doing systematic surveys of birds on Napatree since 1963. He conducts his surveys one to two times a month and reports his results on eBird, an online repository of bird survey data managed by the Cornell Lab of Ornithology. These data are used by scientists for many applications. |
| Nanotag Tower | Monitor the flight paths of sea birds and other wildlife species that have been fitted with small radio transmitters. These data are used to plan the placement of offshore wind turbines so that they do not interfere with the flight paths of migratory birds. | The USFWS oversees this project. The Napatree antenna is part of the MOTUS network (https://motus.org/), which consists of similar antenna towers in Canada, the United States, the Caribbean, Europe, and Africa. The Bureau of Offshore Energy Management use these data for planning sites for offshore wind turbine projects. |
| Bat Monitoring | Record seasonal and annual changes in bat activity on Napatree. | Results are provided to RI DEM for inclusion into their statewide bat monitoring network. |
| Shoreline and Dune Monitoring | Measure dune height and last high tide swash line. | Napatree is a storm-driven ecosystem. The quarterly monitoring of dune heights and high tide lines will provide us a baseline to assess impacts of the next large storm and how long it will take Napatree to recover. These data are managed by Dr. Bryan Oakley at Eastern Connecticut State University. |
| Camera Trap Monitoring | Record changes in predator and prey abundance on Napatree. | There are several species of mammalian predators of ground nesting birds on Napatree. Our camera trap surveys allow us to determine if there are major changes in predator populations. These data are reported to USFWS. |
| Marine Life Monitoring in the Napatree Lagoon | Record the diversity and abundance of fish, crabs, and algae in the Napatree Lagoon. | The Napatree Lagoon is the most dynamic and ever-changing region on Napatree. Over the years, the entrance to the lagoon (where it connects to Little Narragansett Bay) has radically changed. As the lagoon entrance changes, so does the inflow of algae into the lagoon. As algae decomposes, dissolved oxygen decreases, and this impacts fish in the lagoon. The next large storm will likely change lagoon hydrodynamics, so we will be able to assess the impact of these changes with the baseline data we have collected. These studies are conducted by the URI Coastal Institute, URI Biology Department, and Eastern Connecticut State University. |
| Vegetation Monitoring | Monitor the status of invasive plants, restored vegetation, and shrub patches. | The invasive monitoring on Napatree allows us to control invading species before they become established. Over the past few years, well over 2,500 plants of twenty species that are high value to pollinators have been planted on old dune trails. Our monitoring of the survivorship of these plants helps us understand which plants thrive in this habitat and should be used in future restoration projects. These data are managed by the RI Natural History Survey, RhodyNative, and they are used by the CRMC. |
| Visitor Counts | Record the number of people on the Napatree beach and boats anchored on the bayside of Napatree. | Napatree Naturalists record visitor and boat numbers during their summer patrols. These data allow us to monitor changes in the use of Napatree and to predict when visitor activity will be high. They also allow us to monitor compliance in using approved paths on Napatree. |

*Table 1. Description of data collection activities along with associated partners and how the data are used to support stewardship of Napatree.*

## SHARING INFORMATION

Many organizations collaborate with us and use the data collected on Napatree (Table 1, Figure 1). The data support management and stewardship activities in the NTPCA, public education through the *State of Napatree Report*, scientific publications, public presentations, student theses and dissertations, statewide data syntheses (e.g., Watershed Watch), and outreach materials.

*Figure 1. Users of data collected on Napatree.*

## FUNDING STEWARDSHIP DATA COLLECTION

Scientific data collection is guided by the Napatree Science Advisors and coordinated by the Napatree manager. All the advisors have and continue to actively participate in data gathering to support management activities. They volunteer their time to The Watch Hill Conservancy for this work. In 2018, the total time provided by Napatree scientists and their students and assistants was 1,574 hours. Using established rates of the value of volunteer time for Rhode Island ($26.17, Independent Sector annual compilation), this equals $41,191 or 0.9 FTE (full-time equivalent) staff effort. In 2017, 1,106 volunteer hours were provided by scientists. In 2016, 739 hours were provided by volunteers.

The Napatree scientists have been very active in seeking extramural grant funding to support their stewardship activities. Since 2014, twenty-two grant proposals were submitted to funding agencies, nineteen were awarded to support Napatree stewardship for a total of $218,233.

Coordination of stewardship activities and the logistic support provided by The Watch Hill Conservancy and Napatree Naturalist staff are essential to the success of our management programs. Coordination ensures that teams of scientists do not interfere with each other's work. Almost all the projects contributing to Napatree stewardship require logistic support in the form of ATV transportation of equipment and personnel. Frequently, Napatree Naturalists oversee data collection (e.g., horseshoe crab surveys, visitor counts) or assist scientific teams (e.g., plant restoration).

## ACKNOWLEDGMENTS

Special thanks to The Watch Hill Conservancy and the Watch Hill Fire District for their ongoing commitment to the stewardship of the NTPCA. We are especially grateful to the Napatree Naturalist staff for their field assistance and the Napatree Science Advisors for their continual guidance on stewardship of Napatree. The following organizations have generously funded our work: the Alfred M. Roberts Foundation, the URI Coastal Institute, the Coastal Resources Management Council, the USDA RREA program, the Sounds Conservancy, the Rhode Island Foundation, the Lattner Family Foundation, and the Mystic Garden Club.

*Napatree from the air.*
*Photo Credit: Kevin Rogers*

CHAPTER 2

# The Geography of the Napatree Point Conservation Area (2020 Update)

*Peter August,[1] Janice Sassi,[2] Kevin Rogers[2] & Jessica Cressman Greene[1,2]*

[1] Department of Natural Resources Science, University of Rhode Island

[2] Napatree Point Conservation Area, The Watch Hill Conservancy

## INTRODUCTION

The purpose of this chapter is to describe the physical and ecological geography of the Napatree Point Conservation Area (NTPCA). We hope it will serve as the single go-to reference for Napatree facts and figures. Note, however, that Napatree is always changing. The statistics we present here are a snapshot of conditions on the date the source data were recorded. These results are a baseline from which future comparisons can be made.

Updates were made in 2020 to reflect changes in the ownership of parcels on Napatree. The bathymetric data have been updated to reflect the changes reported in Oakley *et al.* (2018). The remaining data reported here are the same as originally presented in August *et al.* (2016).

## METHODS

The maps and measurements reported here were made with ArcGIS (v 10.3.1) geographic information system (GIS) software (Environmental Systems Research Institute, Redlands, California). We used the best available data for each of the categories reported on. Some of the geospatial data used in our tabulations were obtained from the Rhode Island Geographic Information System (RIGIS) database (www.rigis.org). Other data were developed by Napatree staff or scientists. This compilation will be a dynamic document and will be updated in future years as new data become available or significant changes occur on Napatree.

For each theme, we describe the feature measured, where we obtained the source data, and how the feature was measured using the GIS. For each theme, we also present a map and a statistical summary. Measurements are given in both metric and English units.

# RESULTS

## Size of the Napatree Point Conservation Area

*Source Information*: Area and length measurements were taken from the 2016 shoreline of the NTPCA. The shoreline was heads-up digitized from the digital orthophotography obtained for the 2016 eelgrass mapping project. The imagery is true color with a 0.5 m (1.64 feet) pixel size and was obtained in June 2016. It is available from RIGIS. The last high tide swash line was considered the shoreline in the imagery. This is consistent with the shoreline definition used by Napatree geologist Dr. Bryan Oakley (chapter 11). A number of factors affect shoreline position: tidal state when the imagery was obtained, deposition or erosion of sand, and recent storms.

On the north shore of NTPCA, the bayside shoreline measurements extended from the entrance to the conservation area to the eastern border of the lagoon. The lagoon shoreline circumnavigated the lagoon. The cobble shoreline north and west of the lagoon extended from the western entrance to the lagoon to the western rocky shore of Napatree. The ocean beach extended from the western rocky shoreline to the jetty on the eastern end of the conservation area. The eastern border extended from the entrance of Napatree to the jetty on the oceanside. To calculate the area of the lagoon, a polygon was formed by closing off the lagoon at the current breach where it connects with Little Narragansett Bay.

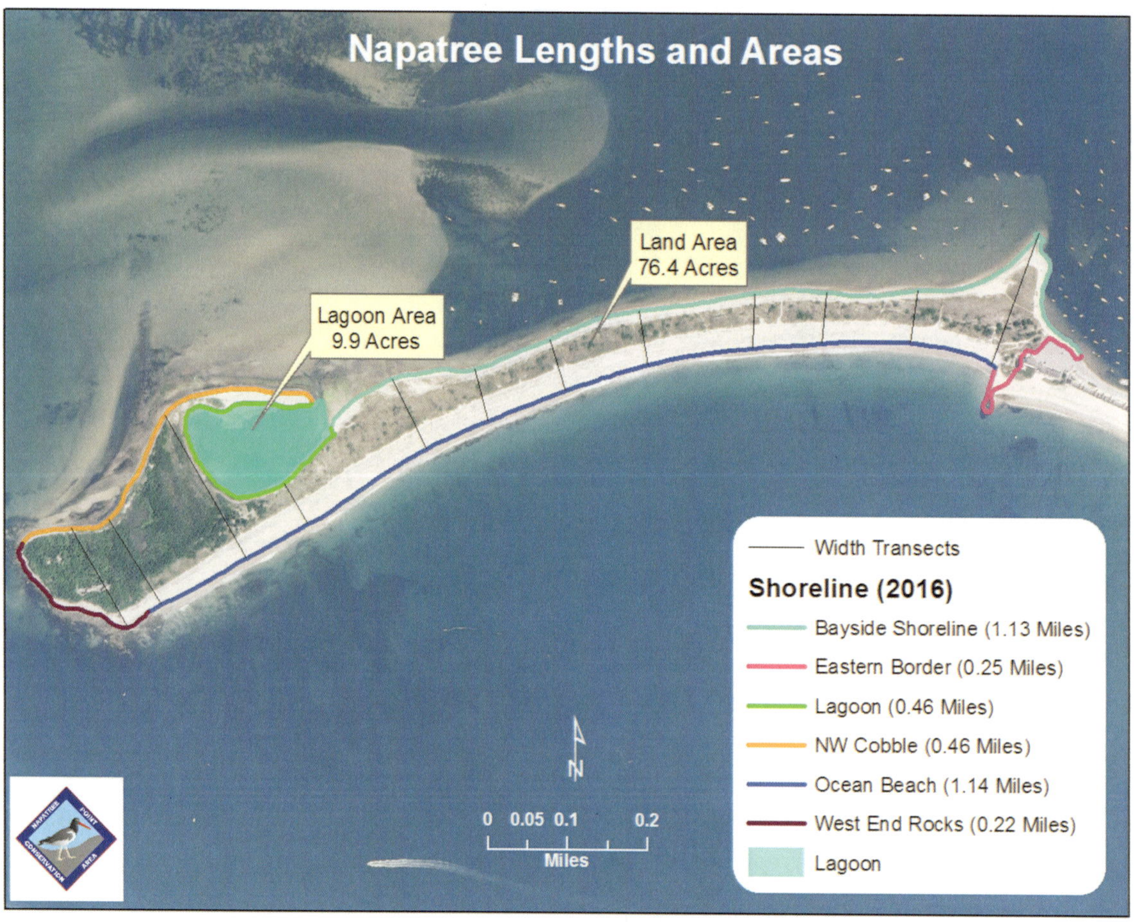


*Figure 1. Areas and distances in the Napatree Point Conservation Area.*

## Statistical Summary

| Land Area | 76.35 acres | 30.9 hectares |
|---|---|---|
| Lagoon Area | 9.90 acres | 4.02 hectares |
| Total Area (Land and Lagoon) | 86.25 acres | 34.92 hectares |
| | | |
| Bayside Shoreline | 1.13 miles | 1.82 kilometers |
| Lagoon Shoreline | 0.46 miles | 0.74 kilometers |
| NW Cobble Shoreline | 0.46 miles | 0.75 kilometers |
| Rocky Western End | 0.22 miles | 0.36 kilometers |
| Ocean Beach | 1.14 miles | 1.84 kilometers |
| Eastern Border | 0.25 miles | 0.39 kilometers |
| Total Perimeter of NTPCA | 3.67 miles | 5.90 kilometers |
| | | |
| Average Width of Barrier (12 Cross Sections along the Barrier) | 517.4 feet | 157.7 meters |
| Width at Fort Mansfield | 734.8 feet | 223.9 meters |
| Width in Salt Marsh West of Lagoon | 1,061.4 feet | 323.5 meters |
| Width From Bayside Jetty (East End) to Ocean | 892.9 feet | 272.2 meters |

## Elevation and Depth of Napatree Point Conservation Area

*Source Information*: Land elevations were obtained from the 2014 NOAA Post-Sandy Topobathymetric Lidar digital elevation model obtained from RIGIS. Lagoon depths were taken from the bathymetry dataset reported by Oakley *et al.* in 2018. The bathymetry of the Napatree Lagoon has changed dramatically since the original 2014 mapping survey. This is likely a result of circulation changes in the lagoon resulting in the shift in the location of the opening with Little Narragansett Bay.

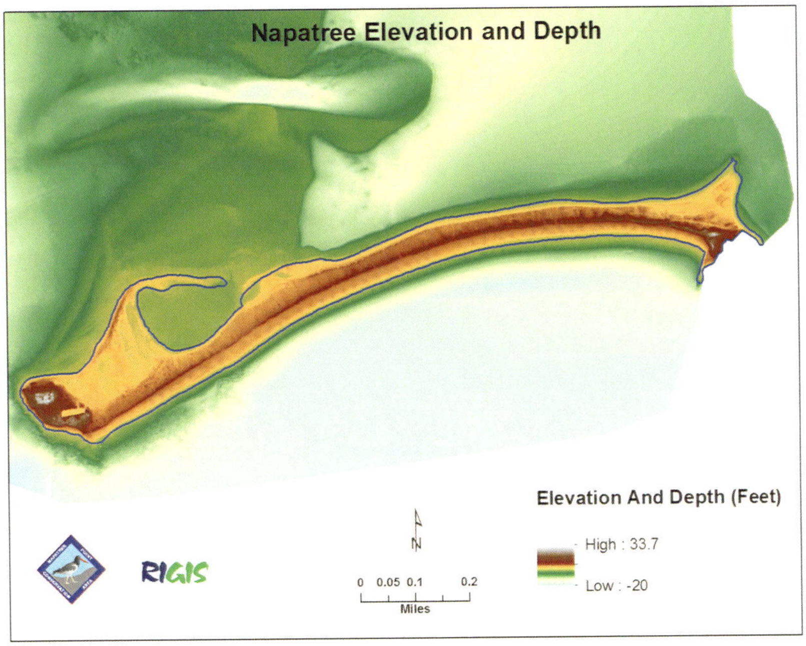


*Figure 2. Elevations and depths in the Napatree Point Conservation Area.*

## Statistical Summary

| | | |
|---|---|---|
| Maximum Land Elevation | 33.7 feet | 10.27 meters |
| Average Land Elevation | 6.21 feet | 1.89 meters |
| Elevation at Top of Dune on Eastern-most Dune Crossing Path | 28.8 feet | 8.78 meters |
| Dune Crest along Paths | 10 to 13 feet | 3 to 3.9 meters |
| Average Depth of Lagoon (MLLW) | -1 to -2 feet | -0.3 to -0.6 meters |
| Maximum Depth of Lagoon (MLLW) | -3.4 feet | -1.0 meters |

## Habitats

*Source of Information:* Eelgrass (*Zostera marina*) data were taken from the 2017 Long Island Sound eelgrass survey available from USFWS and the URI Environmental Data Center. All eelgrass patches in Rhode Island waters between Napatree Point and Sandy Point are included in the summary. Mussel beds (*Mytilus edulis*) were delineated in the field (2013) using a handheld GPS receiver (Garmin, WAAS-enabled) at low tide by Kevin Rogers. The minimum mapping unit was 0.46 sq m (5 square feet). Patches of sparse mussel density were not included in the mapping. Natural community data were photo-interpreted from the 2011 RIGIS orthophoto imagery (available on RIGIS) by Kevin Rogers using the RI Ecological Communities Classification System (Enser *et al.*, 2011, available at http://rinhs.org/wp-content/uploads/2020/04/RI-Ecological-Community-Classif_Enser-etal-2011_FINAL.pdf). The minimum mapping unit was approximately 0.10 ha (0.25 acres).

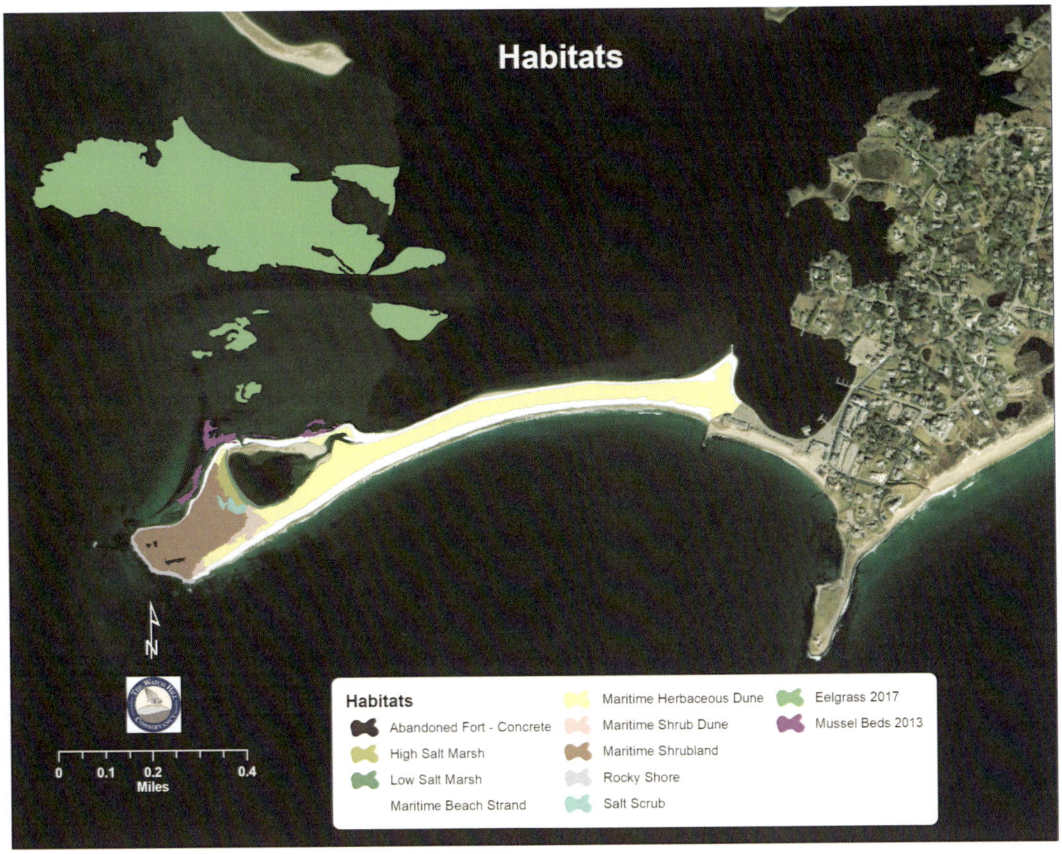


*Figure 3. Habitats in the Napatree Point Conservation Area.*

## Statistical Summary

| Terrestrial and Shoreline Habitats | | |
|---|---|---|
| Concrete (Abandoned Fort) | 0.30 acres | 0.12 hectares |
| High Salt Marsh | 1.34 acres | 0.54 hectares |
| Low Salt Marsh | 0.77 acres | 0.31 hectares |
| Maritime Beach Strand | 18.74 acres | 7.58 hectares |
| Maritime Herbaceous Dune | 28.99 acres | 11.73 hectares |
| Maritime Shrub Dune | 1.81 acres | 0.73 hectares |
| Maritime Shrubland | 14.27 acres | 5.77 hectares |
| Rocky Shore | 1.26 acres | 0.51 hectares |
| Salt Scrub | 0.7 acres | 0.28 hectares |
| **Submerged Habitats** | | |
| Mussel Beds | 2.24 acres | 0.91 hectares |
| Eelgrass Beds | 92.7 acres | 37.5 hectares |

## Property Ownership

*Source of Information:* Parcel boundaries and ownership data were obtained from the Town of Westerly parcel database via website download in September 2020. Area values were taken from the GIS calculation of the size of each parcel. The data were clipped to the NTPCA eastern boundary. The total land area of the parcels' dataset will slightly differ from the total area presented in the area/length data because of different shorelines used to delineate the outline of Napatree.

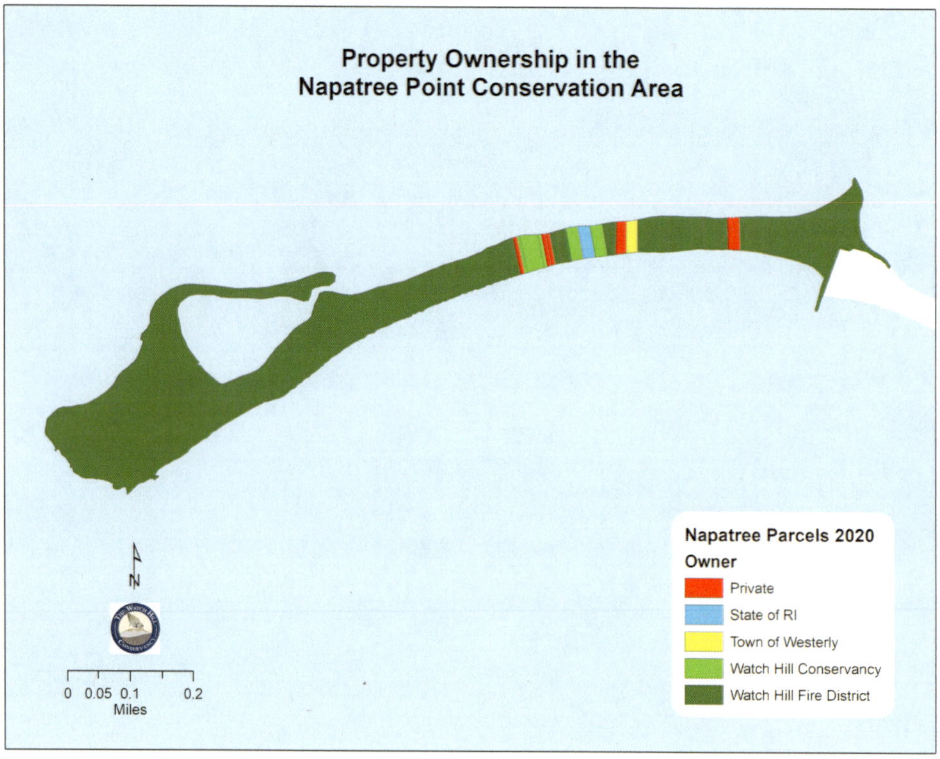


*Figure 4. Property ownership in the Napatree Point Conservation Area.*

## Statistical Summary

| | | |
|---|---|---|
| Total Privately Owned Parcels (5 parcels) | 1.76 acres | 0.71 hectares |
| Parcels Owned by Watch Hill Fire District (10 parcels) | 66.87 acres | 27.06 hectares |
| Parcels Owned by The Watch Hill Conservancy (5 parcels) | 2.55 acres | 1.03 hectares |
| Parcels Owned by State of RI (1 parcel) | 0.60 acres | 0.24 hectares |
| Parcels Owned by the Town of Westerly (1 parcel) | 0.54 acres | 0.22 hectares |
| Total Area of All Parcels (22 parcels) | 72.32 acres | 29.27 hectares |

## Trails (Paths)

*Source of Information:* Path data were collected in 2012 by Jessica Cressman Greene by walking all trails with a handheld GPS. Path pole locations were collected by Peter August in November 2016 using a handheld GPS receiver (Garmin eTrex, WAAS-enabled). The 2012 trails were categorized into two categories: cross dune trails, which extended from bay to ocean, or lateral trails, which ran east-west and parallel to the shoreline or extended a short distance into the dune to connect to a lateral trail. The 2016 trails were those approved trails as of November 2016.

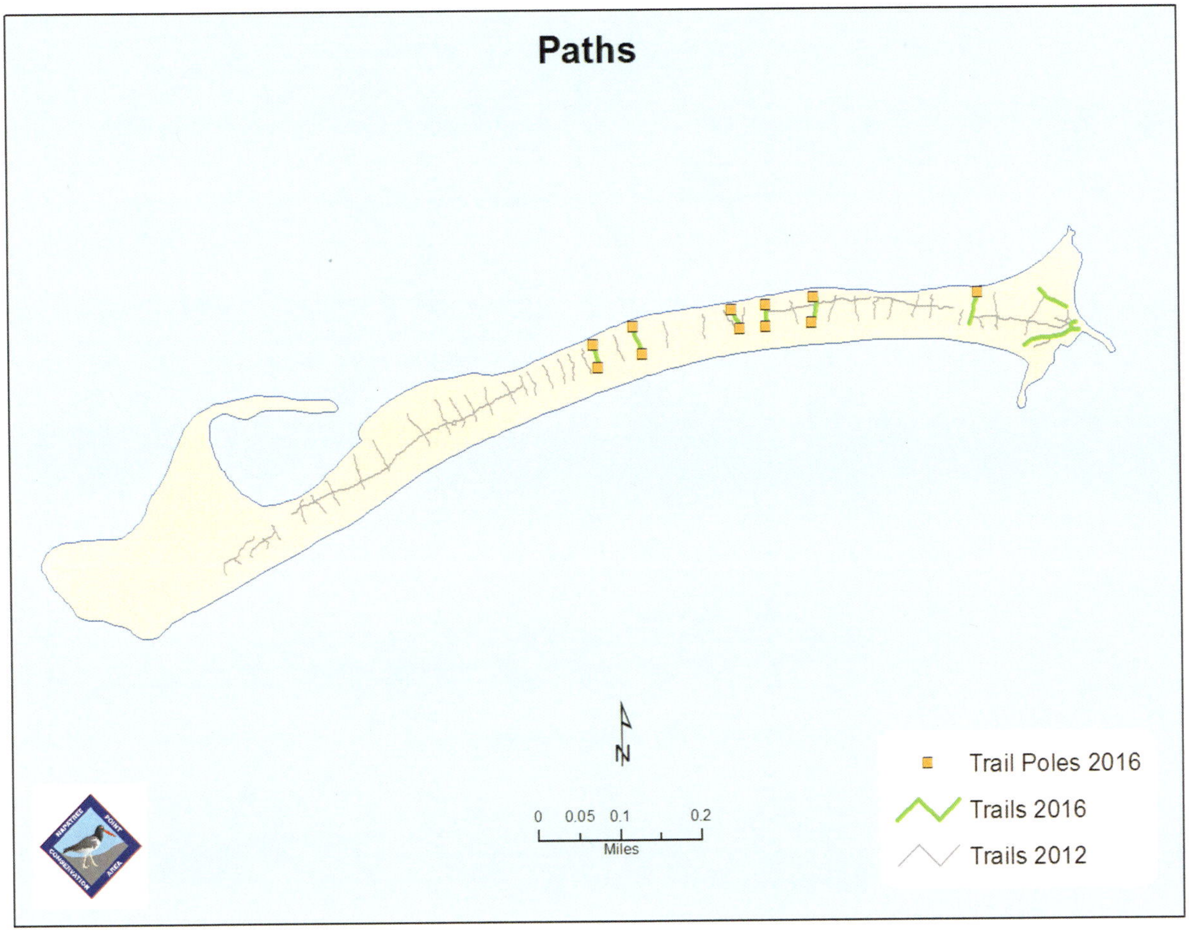


*Figure 5. Trails in the Napatree Point Conservation Area.*

## Statistical Summary

| | | |
|---|---|---|
| 2012 Cross Dune Trails (40) | 1.43 miles | 2.30 kilometers |
| 2012 Lateral Trails (24) | 0.99 miles | 1.59 kilometers |
| 2012 Total Trails (64) | 2.42 miles | 3.89 kilometers |
| | | |
| 2016 Cross Dune Trails (7) | 0.28 miles | 0.45 kilometers |
| 2016 Lateral Trails (1) | 0.04 miles | 0.06 kilometers |
| 2016 Total Trails (8) | 0.32 miles | 0.51 kilometers |

# CONCLUSIONS

As new data are collected on Napatree, or existing data are revised, we will update this descriptive geography. Until then, we recommend using these data as standard facts and figures describing the NTPCA.

## ACKNOWLEDGMENTS

This project was supported by the University of Rhode Island Coastal Institute, the USDA Renewable Resources Extension Act, and the URI College of the Environment and Life Science Extension Program in Natural Resource Conservation and Management, The Watch Hill Conservancy, and the Watch Hill Fire District.

*Photo Credit: Janice Sassi, The Watch Hill Conservancy*

SECTION TWO

# THE HUMAN DIMENSION

*Investigators using a seine net to discover marine life.*
*Photo Credit: The Watch Hill Conservancy*

CHAPTER 3

# Napatree Point Children's Education Program: Investigators 2022

*Stephen Brown, Hugh Markey, Emily Bodell, Maddie Smith & Emily Watling*

Napatree Point Conservation Area, The Watch Hill Conservancy

## INTRODUCTION

The Investigator program is a free, hands-on learning experience for children ages seven to fourteen. The children spend two hours on the beach observing natural biotic and abiotic elements, as well as analyzing or researching their finds using age-appropriate picture pamphlets. The program goals are to foster a curiosity for and appreciation of the fauna and flora of the various habitats and to learn facts about the ecosystems of the Napatree Point Conservation Area (NTPCA). While guiding the Investigators, Napatree Naturalists model proper beach conservation, such as picking up trash, staying on trails, respecting all forms of life, and avoiding climbing dunes. At the same time, Naturalists discuss how these behaviors foster the preservation of Napatree, encouraging the children to become lifelong protectors of beach habitats.

## OUTREACH

We advertise the program in various ways. We email all families who have previously attended the program, provide a sign-up sheet in the children's area of the Westerly Library, send digital flyers to the Stonington School District, and post on the Westerly Schools Backpack website. Registration is done online via a Google form, which is filled out by parents or caregivers of the children. These data are compiled on a Google spreadsheet (see Results). Families must sign a consent form that includes a photo and video release, an acknowledgment of the child's physical condition, and a liability release for The Watch Hill Conservancy. Parents may sign up for one of three sessions offered each week.

## PROGRAM FORMAT

Each session begins at the gazebo in Watch Hill with Naturalists taking attendance, answering parent questions, conducting introductions, and exploring the theme of the day's program. The latter part may involve the Naturalists accessing prior knowledge through questioning and/or visual aids.

Our target enrollment is twenty students per session. The program is two hours long and focuses on a specific theme each week, but students are encouraged to share any findings made on the beach. The themes include wrack line, horseshoe crabs, fish, crabs, mollusks and shells, rocks, and seaweeds. The summer's program concludes with the opportunity for students to collect small abiotic items in a specimen box that they can keep for future reference. Other topics covered informally include jellyfish, birds, tides, conservation, food chains, predators and prey, camouflage, minerals, temperature impacts, adaptations, and wildlife tracking. Each program ends with the group sharing their findings and what they have learned during the session. Families pick up their children at the gazebo.

## RESULTS

We educated sixty-three different children over seven weeks in the summer of 2022, which was ten more children than in 2021 (Figure 1). There was a total of 396 *registered* child hours on Napatree Point. Because of two cancelled classes due to inclement weather as well as children being absent, we had 274 *actual* child hours (registered students not included in cancellations). The percentage of students who actually attended the program after registering was 70 percent. This was a decrease of 8 percentage points from 2021.

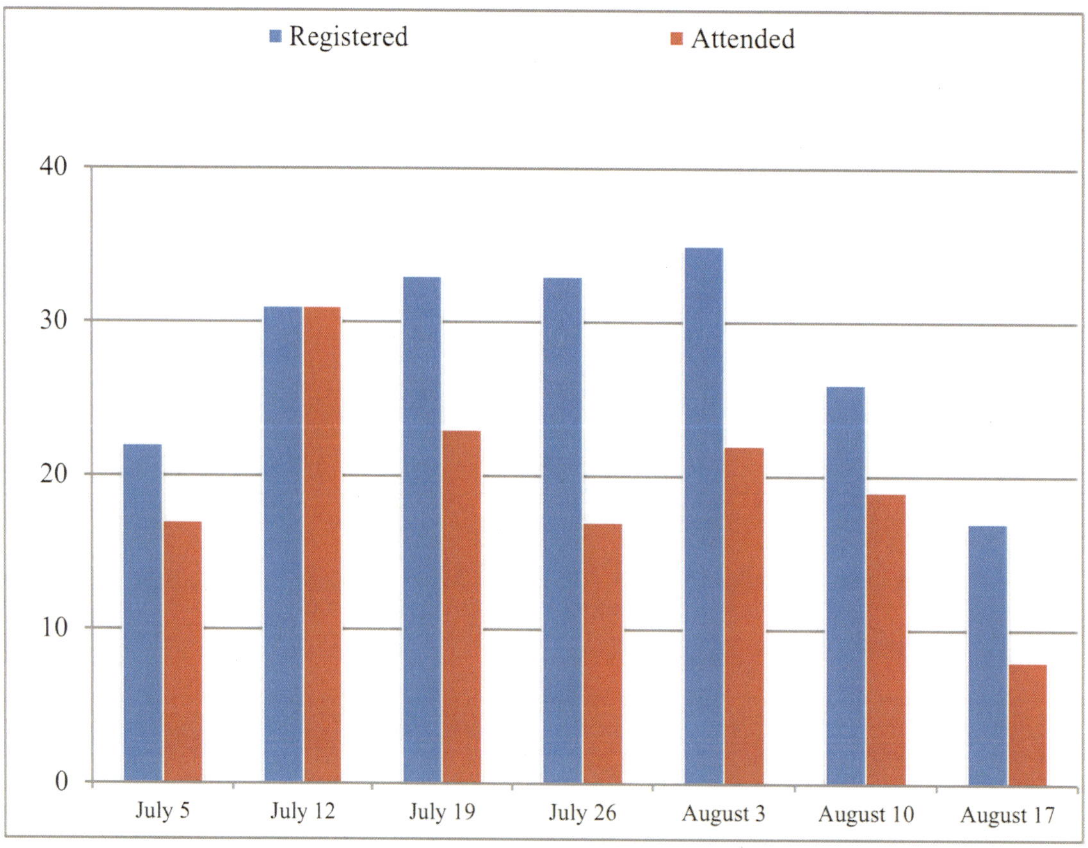


*Figure 1. Weekly totals of the number of registered students and students actually attending the Investigators program during the summer of 2022. Two class days were cancelled due to rain.*

## CONCLUSIONS

Due to the high registration numbers, we suggest that a team of three Naturalists be employed each year, with both genders represented. In addition, we had three high school students help out. Having them was invaluable, especially when the actual numbers were high and/or when we split the groups up for lessons. If we want to increase the number of students who attend the program, we may want to reach out through social media platforms. Most parents get their information via their phones.


## ACKNOWLEDGMENTS

We are grateful for the excellent help we received from 2022 Napatree Investigator Program Assistants Sofia Chronister, Wick Ackerman, Kate Ghaffari, and Penelope Fillipponi.

*Fourth of July weekend at Napatree.*
*Photo Credit: Emily Bodell*

CHAPTER 4

# Visitor Activity on Napatree: 2022

*Daniel Cole, Janice Sassi, Emily Bodell & Tom Pappadia*
Napatree Point Conservation Area, The Watch Hill Conservancy

## INTRODUCTION

Historically, visitor use of the Napatree Point Conservation Area (NTPCA) was inconsistently documented. This made measuring changes in patterns of visitor use difficult. Beginning in 2013, a standardized data sheet was used to record the number of visitors on NTPCA as well as boats anchored along the shore or pulled up onto the beach. Using a standardized data sheet allows for comparisons of visitor data from year to year, which is imperative for understanding trends over time. In 2022, the analog data sheet was converted to an online digital form, thus allowing Naturalists to submit count data and any additional notes from the beach during a shift. In addition, the greeter, Tom Pappadia, stationed at the Napatree main entrance in the summer season, records visitors entering Napatree. His data are a critically important addition to our understanding of visitor traffic in the NTPCA.

## METHODS

We used two sources of information to record visitor usage on Napatree: counts made by Napatree Naturalists of people on the beach and the bay shore and counts of people entering Napatree at the main entrance. In this report, weekends are considered Saturdays and Sundays; weekdays are Monday through Friday.

### Naturalist Count Data

From mid-May through October 2022, visitor use data were collected each day the NTPCA Naturalists were working. The counts documenting the visitor and boat totals were collected as close to 1:00 PM as possible. Data collection on weekdays was done opportunistically whenever a Naturalist was present.

Using the standardized data sheet, Naturalists tallied the number of visitors which included individuals on boats anchored along the beach that were likely to come ashore. Those boats, as well as kayaks, jet skis, and dinghies on the beach (tenders for larger boats), were also counted.

### Main Entrance Count Data

Tom Pappadia counted visitors entering Napatree at the main entrance on Thursdays, Fridays, Saturdays,

Sundays, and holidays from 15 April to 24 September. Data were occasionally collected on various Mondays, Tuesdays, and Wednesdays as well. He typically began recording data at 8:00 AM and finished by 4:00 PM. He also recorded the number of dogs that visitors brought (or tried to bring) onto Napatree. He did not count people leaving the conservation area.

## RESULTS

### Naturalist Visitor Data

A total of 108 days of visitor counts were obtained from 21 May 2022 through 10 October 2022. Over these 108 days, 17,320 visitors were recorded at Napatree (Table 1). Weekends were generally busier than weekdays, with an average of 273 visitors per weekend day. The highest use occurred in July, with an average of 315 visitors per weekend day (Figure 1). The highest visitor count occurred on Monday, 4 July, with 812 people, while the highest boat count occurred on Sunday, 3 July, when 312 boats were tallied.

We tallied the number of boats anchored off Napatree every day for the 108 days sampled (Table 1). Boat numbers were higher on weekends compared to weekdays, with a season-long average of 107 boats per day on weekends and just 29 boats per day on weekdays. July had the highest average number of boats per day, with an average of 48 boats per weekday and 174 boats per weekend day (Figure 2).

| Metric/Time Period (n) | Total Counted | Daily Average | Maximum Count |
|---|---|---|---|
| Visitors | | | |
| All Days (108) | 17,320 | 160 | 812 |
| Weekends (37) | 10,097 | 273 | 742 |
| Weekdays (71) | 7,223 | 102 | 812 |
| | | | |
| Boats | | | |
| All Days (108) | 5,942 | 56 | 312 |
| Weekends (37) | 3,918 | 107 | 312 |
| Weekdays (71) | 2,024 | 29 | 273 |
| | | | |
| Main Entrance | | | |
| All Days (100) | 23,338 | 234 | 509 |
| Weekends (47) | 14,731 | 313 | 509 |
| Weekdays (53) | 8,657 | 163 | 477 |

*Table 1. Visitor and boat count data for 2022 from all sources.*

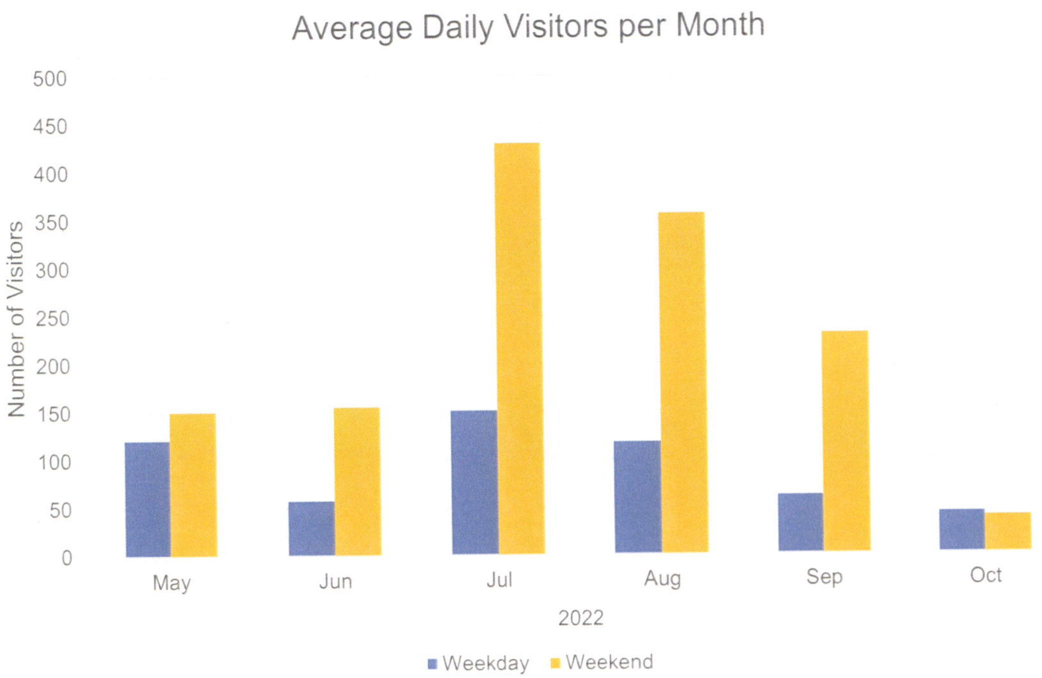


*Figure 1. Average number of people observed per day during Naturalist counts on NTPCA from May through October.*

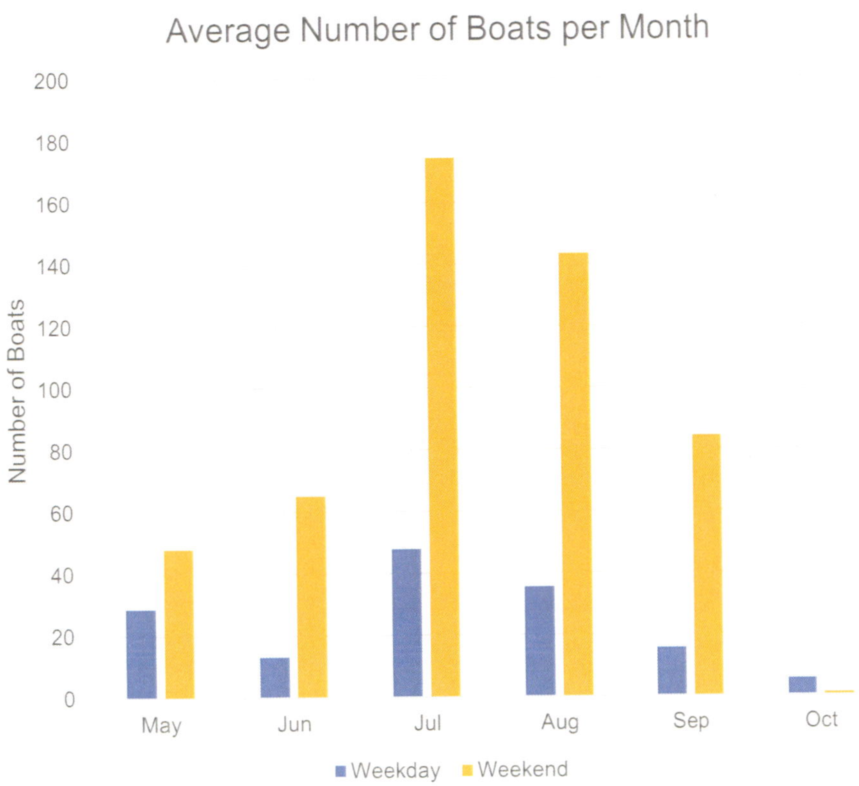


*Figure 2. Average number of boats observed per day during Naturalist counts on NTPCA from May through October.*

### Main Entrance Count Data

A total of 23,338 visitors were counted at the entrance in over one hundred days between 15 April and 24 September (Table 1). The patterns observed with the Naturalist counts were similar for main entrance counts—weekends were busier than weekdays, and July was the busiest month of the summer with 6,159 visitors counted at the main entrance (Figure 3).

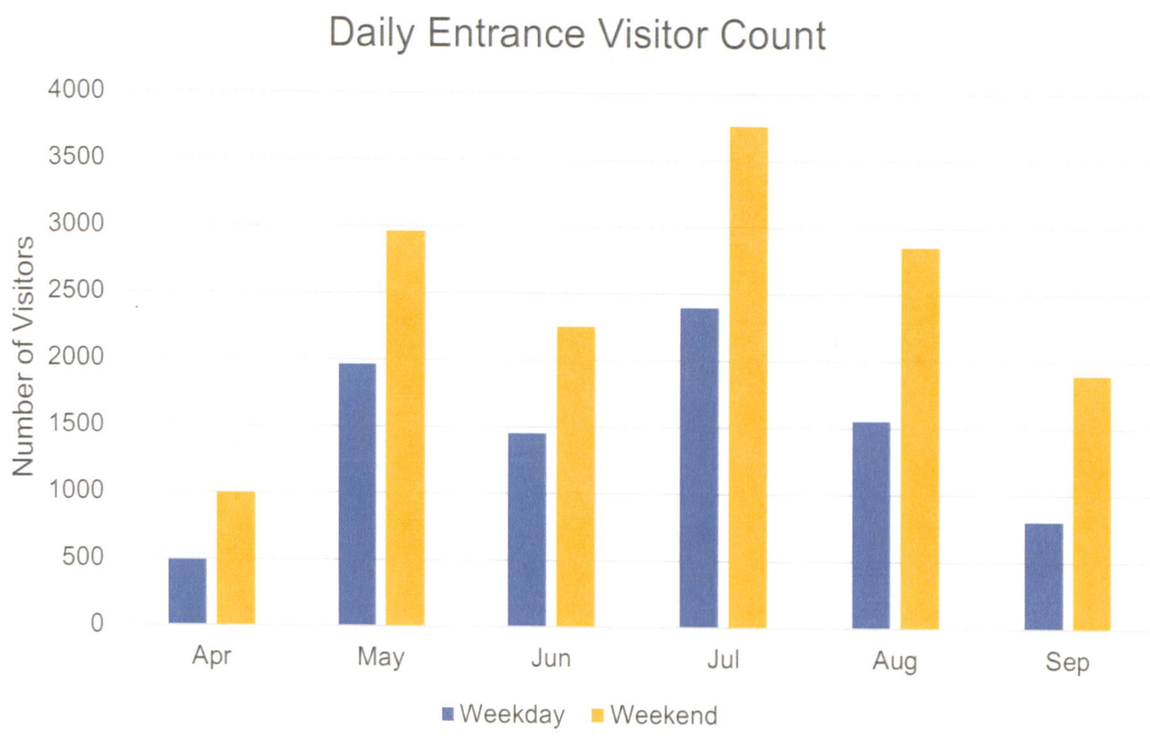


*Figure 3. Total number of visitors counted at the Napatree main entrance per month from April through September.*

Visitors walking their dogs on Napatree typically enter at the main entrance, especially early and late in the season when most boats are not in the water. According to Westerly Ordinance 76-8, dogs are not allowed in the NTPCA from 8:00 AM to 6:00 PM from 2 May to Labor Day (5 September 2022). This is clearly posted at the Napatree entrance as well as on signage placed along the bayside of the beach. Violators on the beach are advised of the ordinance by on-duty Naturalists, and visitors walking onto the beach via the entrance are advised by Napatree staff prior to entry. From the day after Labor Day to 2 May, leashed dogs are allowed on Napatree at any time. The frequency with which visitors with dogs enter Napatree or are turned away by the Naturalist at the main entrance clearly reflects when Naturalists are on duty and when dogs are and are not permitted in the NTPCA (Figure 4).

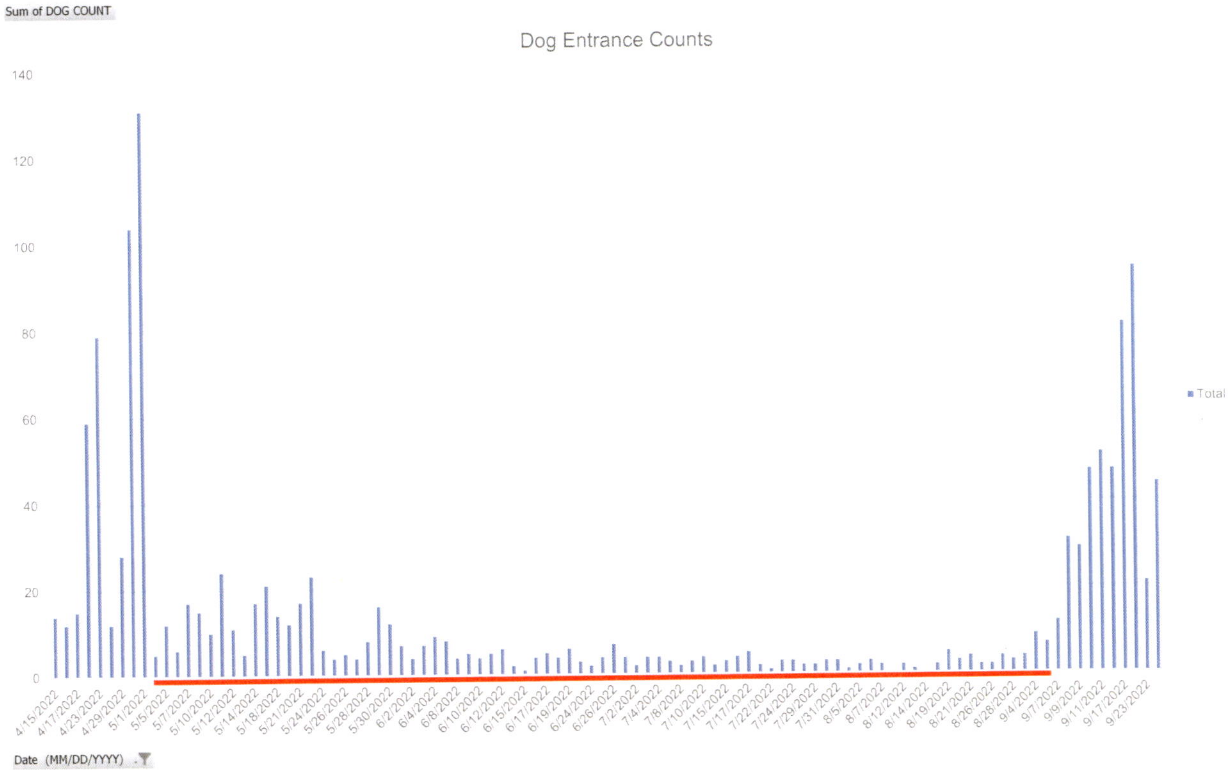


*Figure 4. Dogs counted at the Napatree entrance. By order of Westerly Town Ordinance, dogs are not permitted in the NTPCA between 8:00 AM and 6:00 PM from 2 May to Labor Day (5 September 2022). Red bar indicates dog counts in the prohibited time period when they are not allowed on Napatree (in season).*

## DISCUSSION

Average daily visitation to Napatree in 2022 was the lowest ever recorded (Figure 5). While visitation was only slightly lower than the 2021 season, the 2022 average of 273 visitors per weekend day is a clear departure from 2014–2020 averages, which consistently fell into the 450–800 visitors per weekend day range. Average boat counts followed a similar trend, with just 107 boats per weekend day counted in 2022, the lowest average recorded since 2014, and average boats per weekend day coming in at the second lowest average recorded since 2014. The maximum number of boats recorded in 2022 was also the lowest recorded since 2014, indicating a decline from previous years.

It is difficult to pinpoint which factors may influence visitation trends, though some recent external factors may have played a role in visitation in the past three years. Over the 2020 Napatree season, we saw an increase in visitation and boat use from 2019, which may have been a result of the COVID-19 pandemic. However, visitation is declining for the second year in a row following the 2020 season, perhaps as a result of other recreational opportunities that began to open back up in 2021 and 2022. In addition, gasoline prices soared during the summer, with prices reaching a high in June 2022 not seen in recent years (US Energy Information Administration, 2022). This may have impacted Napatree visitation, as many visitors travel via car or boat to reach the beach. It is worth noting that the 2022 season had the highest number of days collecting visitor data, with 108 days sampled, compared to a previous maximum of 82 days of data in 2019. A higher number of data points would result in a more accurate portrayal of average visitation, underscoring the importance of having Napatree Naturalists on shift to conduct these counts.

As expected, weather affects visitor numbers. Counts are low on rainy days. This has been observed the past few years. However, the weather in 2022 was fine for beach visits (Table 2); it was drier than normal, and temperatures were very typical for the summer.

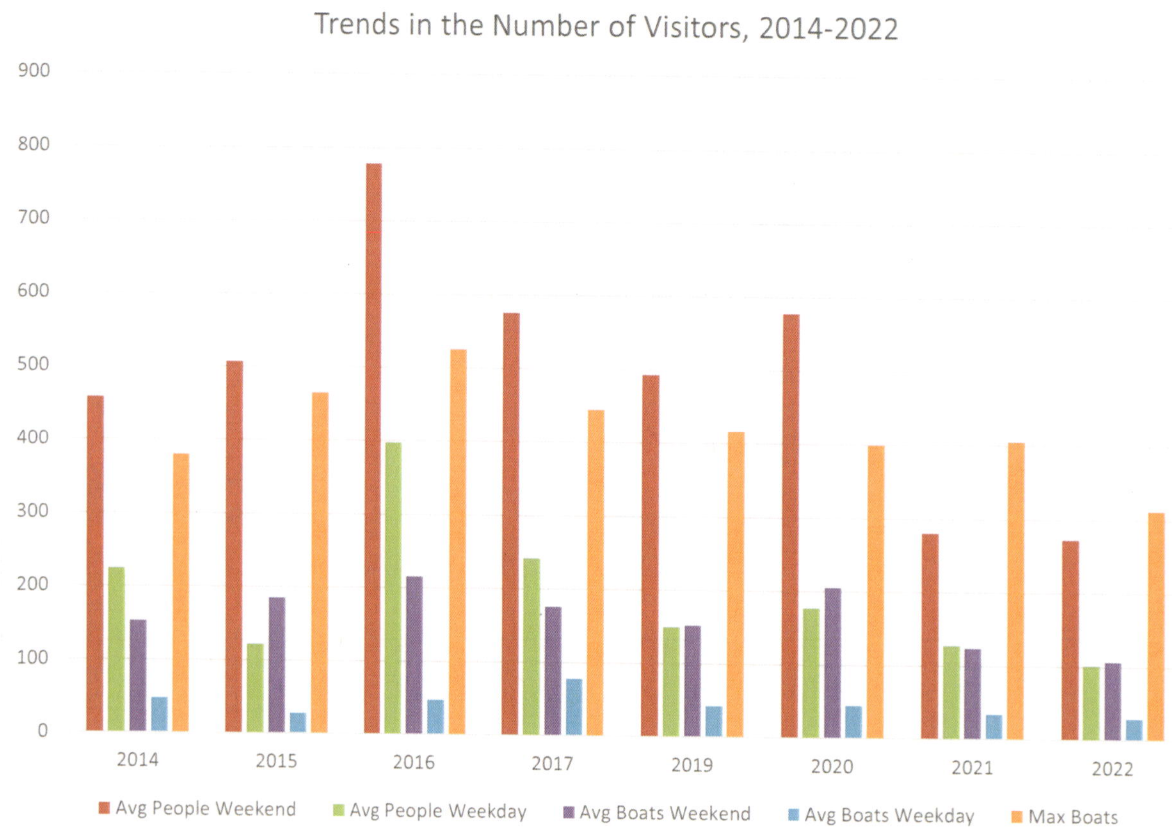


*Figure 5. Eight-year pattern in visitor use of the Napatree Point Conservation Area.*

| | Month | | | |
|---|---|---|---|---|
| | June | July | August | September |
| **Rainfall (Inches)** | | | | |
| Average | 3.1 | 2.8 | 2.4 | 3.8 |
| 2022 | 2.6 | 0.8 | 1.6 | 5.2 |
| **Temperature (Degrees F)** | | | | |
| Average | 83 | 91 | 89 | 83 |
| 2022 | 86 | 90 | 89 | 85 |

*Table 2. Monthly rainfall and temperature for 2022. Data obtained from NOAA NOW data for Westerly, Rhode Island (NOAA 2022). Average values based on the monthly mean of all years from 2014 to 2022. Rainfall is the total rain falling in the month (inches). Temperature is the average daily maximum temperature for each month in degrees Fahrenheit.*

As seen in previous years, visitors were frequently bringing dogs to Napatree. Though leashed dogs are permitted on the beach before 8:00 AM and after 6:00 PM from 2 May through Labor Day, some visitors do arrive at the entrance during "no dog" times, where they are often met by Napatree staff who can relay the town ordinance. During times when dogs are allowed on the beach, Napatree staff consistently need to remind visitors to keep their dogs leashed at all times. In 2022, Napatree Naturalist hours on the beach were expanded by one hour at opening and closing to 8:00 AM to 6:00 PM, which helped to ensure that dogs were not on the beach and/or off leash during the times outlined in the ordinance. Also, Napatree was staffed on Fridays, Saturdays, and Sundays through Columbus Day weekend, which may have deterred off-leash dogs on the beach.

## ACKNOWLEDGMENTS

The information reported here was collected by the 2022 Napatree Naturalists. We are grateful for their diligence in obtaining these important data. Peter August provided a helpful and insightful review of this report.

## REFERENCES

August P., J. Sassi, T. Pappadia, and K. Rogers. 2020. "Visitor Activity on Napatree: 2020." In *The State of Napatree Report: 2020*, edited by J. Sassi. The Watch Hill Conservancy. https://thewatchhillconservancy.org/napatree/napatree-resources/.

NOAA. 2022. "NOAA Online Weather Data (NOWdata)." National Weather Service (website). https://www.weather.gov/climateservices/nowdatafaq.

US Energy Information Administration. 19 December 2022. "New England (PADD 1A) All Grades All Formulations Retail Gasoline Prices (Dollars per Gallon)." https://www.eia.gov/dnav/pet/hist/LeafHandler.ashx?n=PET&s=EMM_EPM0_PTE_R1X_DPG&f=M.

Wilson, J., and G. G. Simmons III. 2021. "Napatree Water Quality 2021." In *The State of Napatree Report: 2021*, edited by J. Sassi. The Watch Hill Conservancy. https://thewatchhillconservancy.org/napatree/napatree-resources/.

*Photo Credit: Janice Sassi, The Watch Hill Conservancy*

SECTION THREE

# PHYSICAL SYSTEMS

*Dr. Oakley and Jonathan Lepire surveying lagoon outlet.*
*Photo Credit: The Watch Hill Conservancy*

CHAPTER 5

# Understanding the Short- and Long-Term Shoreline Change of Napatree Barrier Using RTK GPS Beach Profiles and Mapping of the Last High Tide Swash: 2022 Update

*Bryan A. Oakley*
Department of Environmental Earth Science, Eastern Connecticut State University

## INTRODUCTION

The objective of this project, which began in July 2013, is to examine the short- and long-term response of the Napatree Barrier and headland to storms using beach profiles (Figure 1) and the position of the last high tide swash (LHTS), commonly referred to as the wet-dry line or wrack line (Figure 2). The lack of infrastructure and development on the Napatree Barrier make it an ideal location to examine shoreline change in the absence of the "line in the sand" mentality inherent on more developed portions of the coastline. Storms (both hurricanes and extra-tropical cyclones), not sea level rise, are the driving force in shoreline change (Hayes and Boothroyd, 1987), particularly at the decadal scale (Morton, 2008). Napatree Barrier has migrated more than 60 m (200 feet) landward (toward Little Narragansett Bay) between 1939 and 2014 (Boothroyd and Hehre, 2007; Boothroyd *et al.*, 2015) (Figure 3), via storm surge overwash and deposition of washover fans. This means the landform itself has "rolled over" almost one barrier width since 1939! Much of this measured migration likely occurred during the Great Atlantic Hurricane in 1944, Hurricane Carol in 1954, and the Ash Wednesday Storm of 1962, a significant extra-tropical storm that impacted the mid-Atlantic and southern New England from 6-8 March 1962 (Dolan, 1987). The first vertical aerial photographs in Rhode Island were collected in 1939, so the shoreline position prior to the Great New England Hurricane of 1938 is not known, although a portion of Napatree was captured in 1934 Connecticut aerial photographs. Analysis of historical aerial photographs shows these events overwashed much of the barrier. Superstorm Sandy, a tropical/extra-tropical storm that made landfall in New Jersey in October 2012, affected Napatree with washover fans observed on some parts of the barrier (particularly at the western end) and other sections experiencing extensive frontal erosion. While this study began after Sandy, it provides an excellent laboratory to examine the recovery of the shoreline following a storm event, and some initial work examining trends since Sandy are discussed in this report.

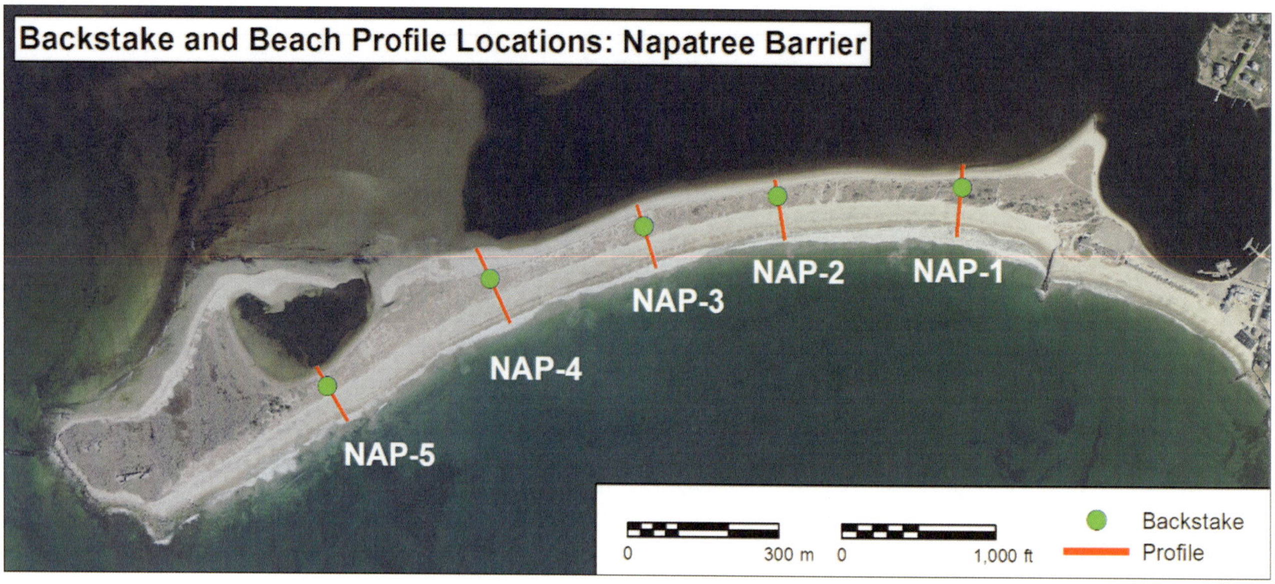


*Figure 1. Benchmark and profile locations on the Napatree Barrier. Basemap is the 2014 USGS orthophotograph downloaded from RIGIS.*

*Figure 2. LHTS on the Little Narragansett Bay shoreline of the Napatree Barrier on 18 December 2013.*

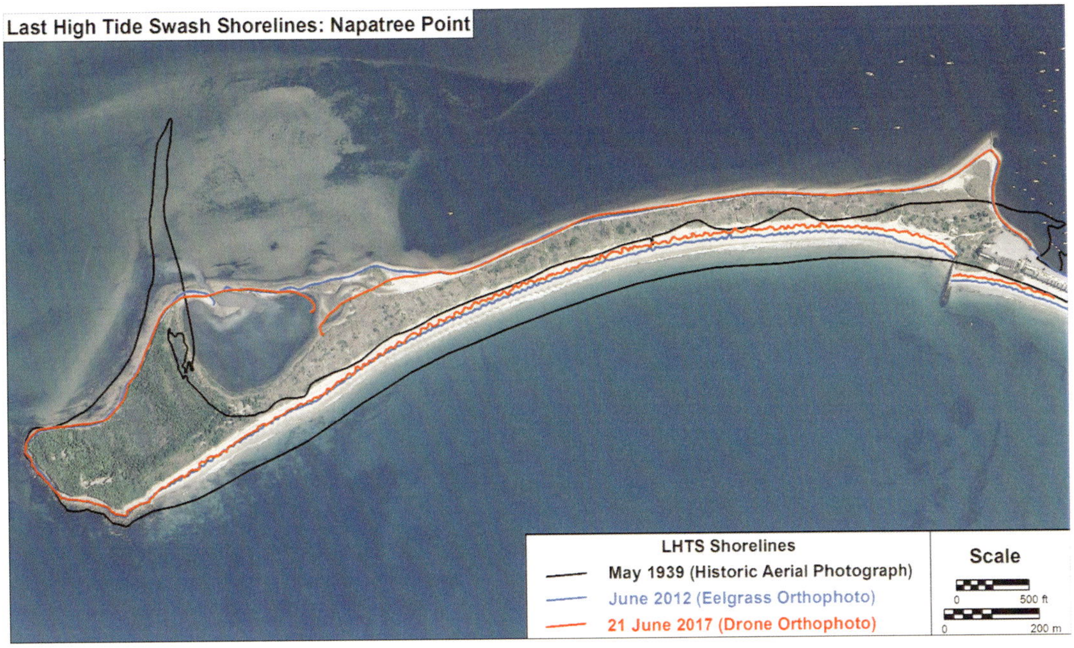


*Figure 3. LHTS, a.k.a. the wet-dry line, derived shoreline positions from 1939 to 2012 (pre-Sandy) (Boothroyd et al., 2016) and the June 2017 orthophotos created from drone imagery (Environmental Data Center, 2017, see Montello et al., 2017).*

## METHODS

### Beach Profiles

Beach profiles are measured quarterly (typically in March, June, September, and December) using RTK GPS, real-time kinematic GPS (Figure 4), which collects an X, Y, Z coordinate—northing, easting relative to Rhode Island State Plane Feet NAD83 and elevation relative to the North American Vertical Datum of 1988 (NAVD88)—for points along the profile. Points are surveyed approximately every 2 m (6.6 feet), as well as at major morphologic features (i.e., the dune crest, berm crest, etc., Figure 5) using three-second occupations (topo points). Elevation is converted to mean lower low water (MLLW) using the V-Datum tool published by the National Oceanic and Atmospheric Administration. Profiles extend as close to MLLW as possible and traverse the barrier from Little Narragansett Bay to Block Island Sound (Figure 1). Profiles were not collected in March 2020 due to the COVID-related travel restrictions.

### Last High Tide Swash (LHTS)

The position of LHTS is mapped using a Trimble 6000XH handheld differential Global Position System and follows the protocol of Psuty and Silveria (2011) developed for the National Park Service on the Atlantic coastline. The position of the shoreline is measured by walking or riding an ATV along LHTS, carrying the handheld GPS. The Trimble 6000XH collects a point every three seconds, with a post-processed reported accuracy of <1 m (3.3 feet) for 75–85 percent of positions recorded, and over 95 percent of positions measured with <2 m (6.6 feet) accuracy. Only portions of the LHTS were resurveyed in 2019, largely around the lagoon spit and adjacent portions of the bayside shoreline. Using LHTS as a proxy for shoreline position, GPS shoreline positions can be directly compared to shoreline positions digitized in historic vertical aerial photographs (discussed later in the report), as well as present and future orthophotographs. The position of the shoreline was also digitized from vertical aerial photographs created from imagery collected using a drone on 21 June 2017 (Environmental Data Center, 2017; Montello *et al.*, 2017) and 2018 Rhode Island Pictometry Orthophotographs. The position of the LHTS will be monitored as needed after storm events.

*Figure 4. Eastern Connecticut State University undergraduate student Joshua Bartosiewicz measuring a beach profile using RTK GPS.*

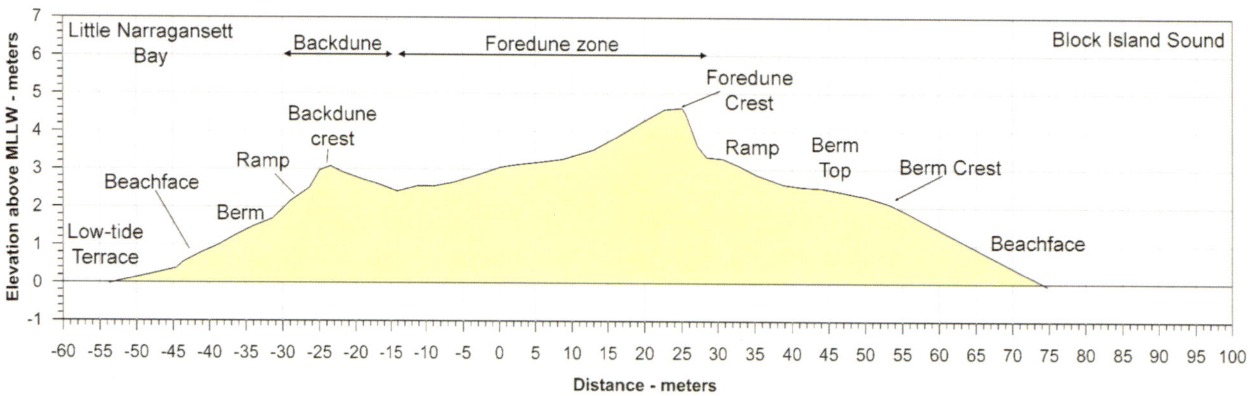


*Figure 5. Major geomorphic features of the Napatree Barrier. Not all profiles show all of the features, and many variations can occur. Profiles NAP-1 and NAP-5 lack an appreciable backdune, and the profiles can show multiple berms on the Block Island Sound shoreline.*

## RESULTS

### Beach Profiles

Profile volumes at the end of 2022 are relatively consistent with those at the end of 2021. NAP-1 and NAP-2 (Figure 1) are slightly higher, while NAP-3, NAP-4, and NAP-5 are down slightly relative to the December profiles. The overall volume of the profiles has generally increased since July 2013 (Table 1; Figure 6). The net accretion can be seen in all five profiles (Figure 7). The dune crest elevation has increased by 0.4–0.6 m (1.3–2.0 feet) on NAP-1, NAP-2, and NAP-5; the elevation gain on the wider profiles (NAP-3, NAP-4) has been less pronounced with an increase of 0.07 m (0.23 feet) and 0.11 m (0.36 feet). The elevation gain can be seen in comparison plots of the minimum volume profiles (December 2013) compared against the 9 December 2022 profiles (Figures 8–12). The trend has begun to flatten out over the last few years, with the profiles maintaining a more consistent dune crest elevation through 2018 with slight increases in some profiles. Much of the volume added to the dunes at all five profiles have occurred on the dune ramp, the area in front of the primary foredune (Figure 5), where incipient foredunes have begun to develop, and in some cases nearly equal the height of the primary foredune (i.e., NAP-5, Figure 12). The development of the incipient dunes is discussed below in the section outlining the recovery since Superstorm Sandy.

| Profile | Volume change ($m^3\ m^{-1}$): July 2013 to 9 Dec 2022 | Percent change in volume: July 2013 to 9 Dec 2022 | Change in foredune elevation (m): July 2013 to 9 Dec 2022 |
|---|---|---|---|
| NAP-1 | 24.8 | 9.5% | 0.41 |
| NAP-2 | 43.5 | 17.7% | 0.54 |
| NAP-3 | 23.1 | 8.4% | 0.07 |
| NAP-4 | -35.2 | -10.4% | 0.11 |
| NAP-5 | 2.1 | 1.0% | 0.6 |

*Table 1. Volume change (expressed as $m^3\ m^{-1}$ and percentage of change) and dune crest elevation increase (m) for the profiles between July 2013 and December 2022.*

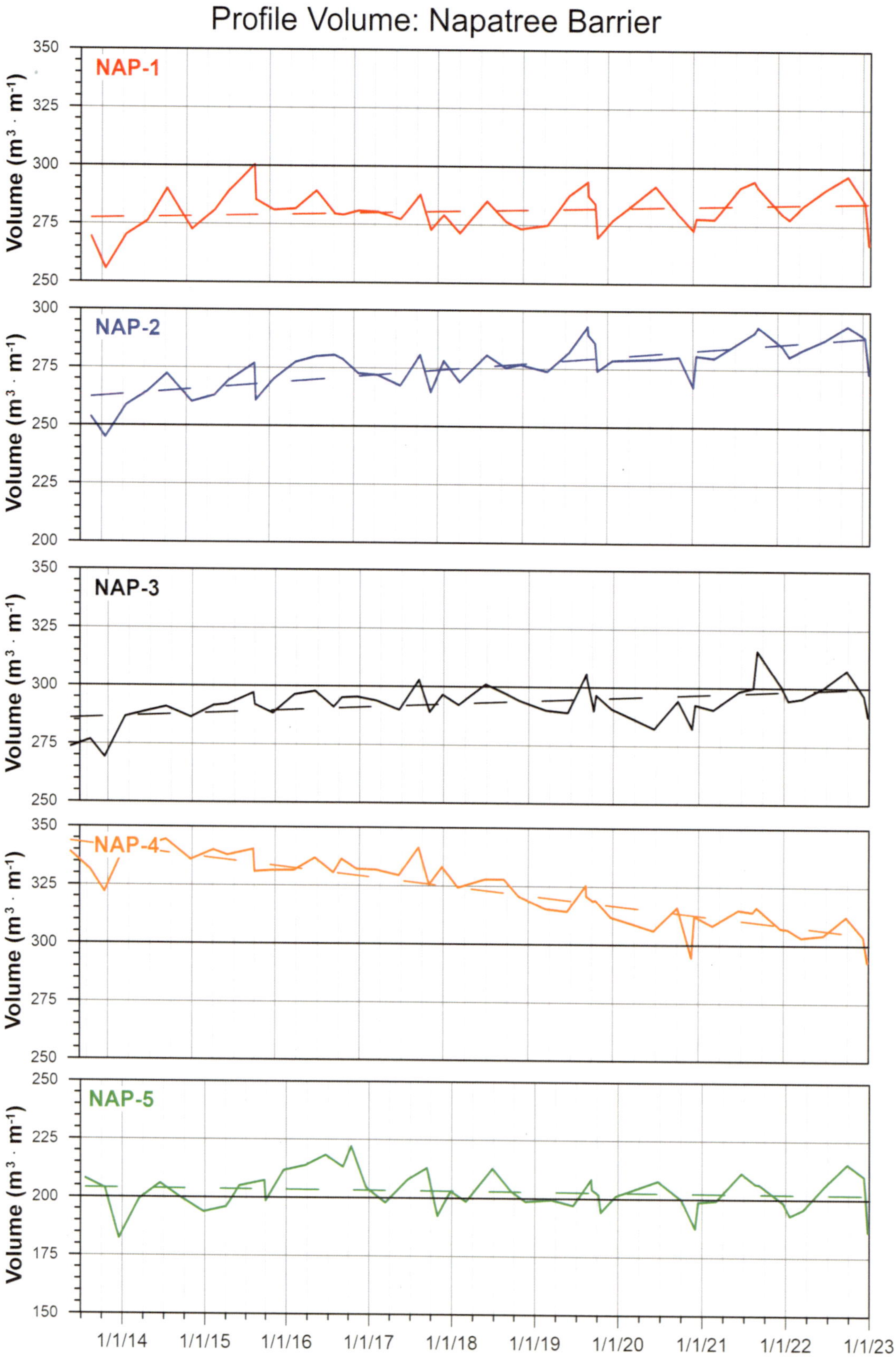


*Figure 6. Profile volume ($m^3 m^{-1}$) for profiles NAP-1 to NAP-5 from July 2013 to December 2022. Dashed lines are the interpolated linear trend. Profiles NAP-1, NAP-2, and NAP-3 show a positive (depositional) trend; NAP-4 shows a slightly negative trend, while NAP-5 shows little net change.*

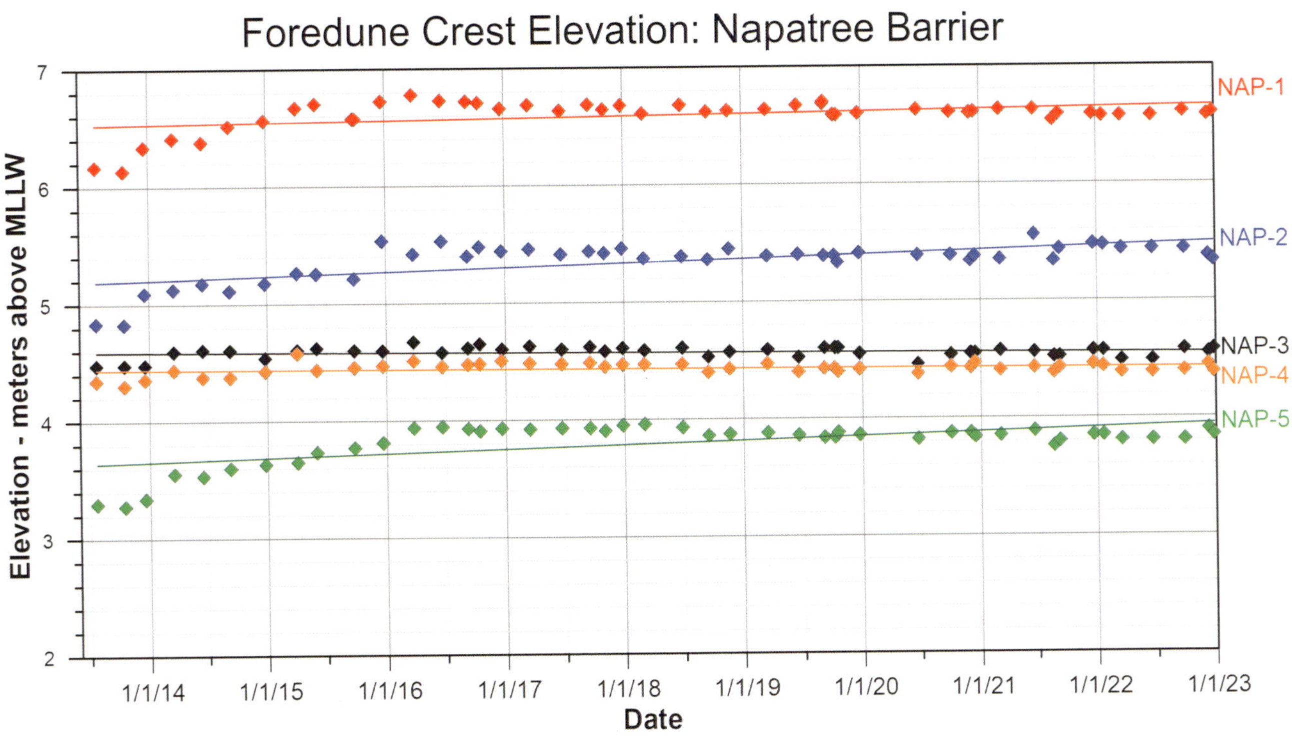


*Figure 7. Foredune crest elevation on measured profiles between July 2013 and December 2022. Solid line is the interpolated linear regression.*

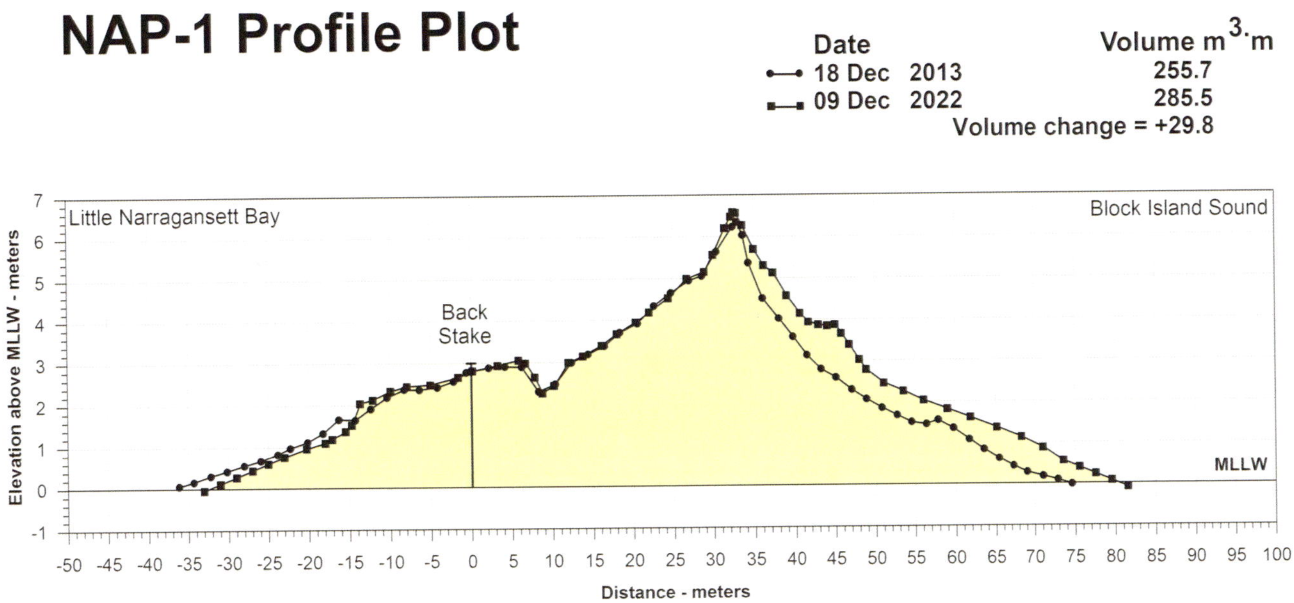


*Figure 8. Plotted profile of NAP-1 comparing the 18 December 2013 and 9 December 2022 profiles; note accretion in the foredune and on the ramp (seaward slope in front of the dune). See Figure 4 for location of the different morphologic zones of the barrier.*

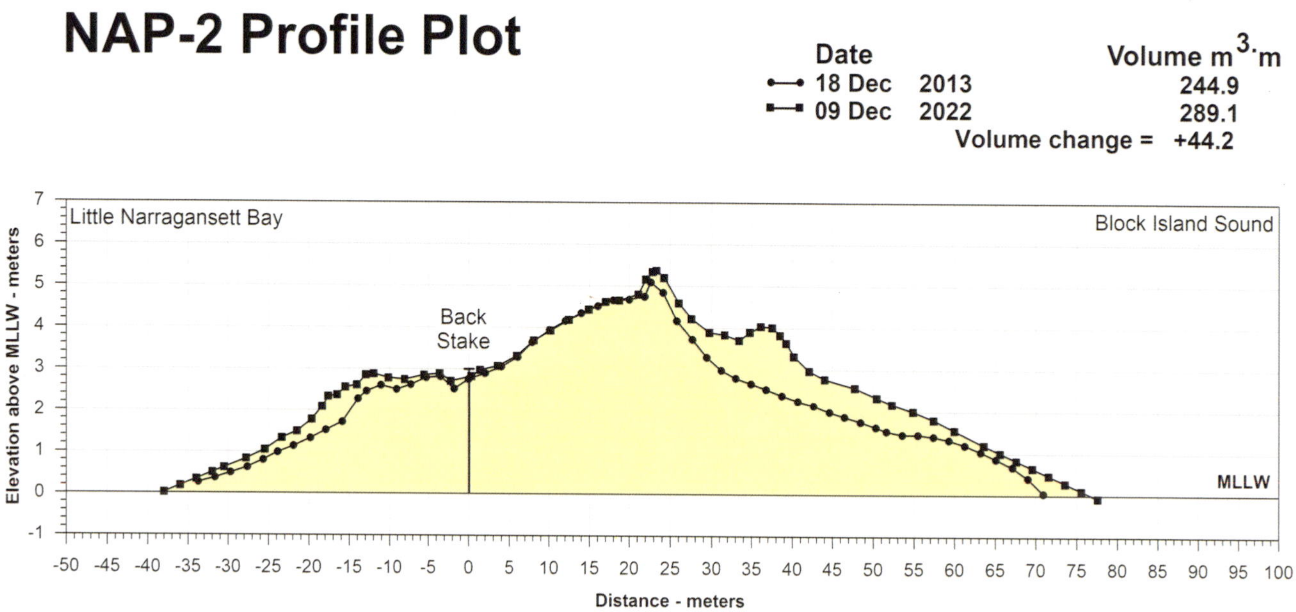


*Figure 9. Plotted profile of NAP-2 comparing the 18 December 2013 and 9 December 2022 profiles; note accretion in the foredune and on the ramp (seaward slope in front of the dune). The backdune (small dune on the Little Narragansett Bay shoreline) has also shown some increase in width and height.*

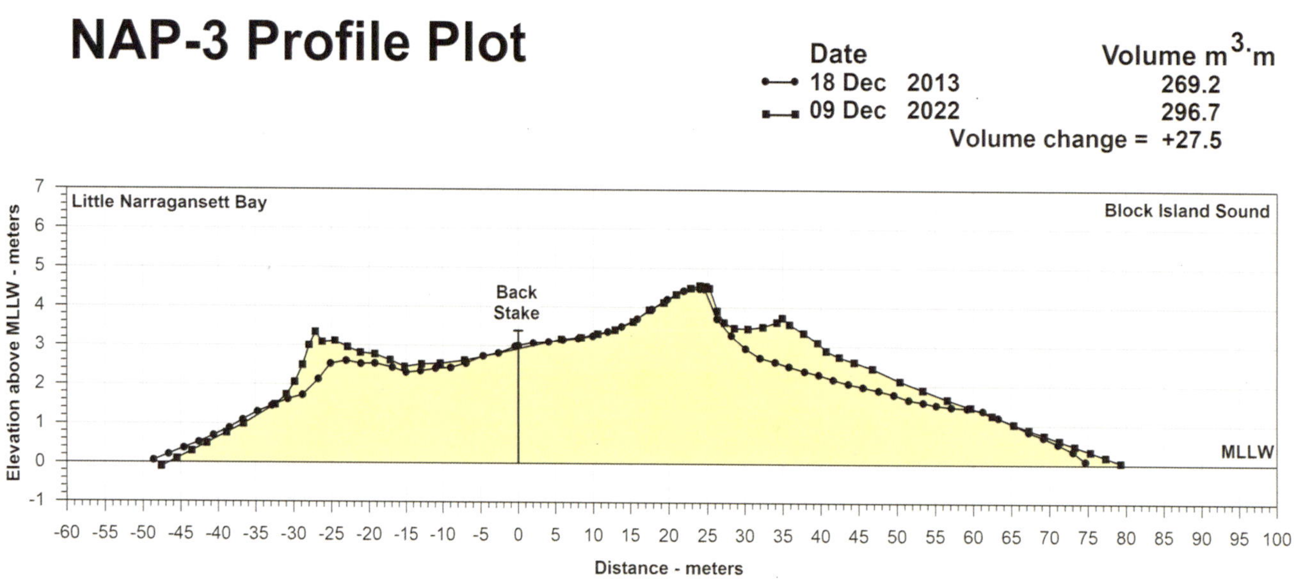


*Figure 10. Plotted profile of NAP-3 comparing the 18 December 2013 and 9 December 2022 profiles; note accretion on the ramp (seaward slope in front of the dune) and the backdune.*

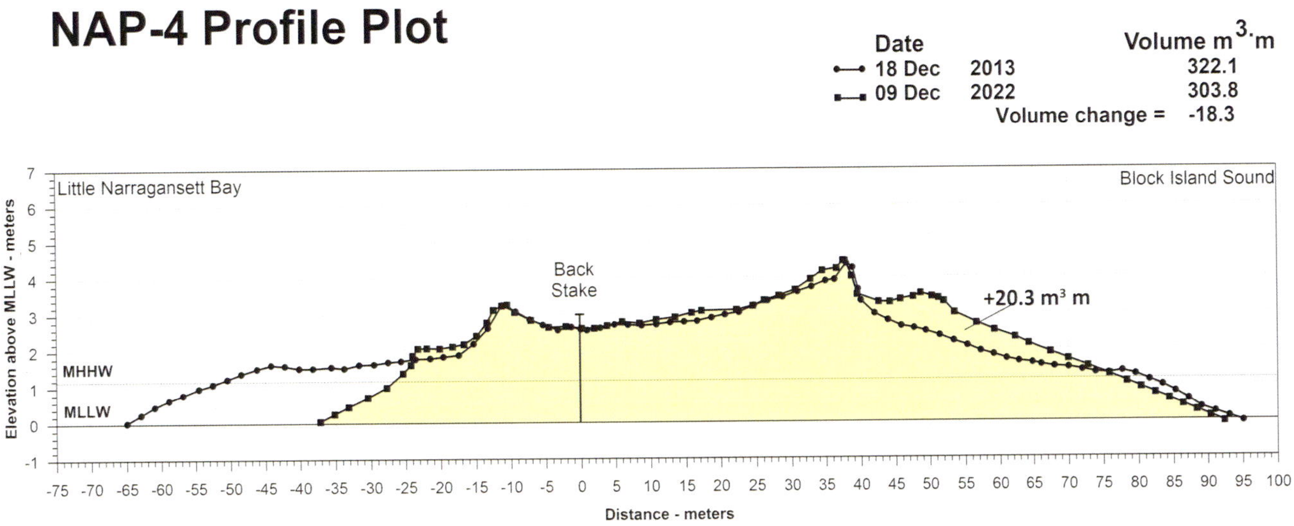


*Figure 11. Plotted profile of NAP-4 comparing the 18 December 2013 and 9 December 2022 profiles. The berm on the Little Narragansett Bay shoreline has retreated significantly, as the welded spit from the eastern end of the lagoon has been eroded. Accretion in the foredune on the oceanside has resulted in an increase there of 20.3 $m^3$ $m^{-1}$.*

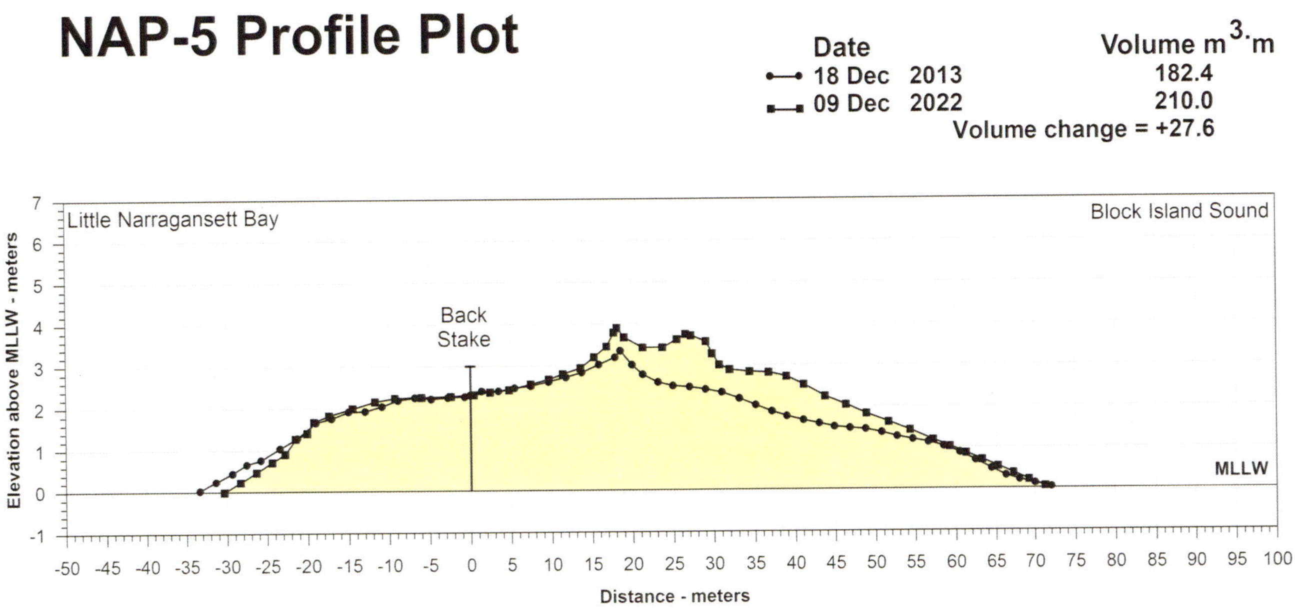


*Figure 12. Plotted profile of NAP-5 comparing the 18 December 2013 and 9 December 2022 profiles. Note the accretion in the foredune and ramp (area just seaward of the dune).*

## DISCUSSION

The source of sediment added to the profiles since 2013 has likely been transported onshore from the portion of the shoreface known as the depositional platform as the barrier recovered from Superstorm Sandy. This sandy geologic habitat mapped along other portions of the Rhode Island south shore is interpreted to represent the sediment actively exchanged between the intertidal beach and the shoreface (Oakley *et al.*, 2019). Onshore sediment transport occurs via the combined action of wave orbital motion and longshore transport to the east (driven by southwesterly sea breezes) during periods of fair weather. Given the dominant longshore transport direction on the adjacent south shore of Rhode Island (to the east) (Boothroyd *et al.*, 1985; Fisher, 1988; Oakley, 2019), it seems likely that if any sediment bypasses around Watch Hill, it will happen during periods of easterly swell and probably does not contribute much (if any) sediment to Napatree. The northward migration of Sandy Point (and the continual dredging of the navigation channel) suggests sediment transport here is to the north, and significant amounts of sediment are probably not being transported from Little Narragansett Bay to the seaward side of the Napatree Barrier.

Sand transported onshore to the berm via wave-orbital motion is then available to be transported by eolian (wind) processes and deposited in the center of the barrier. Southerly winds transport sand from the Block Island Sound shoreline; northerly winds transport sand from the Little Narragansett Bay shoreline. The net prevailing westerly wind direction (northwest and southwest) transports sediment towards the eastern portion of the barrier. The vegetation (and limited sand fencing) in the foredune and backdune zones baffles the wind, and sediment is deposited in the wind shadow. A consistent trend over the last five years has been the deposition of sand on the ramp (front of the dune, Figure 5) on all the profiles, much of this is accompanied by the reestablishment of vegetation. The increase in foredune height and ramp deposition is very apparent when examining the over eight years of data gathered for this project (Figures 8–12). The flattening of the trend over the past few years may represent the reaching of a maximum dune height, controlled by a combination of factors relating to sediment supply, beach profile width/slope, vegetation type/abundance, and other factors (Durán and Moore, 2013; Hesp, 1988; Psuty, 2004).

Barring another major storm, the barrier will continue to see a relatively consistent volume or perhaps a slight increase in height and volume. Small storms (i.e., the late September/early October extra-tropical storm in 2015; Tropical Storm Hermine in 2016; Hurricane Jose in 2017; March 2018 nor'easter; Hurricane Dorian in September 2019, and a December 2020 nor'easter) will continue to erode the berm (followed by subsequent recovery), and during larger storms, low-lying areas of the barrier (i.e., NAP-5, Figure 1) will continue to be overwashed. Hurricane Henri, which made landfall over Westerly in late August 2021, caused little erosion along Napatree, as measured in profiles collected on 7 September 2021, and most profiles showed little net change in volume following that storm (Figure 13). Given the track, intensity, and storm speed coupled with the timing of the storm (peak storm surge occurred near low tide) resulted in little damage along the southern Rhode Island shoreline. Hurricane Fiona, which passed offshore in September 2022, resulted in onshore transport of sediment and debris (as discussed below).

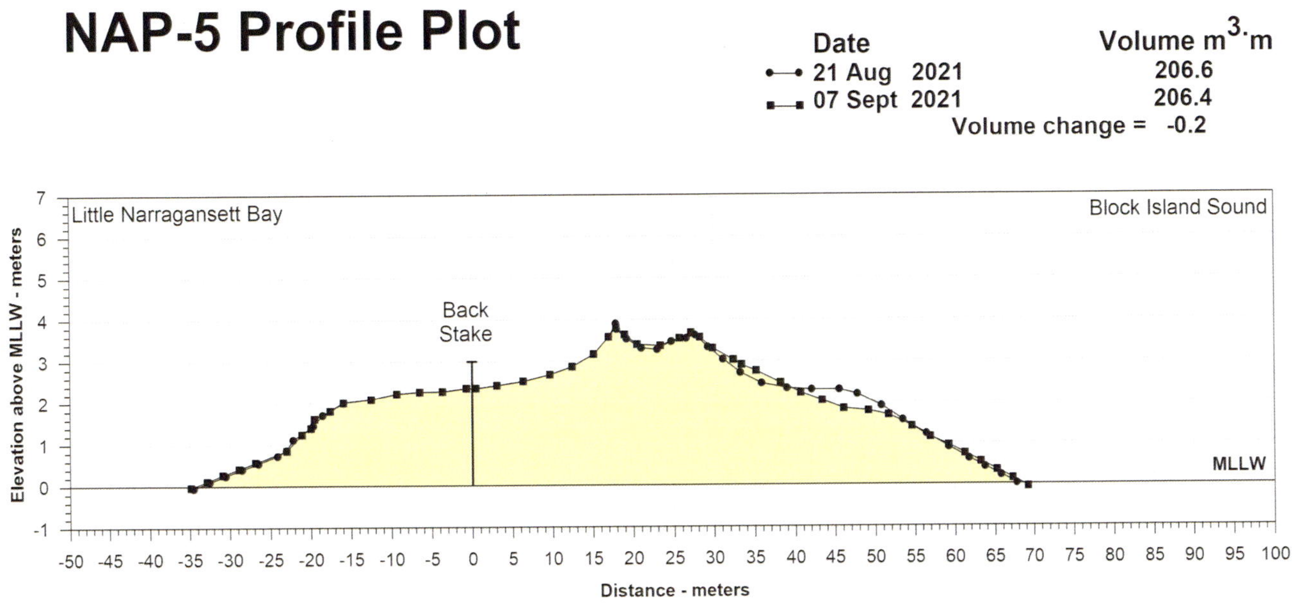


*Figure 13. Plotted profile of NAP-5 comparing the 21 August 2021 (pre-Henri) and 7 September 2021 profiles. Note the minor berm erosion and ramp deposition resulting in little net change in volume.*

Mapping of the shoreline position and field observations made between profile NAP-4 and the lagoon (Figure 1) suggest that landward migration of the shoreline has been occurring as the former spit enclosing the western end of the lagoon has been migrating onshore. This is readily seen by the presence of an erosion scarp (near vertical surface) in the backdune in this area (Figure 11). This process continued largely during extra-tropical storms (nor'easters) in the fall/winter, where northeast and northwest winds can erode the shoreline here. The fate of this sediment is unclear; some may be transported into the lagoon, the rest is likely transported alongshore to the east eventually ending up in Watch Hill Harbor.

While the perception exists that installing sand fences or otherwise creating a more reflective profile is necessary to maintain the barrier (see the ubiquitous KEEP OFF THE DUNES signs along the coastline), geologically this is not correct. Studies on adjacent segments of barriers in the Netherlands have shown that while the addition of sand fences affected morphology of the barrier, they did not directly affect the total volume of sediment in the foredune (De Jong *et al.*, 2014). The importance of leaving Napatree as a natural laboratory cannot be overstated, and a lower, more dissipative profile probably reflects the historical shape of the barrier. Allowing most of the barrier to evolve in the absence of sand fencing or other structures is crucial to helping understand the response of a shoreline to storms. The recently installed cedar split-rail fencing was chosen specifically because of the limited impact on sediment transport.

Overwash, which occurs when the total water level exceeds the elevation of the dune crest during a storm event, should not be viewed as a negative process; the subsequent landward deposition of washover fans create a more dissipative profile and is the natural process that allows the barrier to evolve in response to storms and sea level rise. Napatree has migrated substantially since the 1880s, maintaining a relatively consistent width via washover fan deposition. The results of the shoreline change analysis between 1883 and 2016 are summarized in a recent publication in the journal *Geosciences* (https://www.mdpi.com/2076-3263/11/8/330). The presence of washover fans is readily seen in the 1939, 1951, and 1963 aerial photographs (Figure 14) shoreline positions are summarized in Figure 15 (Oakley, 2021).

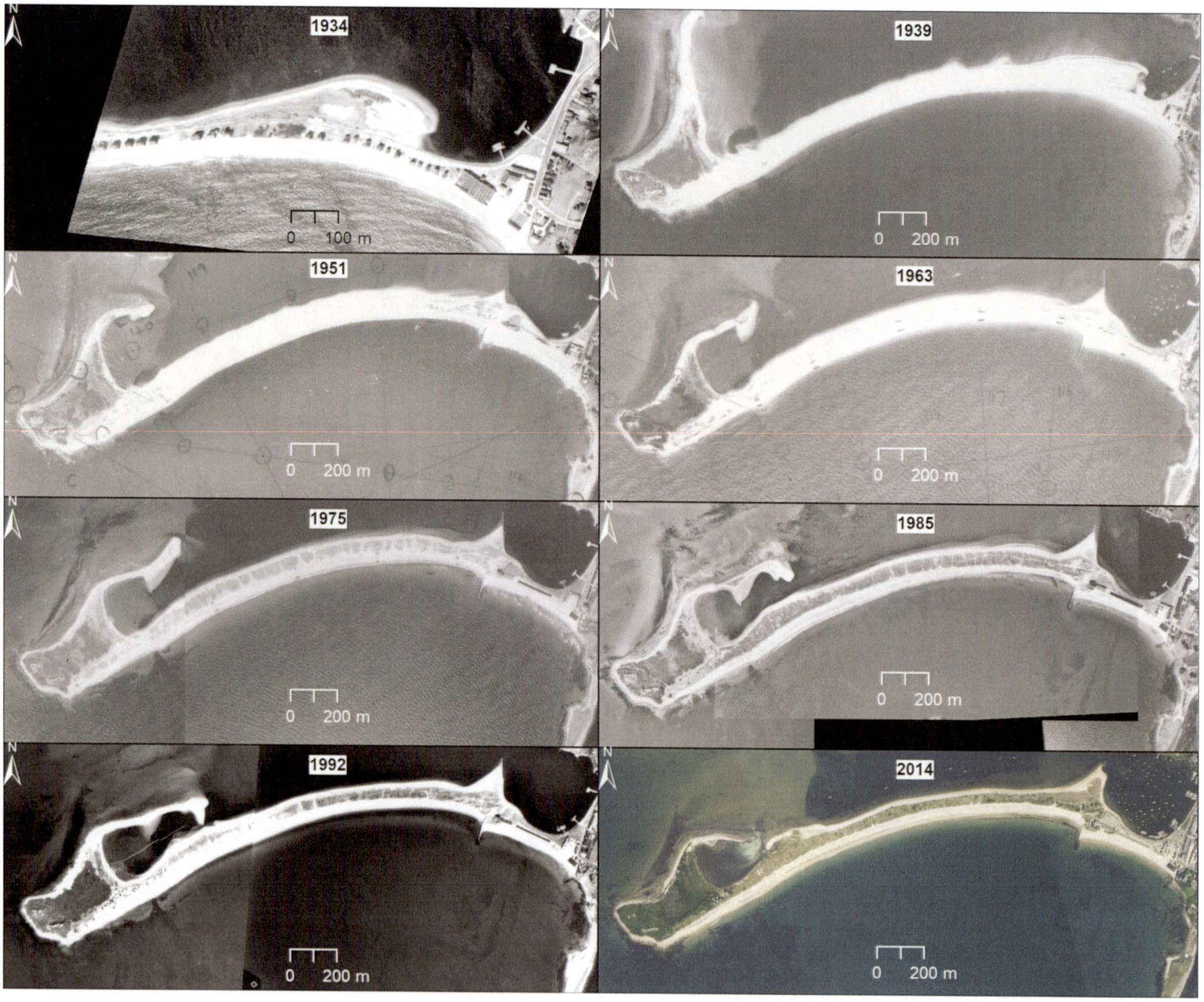


*Figure 14. Napatree Barrier 1934 through 2014. Only the eastern portion of the barrier was imaged in 1934. Note the houses and vegetation visible in 1934. Images shown from 1939, 1951, and 1963 all show evidence of recent washover fan deposition. Images from 1975 to 2014 show vegetation reestablished along the barrier. The 1992 images show some overwash of the western half of the barrier but not full removal of the dunes (modified from Oakley, 2021).*

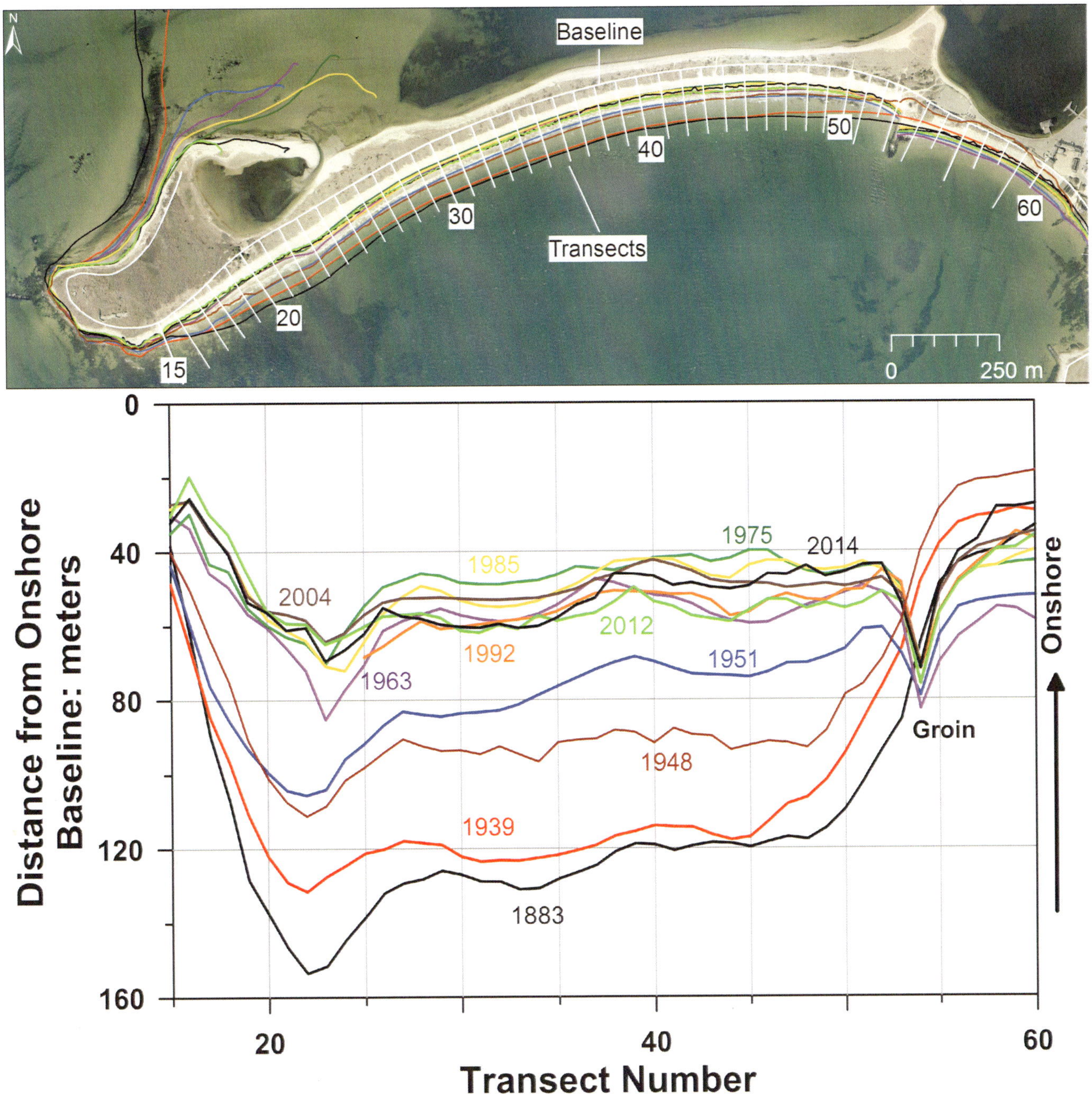


*Figure 15. Top, Digital orthophotograph of the Napatree Barrier showing the location of the transects used in this study. Bottom, Distance from the onshore baseline for transects 15–60 for Napatree Point. The closer to the top of the graph, the further onshore the barrier has migrated. Note the migration of the shoreline between 1939 and 1975 and the lack of change in position between 1975 and 2014. The apparent progradation of the shoreline at transect 54 represents the installation of the small groin between 1948 and 1951. The impact of the groin on shoreline position appears to be limited to transects 51 to 56. Figure modified from Oakley (2021).*

Deposition of washover fans is the primary process that allows the landward portions of the barrier (known as the backbarrier flat and surge platform) to accumulate sediment (Leatherman, 1979). It has been known for some time (Godfrey and Godfrey, 1976; Leatherman, 1979), and recently reinforced (Houser and Hamilton, 2009; Timmons *et al.*, 2010), that overwash and the subsequent deposition of washover fans on the backbarrier is critical for barriers to continue to migrate in response to storms and sea level rise.

This geologic process, which can look "catastrophic" in the immediate aftermath of a storm, is *vital* to the evolution of Napatree in response to future storms and sea level rise. The lower, wider dissipative profile configuration is also generally less erosive for any given storm than a reflective barrier (Dolan, 1972). An added benefit is that these wider, more dissipative barriers promote a more diverse array of habitats than those typically found on many managed shorelines (Dolan *et al.*, 1973).

## Recovery of the Profiles: Superstorm Sandy (October 2012) to Present

One obvious metric the profiles offer is a measurement of the volume of sediment eroded during a given storm. Equally, if not more, important are the details learned of the recovery of the system following a storm event. Superstorm Sandy is the most significant storm (in terms of storm surge elevation) to impact the area since Hurricane Carol in 1954. Profiles extracted every 50 m alongshore from digital elevation models derived from pre– (April 2011) and post-storm (November 2012) LiDAR show that the dune crest retreated up to 13 m (42.6 feet) (mean retreat distance 4 m, 13.1 feet). The elevation of the dune crest lowered by up to 1.4 m (4.6 feet) (mean 0.4 m, 1.3 feet) (Figure 16). Dune volume (measured above 2.5 m (8.2 feet) NAVD88 decreased by up to 23 $m^3$ $m^{-1}$ (mean 11 $m^3$ $m^{-1}$). The foredune was largely removed along the western portion of the barrier (overwash regime) and retreated due to frontal erosion at the eastern half (collisional regime). Washover fan deposition was limited to trail crossings on the eastern portion of the barrier, which acted as surge channels. Erosion/recovery of the berm was not quantified, as the post-storm LiDAR was collected two weeks after Sandy and profiling did not begin until 2013 (eight months after the storm), however, much of the dune response can be quantified.

The foredune crest elevation has increased by an average of 0.4 m (1.3 feet) following Sandy (2012–2018). The dune crest now exceeds the 2011 (pre-storm) elevation at 68 percent of the LiDAR transects. Dune volume (above 2.5 m (8.2 feet) NAVD88) increased by an average of 2.75 $m^3$ $m^{-1}$ $yr^{-1}$ between 2012 and 2018. The dune volume for the barrier in 2018 ($5.9 \times 10^4$ $m^3$ ($\pm$ $8.8 \times 10^3$ $m^3$)) (Figure17) exceeds the 2011 volume (within the uncertainty) suggesting the barrier is largely recovered from Sandy, although the dune crest is still lower in some areas. Volume added within the last three years has particularly focused on eolian transport to the dune ramp, as incipient dunes formed around the reestablished vegetation—largely American beach grass (*Ammophila breviligulata*). Deposition on this portion of the barrier was exemplified following the passage of an extra-tropical storm with southwest (onshore oblique) winds greater than 20 m $s^{-1}$ (65.6 feet $s^{-1}$). Profiles collected the following day show an increase in sediment volume on the dune ramp for four of the five measured profiles (Figure 18). The results indicate that for moderate storm impacts (collisional/overwash regime), the recovery of the NTPCA barrier system is approximately five years. The results of this on-going work were presented online at the American Shore and Beach Preservation Conference in October 2020 (Oakley, 2020).

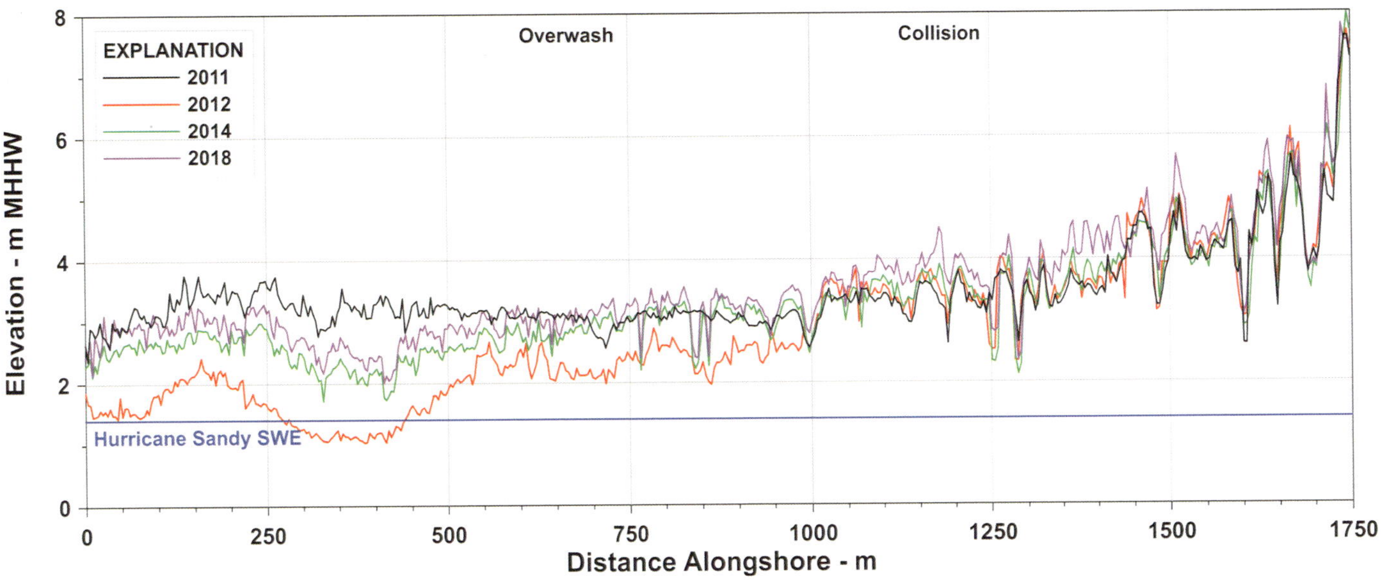


*Figure 16. Dune elevation along the Napatree Barrier from west (left) to east (right). The area in gray was largely overwashed during Sandy and saw the most significant lowering of the dune crest. The purple line represents the dune crest elevation in 2018 and is near or above the dune crest elevation.*

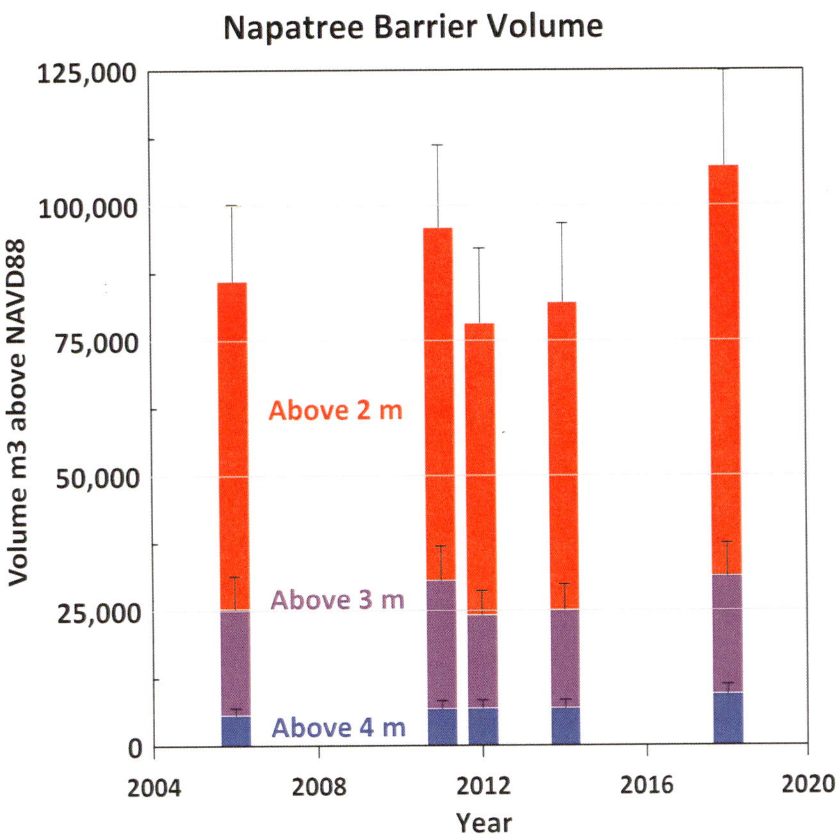


*Figure 17. Volume of the Napatree Barrier above 2 m (6.6 feet), 3 m (9.8 feet), and 4 m (13.1 feet) NAVD88. Note there was little change in overall volume between 2011 and 2012 (post-Sandy) LiDAR; however, there was a reduction in volume of sediment above 2 m (6.6 feet) and 3 m (9.8 feet) NAVD88 following Sandy and the subsequent recovery and increase in volume between 2012 and 2018.*

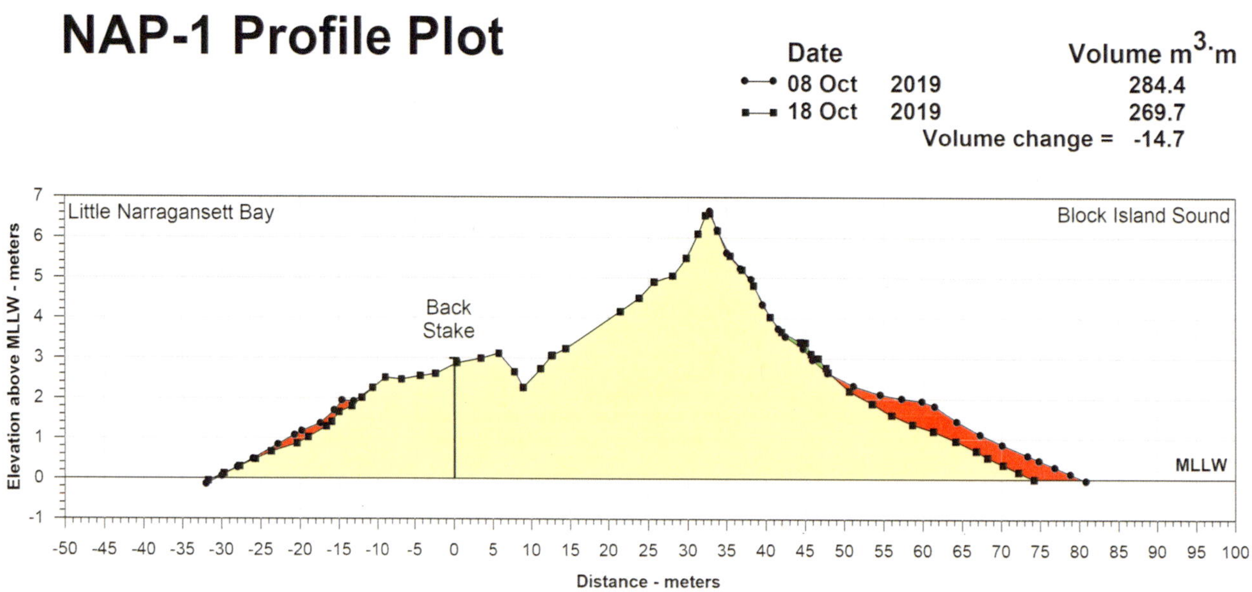


*Figure 18. Profile collected on 8 October 2019 and 18 October 2019 showing the deposition (green) of 0.8 $m^3$ $m^{-1}$ (see green arrow) of sand on the incipient foredune during the wind event on 17 October 2019. Red areas represent minor berm erosion on both sides of the barrier during the storm.*

## Onshore Transport of Concrete Debris during the Passage of Hurricane Fiona

Following the passage of Hurricane Fiona, which tracked east of Bermuda in late September 2022, concrete debris was observed near the western end of the barrier adjacent to profile NAP-5 (Figure 1). Intuitively, this could have been perceived as evidence of erosion of the barrier uncovering remnants of Fort Mansfield; however, time-lapse imagery collected at the Osprey pole documented the onshore transport of these blocks (Figure 20). A post-storm profile (Figure 21) measured at NAP-5 also supports the depositional nature of this event, as both the berm and beachface showed deposition, including a cobble berm (Figure 22).

*Figure 19. Time-lapse image collected on 23 September 2022 at 6:30 PM from the center Osprey pole near profile NAP-5 (Figure 1). In this image, wave swash has overtopped the berm and concrete debris is visible in the images.*

*Figure 20. Concrete debris that was transported onto the Napatree Barrier during the passage of Hurricane Fiona. Note hiking boot (size 11) for scale.*

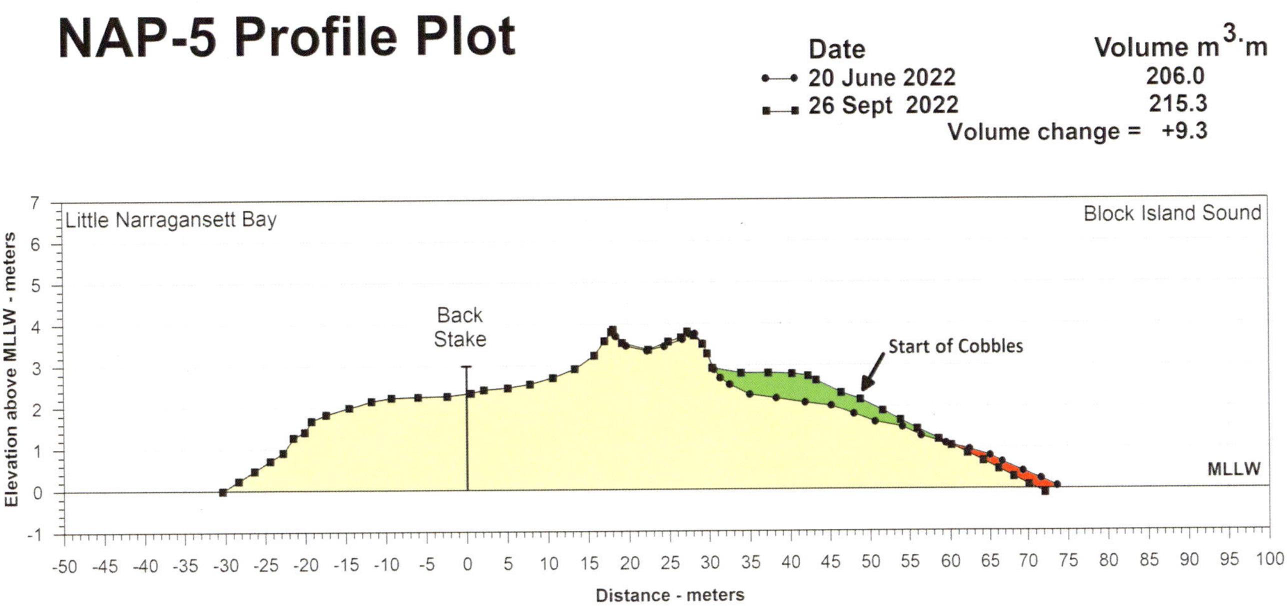


*Figure 21. Profile of NAP-5 collected before and after the passage of Hurricane Fiona. Deposition is noted in green, erosion in red. Note the net deposition of the berm, including a cobble-gravel berm.*

*Figure 22. Time-lapse camera image taken on 24 September 2022 at 8:00 PM. Profile NAP-5 runs through the center of this field of view. Note the cobble-gravel berm deposited by the Hurricane Fiona swell (see Figure 21 for the extent).*

## Erosion during the "Festivus" Storm (Extra-Tropical Storm Elliott): December 2022

Extra-tropical storm Elliott, which impacted Napatree and the rest of southern New England on 23 December 2022, caused extensive flooding along portions of Watch Hill Harbor (Figure 23). The peak water level measured at the Watch Hill Tide Gauge was 0.96 m (3.15 feet) above MHHW. This is the highest water level recorded at the USGS Watch Hill Tide Gage since it was deployed in 2015 (USGS, 2022). Return period analysis at the Newport and New London Tide Gages places this as approximately a ten-year return period water level (NOAA, 2022*A*; NOAA, 2022*B*). Storms of this magnitude show the vulnerability of these areas, as all flooded areas would be inundated at high tide by a three-foot rise in sea level.

Storm impacts along the barrier were largely collisional (Sallenger, 2000), where waves impact the dune face and cause some retreat but did not cause wholesale overwash and migration of the barrier. Overwash was limited to some of the dune crossovers and some low areas of the dune, but as stated previously, overwash of the barrier and washover fan deposition is a key part of the long-term resilience of the barrier! Partial erosion of the foredune does increase the vulnerability in future storm events for more dune erosion and will be monitored in the future.

*Figure 23. Photograph taken at 8:37 AM on 23 December 2022 of water level +0.92 m (4.25 feet) MHHW, showing inundation of the parking lot and adjacent Larkin Square shops.*

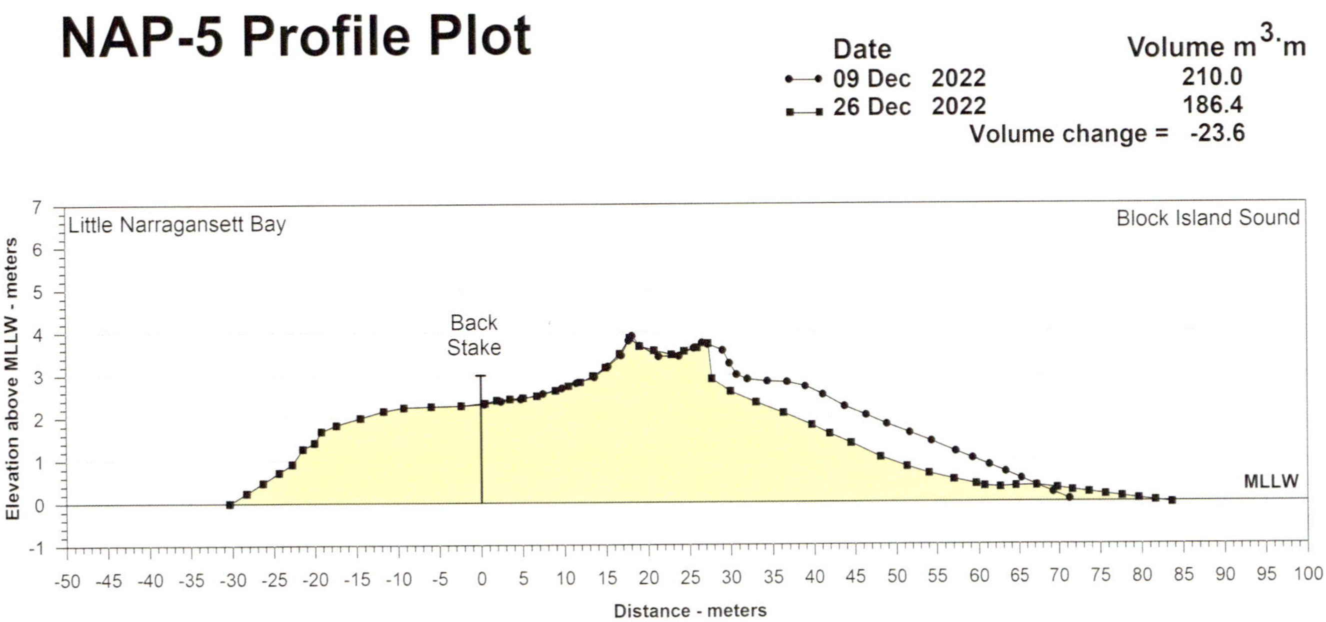


*Figure 24. Profile of NAP-5 collected following the passage of extra-tropical storm Elliott on 23 December 2022. Note the erosion of the foredune and berm. The low, convex feature at the seaward edge of the profile on 26 December 2022 is a swash bar, which is a constructive feature common during storm recovery as sediment is transported onshore.*

*Figure 25. Photograph taken on 23 December 2022 showing overwash and wave run-up on the incipient foredune.*


## ACKNOWLEDGMENTS

Fieldwork has been assisted by Peter August, Janice Sassi, Laura Craver-Rogers, Kevin Rogers, Melissa Cote, Braden Fleming, Emily Bodell, Noah Barreto, and Daniel Cole (The Watch Hill Conservancy and the Watch Hill Fire District); Alyson Augenstien, Joshua Bartosiewicz, Tim Ciskowski, Jennifer Croteau, Cody Murphy, Brandan Sumeersarnauth, and Madie Varney (Eastern Connecticut State University undergraduate students); and Drew Hyatt (Eastern Connecticut State University faculty). This project has been supported by grants from The Watch Hill Conservancy and the Watch Hill Fire District, Rhode Island Bays Rivers Watershed Coordination Team, the URI Coastal Institute Demonstration Site Project, and the Eastern Connecticut State University Exemplary Program Fund. Pete August provided constructive edits of this document.


## REFERENCES


Boothroyd, J. C., N. Friedrich, and S. McGinn. 1985. "Geology of Microtidal Coastal Lagoons; Rhode Island." *Marine Geology*, 63:35–76.

Boothroyd, J. C., R. Hollis, B. Oakley, and R. Henderson. 2016. "Shoreline Change Maps of the Rhode Island South Shore, Scale 1:2,000." Rhode Island Coastal Resources Management Council (website). Accessed 5 January 2024. http://www.crmc.ri.gov/maps/maps_shorechange.html.

De Jong, B., J. Keijsers, M. Riksen, J. Krol, and P. Slim. 2014. "Soft Engineering vs. a Dynamic Approach in Coastal Dune Management: A Case Study on the North Sea Barrier Island of Ameland, The Netherlands." *Journal of Coastal Research*, 30(4):670–684.

Dolan, R. 1972. "Barrier Dune System Along the Outer Banks of North Carolina: a Reappraisal." *Science*, 76:286–288.

Dolan, R., P. Godfrey, and W. Odum. 1973. "Man's Impact on the Barrier Islands of North Carolina." *American Scientist*, 61:152–162.

Dolan, R. 1987. "The Ash Wednesday Storm of 1962: 25 Years Later." *Journal of Coastal Research*, 3(2): ii-v.

Durán, O., and L. Moore. 2013. "Vegetation Controls on the Maximum Size of Coastal Dunes." *Proceedings of the National Academy of Sciences*, 110:17217–17222.

Environmental Data Center. 2017. "Aerial Imagery of Napatree Point Rhode Island from June 21, 2017." Digital raster dataset, University of Rhode Island, unpublished work.

Hayes, M. O., and J. Boothroyd. 1987. "Storms as Modifying Agents in the Coastal Environment." In *Beaches and Nearshore Sediments and Processes*, edited by Davis, R.A., and R. Ethington. Volume Reprint Series No. 12: 25–39. Tulsa, OK, SEPM.

Hesp, P. A. 1988. "Surfzone, Beach, and Foredune Interactions on the Australian South East Coast." *Journal of Coastal Research*, Special Issue 3:15–23.

Houser, C., and S. Hamilton. 2009. "Sensitivity of Post-Hurricane Beach and Dune Recovery to Event Frequency." *Earth Surface Processes and Landforms*, 34:613–628.

Leatherman, S. 1979. "Migration of Assateague Island, Maryland, by Inlet and Overwash Processes." *Geology*, 7:104–107.

Godfrey, P. J., and M. Godfrey. 1976. "Barrier Island Ecology of Cape Lookout National Seashore and Vicinity, North Carolina." National Park Service Scientific Monograph Series, Publication No. 9. Washington, DC: US Government Printing Service.

Montello, C., G. Bonynge, C. Damon, P. August, and J. Sassi. 2017. "Mapping Napatree Point with Small Unmanned Aerial Systems." In *The State of Napatree Report: 2017*, edited by J. Sassi. The Watch Hill Conservancy. https://thewatchhillconservancy.org/napatree/napatree-resources/.

Morton, R. A. 2008. "Historical Changes in the Mississippi-Alabama Barrier-island Chain and Roles of Extreme Storms, Sea Level and Human Activities." *Journal of Coastal Research*. 24:1587–1600.

NOAA. 2022*A*. "Annual Exceedance Probability Curves 8452660 Newport, RI." Accessed 29 December 2022. https://tidesandcurrents.noaa.gov/est/est_station.shtml?stnid=8452660.

NOAA. 2022*B*. "Annual Exceedance Probability Curves 8461490 New London, CT." Accessed 5 January 2024. https://tidesandcurrents.noaa.gov/est/curves.shtml?stnid=8461490.

NOS. 2018. "Meteorological Observations at Station 8452660 Newport, RI." Accessed 19 September 2018. https://tidesandcurrents.noaa.gov/met.html?id=8452660.

Oakley, B. A. 2020. "Response and Recovery of Welded Barrier: Napatree Point Conservation Area Following 'Superstorm' Sandy Measured Using LiDAR and Field Surveys." Presentation at the American Shore and Beach Preservation Association Annual Meeting (Online). 14 October 2020. https://asbpa.org/conferences/.

Oakley, B. A. 2021. "Storm Driven Migration of the Napatree Barrier, Rhode Island, USA." *Geosciences*, 11, 330:1–22.

Oakley, B. A., C. Murphy, M. Varney, and R. Hollis. 2019. "Spatial Extent and Volume of the Shoreface Depositional Platform on the Upper Shoreface of the Glaciated Rhode Island South Shore." *Estuaries and Coasts*, 45:1123–1142.

Psuty, N. P. 2004. "The Coastal Foredune: A Morphological Basis for Regional Coastal Dune Development." In *Coastal Dunes: Ecology and Conservation*, edited by Martinez, M., and N. Psuty. Berlin: Springer-Verlag, 11–27.

Psuty, N. P., and T. Silveira. 2011. "Monitoring Shoreline Change along Assateague Barrier Island: The First Trend Report." *Journal of Coastal Research*: Special Issue, 64:800–804.

Timmons, E. A., A. Rodriguez, C. Mattheus, and R. DeWitt. 2010. "Transition of a Regressive to a Transgressive Barrier Island due to Back-barrier Erosion, Increased Storminess, and Low Sediment Supply." *Marine Geology*, 278:100–114.

USGS. 2022. "Watch Hill Cove Tide Gage, Westerly, RI." Accessed 29 December 2022. https://waterdata.usgs.gov/monitoring-location/411838071513000/#parameterCode=00065&period=P7D.

*sUAS image of Napatree Lagoon, 2017. See Montello et al. (2017) for details.*
*Photo Credit: URI Environmental Data Center*

CHAPTER 6

# Tidal Characteristics of the Napatree Lagoon*

*Scott A. Rasmussen*
Environmental Data Center, Department of Natural Resources Science,
University of Rhode Island

## INTRODUCTION

A basic element necessary for understanding the physical dynamics of the Napatree Lagoon is an accurate measure of tidal datums relative to a fixed vertical orthometric datum (e.g., NAVD88) that is used as the reference for vertical elevation data on land (NOS, 2013). Due to spatial variability in the oceanic tidal response and in non-tidal currents, the vertical height of a particular tidal datum (e.g., mean sea level (MSL) or mean high water, or MHW relative to an orthometric datum varies geographically. To address this, a tide gauge has been deployed in the Napatree Lagoon since 24 March 2016. Field methods and results of that deployment as of 19 October 2017 are presented here.

This work is a continuation of a long-term assessment of the physical properties of the Napatree Lagoon beginning in 2014. Previously, the bathymetry of the lagoon was mapped and depths relative to MHW and MLW were calculated (Rohr *et al.*, 2014) based on NOAA tidal parameters of Watch Hill Point (analysis period 1–31 August 1962). A longer and current record of water elevations inside the lagoon was previously developed by Rasmussen (2016) for a six-month water logger deployment period. These data improved depth calculations as well as better prediction of the timing and extent of tides within the lagoon. With the addition of data collected from this longer analysis period (1.5 years), a better estimate of the tidal characteristics of the lagoon can be calculated.

## METHODS

A HOBO U20-001-04 water level logger was used to record water levels in the Napatree Lagoon. The HOBO model measures pressure and converts to water height using the HOBOware-Pro software package to within a typical accuracy of 0.3 cm (0.1 inch) (Onset, 2016). The meaning of *accuracy*, as it applies to these

---

* Originally published in 2017

sensors, is that the indicated pressure will conform to true pressure to within ± *maximum error* (expressed as equivalent depth) throughout the sensor's operating range. The logger was calibrated by the manufacturer and manually tested in an office setting prior to deployment. Measurements were programmed to record the water height above the sensor every fifteen minutes.

In order to account for atmospheric pressure, a second Hobo device (U20L-04) was deployed in a nearby tree. Measurements of the atmospheric logger were programmed to begin at the same time and frequency as the water level logger. At fifteen-minute measurement intervals, the internal memory for these devices has capacity to store data for at least six months.

Real-time kinematic (RTK) GPS measurements were performed to establish a geodetic relationship to tidal datums. This method of surveying can provide 2–5 cm (0.8–1.9 inches) of accuracy by using a reference station (base receiver) that sends position corrections to the rover receiver in real time. The corrections resolve errors in satellite signals resulting from atmospheric interferences. A base receiver was set up on a nearby brass disc benchmark (NAP-5, see Oakley, 2017), which in turn was surveyed for 4+ hours. A second benchmark (NAP-4) was used as a "check-in" to validate the accuracy of the survey (Figure 1). All elevations were recorded in NAVD88 (Geoid 12A) feet and referenced to RI State Plane NAD83 (2011) coordinates.

*Figure 1. Top, Locations of water level logger, atmospheric pressure logger, RTK GPS base receiver, and survey check-in location. Blue lines represent RTK baselines from the base receiver to the rover receiver. Bottom left, RTK GPS base station set-up. Bottom middle, HOBO survey 23 October 2016. Bottom right, HOBO resurvey 20 October 2017.*

The HOBO was placed in a PVC container, which was mounted to a 25 kg (55 pounds) cinder block with two stainless steel rods driven into the lagoon bottom. A survey target was affixed to the PVC housing cap so repeated surveys could be performed on the same location in relationship to the logger sensor (Figure 2). The HOBO is attached to the end of the housing cap with galvanized steel wire. This method provides easy access to the unit for downloading data without disturbing the submerged mount system. The distance from the target face to the end of the HOBO sensor was measured to 19.4 cm (7.64 inches) (Figure 2); therefore, -19.4 cm (-7.64 inches) are added to the RTK GPS measurement to derive the sensor elevation. Water heights above the sensor are manually checked at the time of deployment to serve as an accuracy check and to account for movement and pressure sensor drift. Additional water heights were performed periodically throughout the deployment period.

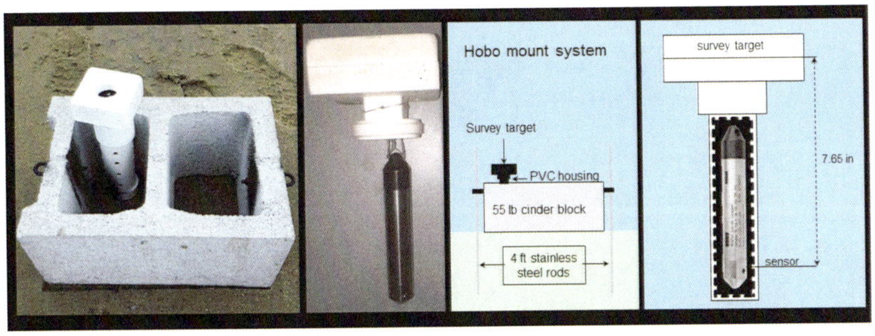


*Figure 2. Mount system for water level logger.*

Since the original deployment of the HOBO logger, data were downloaded in six-month increments: 23 October 2016; 15 April 2017; and 20 October 2017. Water depths were exported to a Microsoft Excel document via the HOBOware-Pro software package. Tidal datums such as MHW and MLW for the deployment period were extracted by an Excel command and manually verified as defined by National Ocean Service (NOS, 2003). Values were converted to NAVD88 based on averaging RTK GPS measurements performed on the mount target at the times of deployment and retrieval. Final tidal datum calculations used in this report are "standalone." They represent conditions from the time of logger deployment and have not been tied into a long-term control station.

## RESULTS

Comparisons of the Napatree Lagoon to that of the NOS station 8510560 at Montauk, New York, during the deployment period show similar tidal patterns (Figure 3) as found in the previous analysis (Rasmussen, 2016). The extent of high tide elevations has no restrictions and is slightly higher than Montauk, while low water levels are lower than Montauk but are also restricted from the narrow width and elevation of the sill of the tidal inlet connecting the lagoon to Little Narragansett Bay. The basin cannot completely drain before the overpowering tide floods the lagoon. This restriction also delays the timing of high and low tide with respect to Little Narragansett Bay. The average delay of MHW and MLW for the entire deployment period was one hour later than the Montauk tide gauge and fifty-five minutes later than at Watch Hill Point.

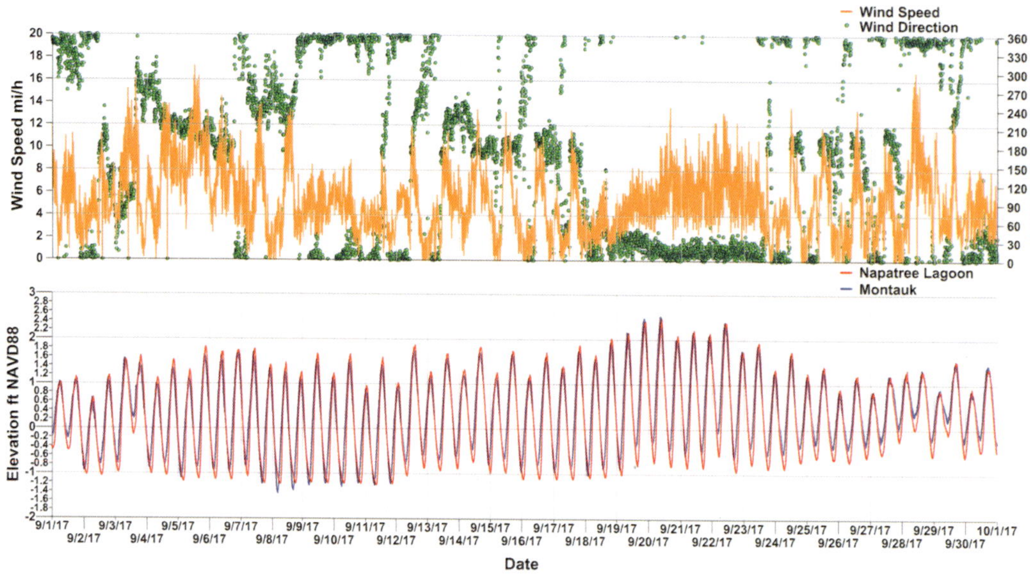


*Figure 3. Concurrent tide graph sample from Napatree Lagoon and Montauk, New York.*

The great diurnal range (GT) of the lagoon was slightly higher than Montauk for the deployment period, with 0.79 m (2.6 feet) from the lagoon and 0.75 m (2.5 feet) from Montauk. The highest and lowest elevation of the tide from the lagoon was 0.99 m (3.25 feet) and -0.80 m (-2.6 feet) NAVD88, respectively. Based on the lagoon tidal datums and previous bathymetry work (Rohr *et al.*, 2014), the greatest depth of the lagoon is approximately 0.94 m (3.1 feet) relative to MSL. The maximum depth at MLW is 0.64 m (2.1 feet) and 1.34 m (4.4 feet) at MHW (Figure 4).

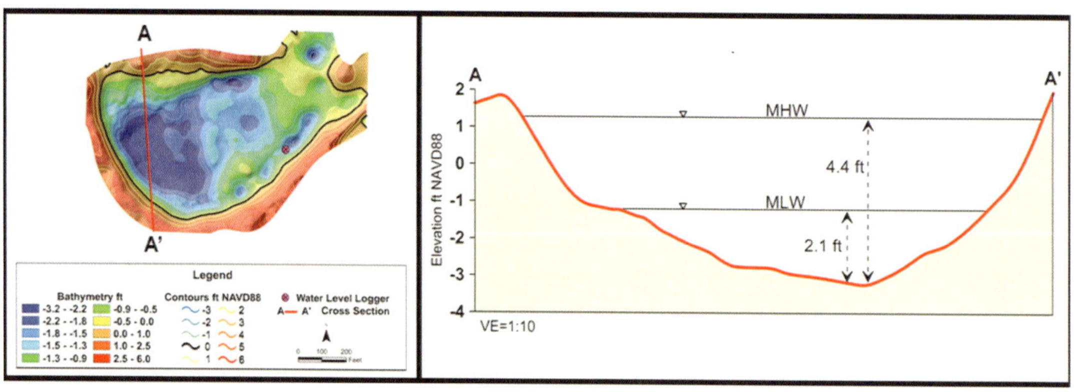


*Figure 4. Left, Bathymetry of Napatree Lagoon (adapted from Rohr et al., 2014). Right, Cross section of deepest area of Napatree Lagoon derived from bathymetric data collected in 2014. MHW and MLW values based on water level logger deployment period.*

## DISCUSSION

Observations at short-term stations are not usually sufficient for a precise independent determination of tidal datum. Results with better accuracies may be obtained by comparison with simultaneous observations at a suitable control tide station such as Montauk (NOS, 2003). At long-term National Water Level Observation Network (NWLON) stations, where the required nineteen-year dataset is available, datums are computed directly by averaging monthly mean values over the entire national tidal datum epoch (NTDE). However, water levels in estuaries can be affected by restrictive channels and dominant wind directions, and therefore, caution should be used when tying to a control station that falls outside this zone (NOS, 2014). Final tidal datum calculations used in this document are "standalone;" they represent conditions from the time of logger deployment and have not been tied into a long-term control station.

To see the effect of wind on the Napatree Lagoon, wind data (speed and direction) from a USGS weather station located in Watch Hill Cove was downloaded for the deployment period and compared to concurrent tidal data from the Napatree Lagoon and Montauk. In the previous SoN report (Rasmussen, 2016), strong north winds were thought to create a regional set-down (lower observed tides than predicted) for the coast and keep the water from entering Little Narragansett Bay and therefore limiting the water supply to enter the lagoon. However, during late September 2017, post-tropical Hurricane Jose approached New England and sat off the coast, producing a strong north wind for about seven days straight (18–24 September). It was expected that the water levels in the lagoon would be "trapped" from the strong north wind and decrease the tidal prism during this event. However, tide patterns remained virtually unaltered in the lagoon during this period (Figure 5).

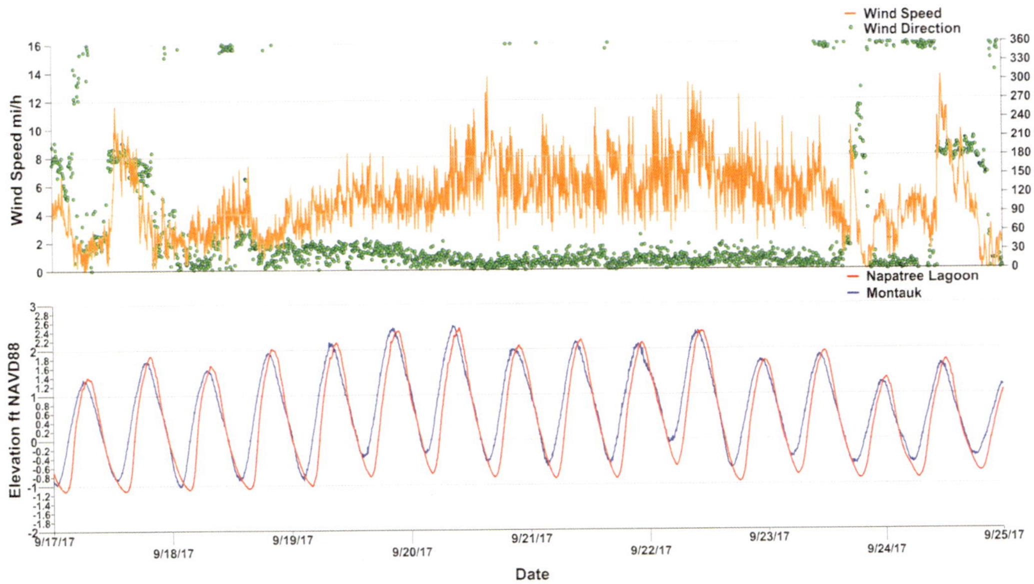


*Figure 5. Simultaneous comparison of tide levels from Napatree Lagoon and Montauk combined with wind speed and direction from a local USGS weather station during post-tropical Hurricane Jose.*

Tools such as NOAA's VDatum can be used to calculate tidal datums in areas that do not possess a long-term control station. However, consideration of a variety of factors should be made over the accuracy of these calculations, including variations in tidal range, bathymetric and coastal features, the density and proximity of nearby geodetic and tide stations used in the corrections (NOAA, 2012). NOAA (2012) further states that larger errors in the calculations are more likely to appear in upstream river environments and marshes.

## CONCLUSIONS

The calculated average tidal datum elevations and great diurnal range of the Napatree Lagoon and the

Montauk gauge were remarkably similar for the deployment period of March 2016 to October 2017 (Figure 6). Therefore, tidal extents and timing from the Montauk gauge is an adequate proxy for those in the lagoon when a one-hour delay offset is applied to Napatree Lagoon tidal conditions.

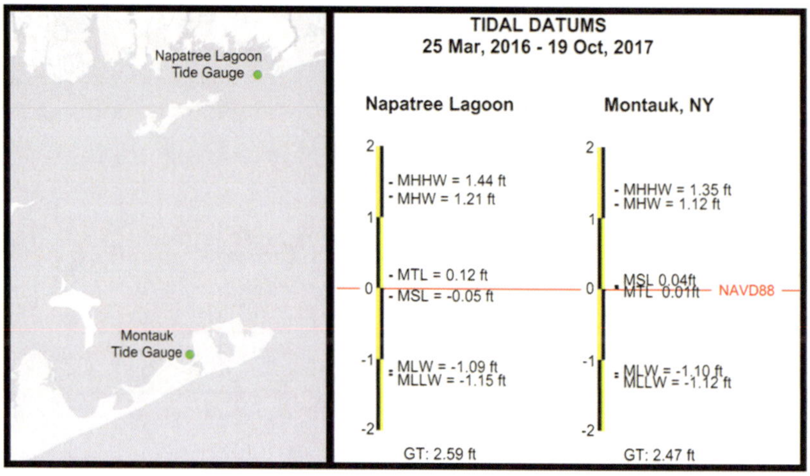


*Figure 6. Locations and tide staffs for Napatree Lagoon and Montauk (NOS, 2017A).*

Mean sea levels can be influenced by regular fluctuations in coastal temperatures, salinities, winds, atmospheric pressures, and ocean currents (NOS, 2016). It is for this reason the HOBO logger was redeployed for an additional six months to capture a full year of data and account for seasonal variability. Figure 7 shows the relationship of MSL in the lagoon for a one-year period (1 April 2016 through 31 March 2017). The average seasonal variation of MSL from this time period was 48 cm (18.9 inches). MSL's seasonal variation at Montauk for this time period was approximately 80 cm (31.5 inches, NOS, 2017*B*). These values are higher than the accepted variation in MSL for Montauk. A longer time series would average out some of the monthly outliers—for instance, the extremely low MSL of 0.28 cm (0.11 inches) below average during March 2017.

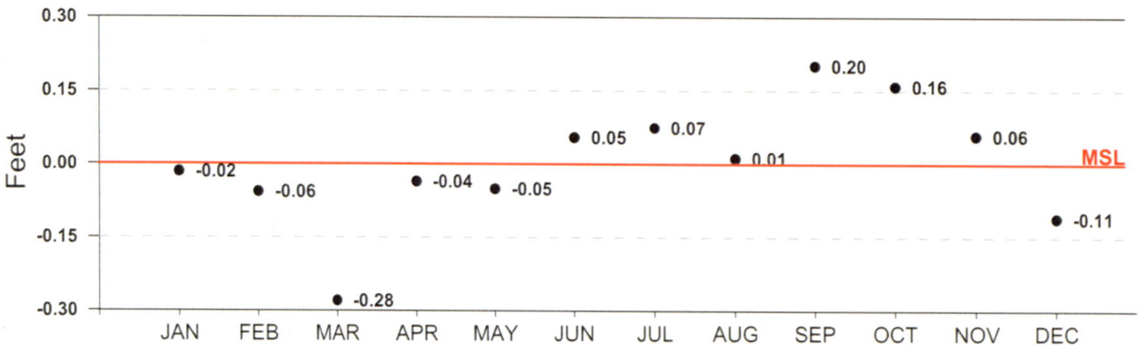


*Figure 7. Seasonal variability of MSL of the Napatree Lagoon. Values represent monthly averages above or below MSL for a one-year period.*

## ACKNOWLEDGMENTS

A grant-in-aid of research from the URI Coastal Institute allowed us to purchase the two HOBO loggers and material for the mount system. The URI EDC provided the use of the RTK GPS equipment and processing software.

## REFERENCES

Montello, C., G. Bonynge, C. Damon, P. August, and J. Sassi. 2017. "Mapping Napatree Point with Small Unmanned Aerial Systems." In *The State of Napatree Report: 2017*, edited by J. Sassi. The Watch Hill Conservancy. https://thewatchhillconservancy.org/napatree/napatree-resources/.

National Ocean Service. 2003. *Computational Techniques for Tidal Datums Handbook*. NOAA Special Publication NOS CO-OPS 2, National Ocean Service, NOAA. http://tidesandcurrents.noaa.gov/publications/Computational_Techniques_for_Tidal_Dat ums_handbook.pdf.

NOAA. 2012. "Estimation of Vertical Uncertainties in VDatum." http://vdatum.noaa.gov/docs/est_uncertainties.html.

NOS. 2013. "Tidal Datums." NOAA (website). https://tidesandcurrents.noaa.gov/datum_options.html.

NOS. 2014. "A Network Gaps Analysis for the National Water Level Observation Network–Updated Edition." NOAA Technical Memorandum NOS CO-OPS 0048, National Ocean Service, NOAA.

NOS. 2017*A*. "Datums for 8510560, Montauk, NY." NOAA (website). https://tidesandcurrents.noaa.gov/datums.html?id=8510560.

NOS. 2017*B*. Data from Montauk, New York Station (8510560). https://tidesandcurrents.noaa.gov/sltrends/.

Oakley, B. 2017. "Understanding the Short- and Long-Term Shoreline Change of NapatreeBarrier Using RTK GPS Beach Profiles and Mapping of the Last High Tide Swash: 2017 Update." In *The State of Napatree Report: 2017*, edited by J. Sassi. The Watch Hill Conservancy. https://thewatchhillconservancy.org/napatree/napatree-resources/.

Onset. 2016. https://www.onsetcomp.com/products/data-loggers/u20-001-04.

Rasmussen. 2016. "Tidal Characteristics of Napatree Lagoon." In *The State of Napatree Report: 2016*, edited by J. Sassi. The Watch Hill Conservancy. https://thewatchhillconservancy.org/napatree/napatree-resources/.

Rohr, N., J. Turenne, M. Bradley, P. August, and J. Sassi. 2014. "An Ecological Reconnaissance of the Napatree Lagoon: Physical Properties." In *The State of Napatree Report: 2014*, edited by J. Sassi. The Watch Hill Conservancy. https://thewatchhillconservancy.org/napatree/napatree-resources/.

*Grant Simmons and Marco Greco-Byrne monitoring Little Narragansett Bay at sunrise.*
*Photo Credit: The Watch Hill Conservancy*

CHAPTER 7

# Napatree Water Quality 2022

*Marco Greco-Byrne & Grant G. Simmons III*
Napatree Point Conservation Area, The Watch Hill Conservancy

## INTRODUCTION

The 2022 Napatree water quality monitoring season marks the fifteenth year of participation in the URI Watershed Watch (URIWW) program (https://web.uri.edu/watershedwatch/). In 2022, water samples were collected weekly from 6 May through 27 October, aside from the second and third week of October. The team sampled twenty-four times throughout the spring, summer, and fall.

Little Narragansett Bay (Figure 1, "Bay") is located at the mouth of the Pawcatuck River. The Wood-Pawcatuck Watershed, a drainage system that encompasses one-third of the land area of Rhode Island (775 square km, 300 square miles) empties into Little Narragansett Bay by the Pawcatuck River. We monitor three sites: off Napatree Beach on the oceanside of Napatree (ocean site), off the bayside of Napatree (bay site), and in Foster Cove (cove site). The effects of frequent rain events, increased temperatures, and nonpoint source pollution detected in the bay and cove can be compared with those from the ocean site south of Napatree Point. Furthermore, the site within Foster Cove is used to assess the effects of septic systems and lawn fertilizers coming from the residential properties surrounding the cove.

## METHODS

Once a week, between the hours of 7:00 AM and 10:00 AM, the Napatree water quality crew visited each monitoring site using a Zodiac powerboat. Samples were collected from both shallow (1 m, 3.3 feet) and deep (1 m, 3.3 feet from the bottom) depths at each site. These samples were brought back to shore where filtration was performed with a total of four filtered samples per monitoring site. The filtered samples were then kept in a plastic bag in a freezer until given to the URI Watershed Watch program (URIWW) to measure algal density (chlorophyll-A) and to assess nutrient enrichment and eutrophication. The analysis complemented water clarity analysis performed with a secchi disk and view tube (Figure 2A). The disk is lowered into the water and the depth is recorded at the last point the disk is visible. Water clarity depends on the amount of suspended sediment and plant matter present in the water column, as well as the presence of algal blooms.

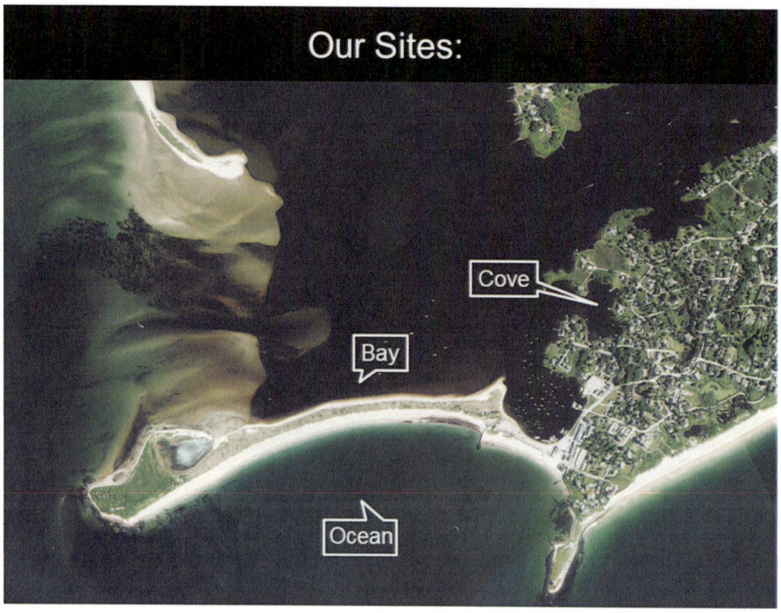


*Figure 1. Napatree water quality monitoring sites. Imagery from US Fish and Wildlife Service, 2018.*

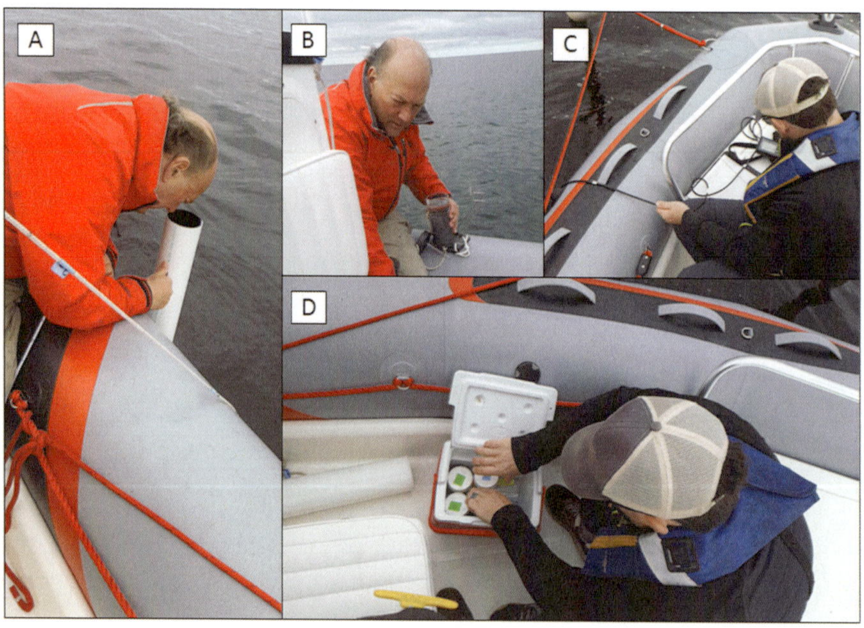


*Figure 2. Water sampling and monitoring. A, Water clarity with view tube and secchi disk. B, Water sampling device. C, Collecting data with YSI meter. D, Storing collected samples in an ice cooler.*

A YSI ProSolo ODO/CT (Figure 3) was used to measure temperature, salinity concentration, and dissolved oxygen (DO) levels at each site. Both temperature and salinity reflect the quality of mixing within the water column, while low levels of DO can indicate stress to aquatic organisms. Readings were taken from both shallow (1 m, 3.3 feet) and deep (1 m, 3.3 feet, from the bottom) at each site. This device was recalibrated every two months. The accuracy of this instrument is given in Figure 3.

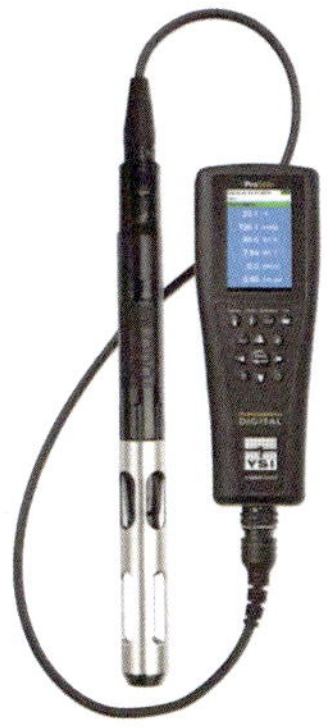

| Parameter | Resolution | Accuracy |
|---|---|---|
| Temperature | 0.1°C | ±0.2°C |
| Dissolved Oxygen | 0.01 mg/L or 0.1 mg/L, auto-adjusts based on range | ±0.1 mg/L |
| Salinity | 0.01 ppt | ±0.1 ppt |

*Figure 3. YSI ProSolo ODO/CT. Used for measuring temperature, dissolved oxygen, and salinity. Specifications obtained from the YSI ProSolo ODO/CT specification sheet found at https://www.ysi.com/File%20Library/Documents/Specification%20Sheets/W145-ProSolo-ODOCT-Specification-Sheet.pdf.*

Additional information was recorded each sampling day, including time of day, stage of tide, light and wind intensity, and the presence of recent rain events.

Once a month during the sampling season, additional water samples were taken to URIWW for further analyses. These included alkalinity and pH, which reflect acidification; total and dissolved phosphorus, ammonia, total and nitrate-nitrogen, which indicate the degree of nutrient enrichment or eutrophication; and bacteria (enterococci and fecal coliform), which is an indicator for suitability for recreational water use and shell fishing. High bacteria levels may indicate sewage contamination. Samples were taken from shallow (1 m, 3.3 feet) and deep (1 m, 3.3 feet, from the bottom) settings and placed in a cooler during transport to URIWW within one hour of sampling.

## RESULTS: 2022 DATA

### Chlorophyll-A

Samples of chlorophyll-A concentrations had not been processed by URIWW at the time this report was written. Results can be obtained upon request to URIWW in late spring 2023. To request data, contact Elizabeth Herron, RM 105 CIK, One Greenhouse Rd., Kingston, RI 02881.

### Temperature

All sites gradually increased in temperature from May through mid-July, peaking late July to early August. By late August temperatures dropped and continued to do so through early October (Figure 4). The cove generally maintained the highest temperature of all sites due to shallow water levels.

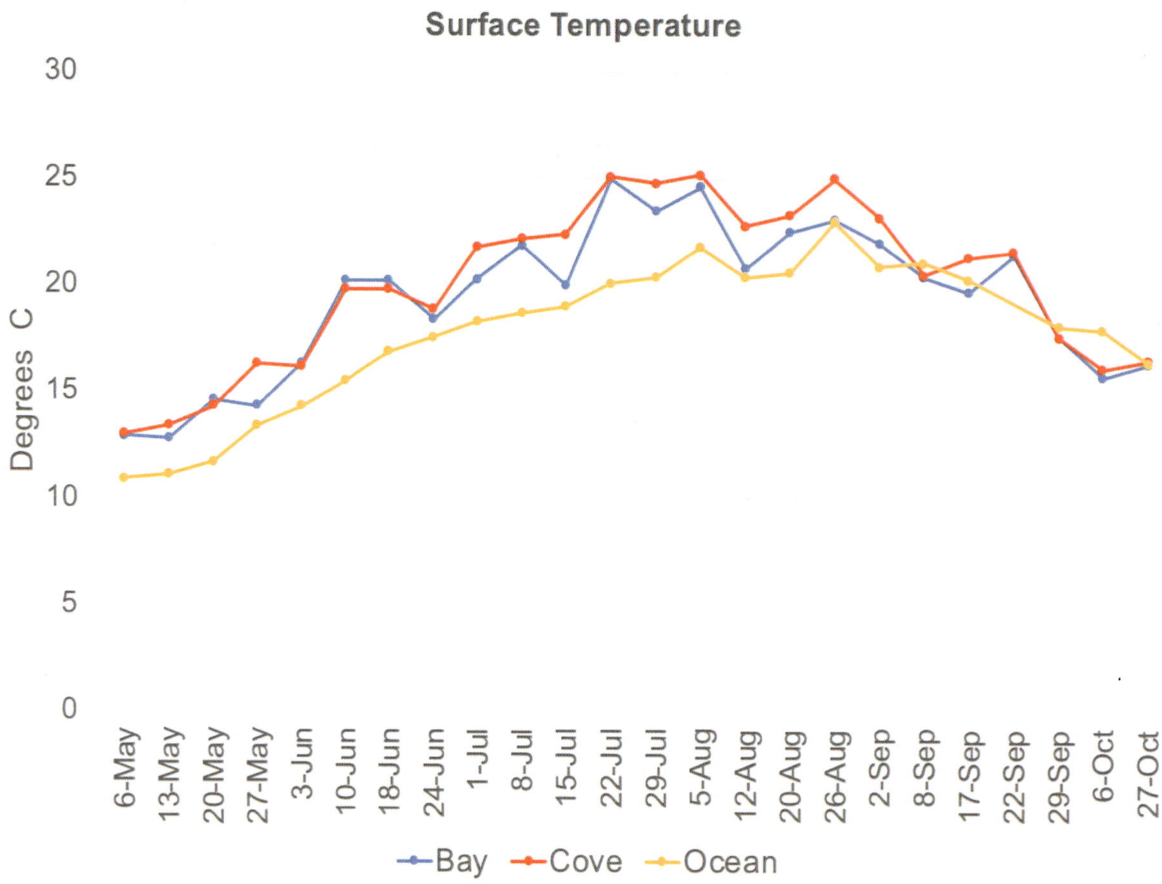


*Figure 4. Surface temperatures in degrees Celsius for all sites of the 2022 monitoring season.*

## Water Clarity

Water clarity as determined by secchi disk visible depth was variable among sites and over the season (Figure 5). The average depth reading was 5.1 m (16.7 feet) for the oceanside, 2.8 m (9.2 feet) for the bay, and 1.9 m (6.2 feet) for the cove. Both the ocean site and bay site experienced higher levels of water clarity in 2022 compared to that observed in 2021.

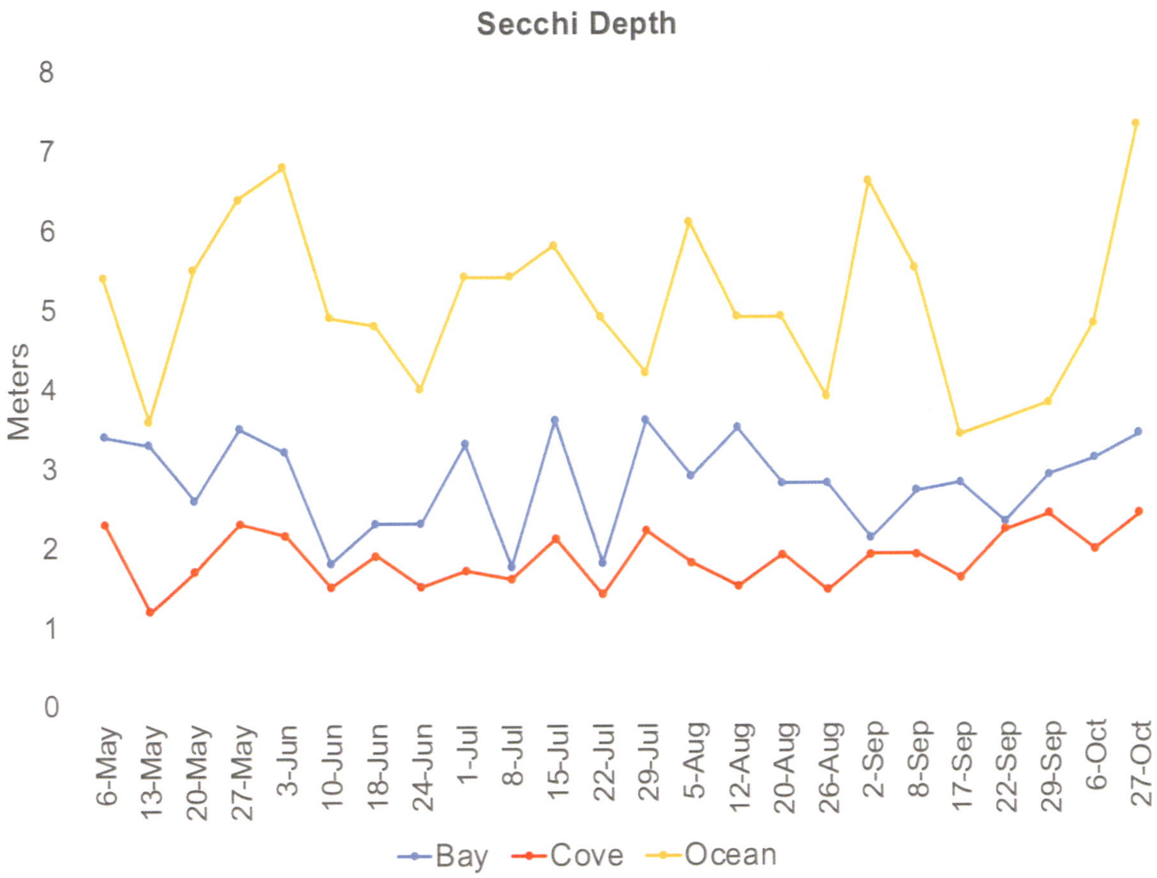


*Figure 5. Water clarity by visible secchi depth in meters of all sites during the 2022 monitoring season.*

## Salinity

The concentrations of salinity at the ocean site had the least amount of variability and the highest levels of salinity between the three sites (Figure 6). The ocean site had an average salinity of 30.7 ppt (parts per thousand) for the shallow sampling site and 30.9 ppt for the deep sampling site. The bay site saw an average salinity concentration of 29.4 ppt for the shallow sampling site and 30.2 ppt for the deep sampling site. The cove saw an average of 29.1 ppt for the shallow sampling site and 30.0 ppt for the deep sampling site.

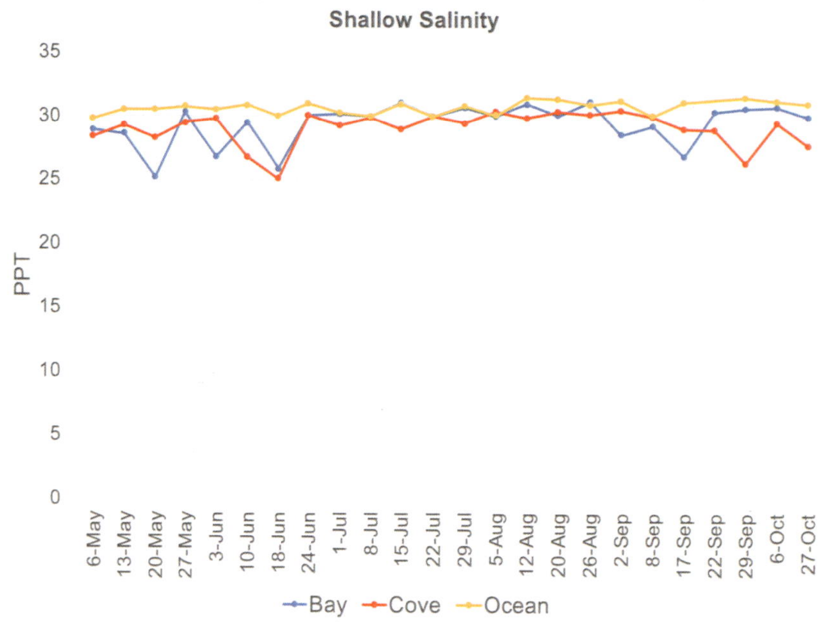


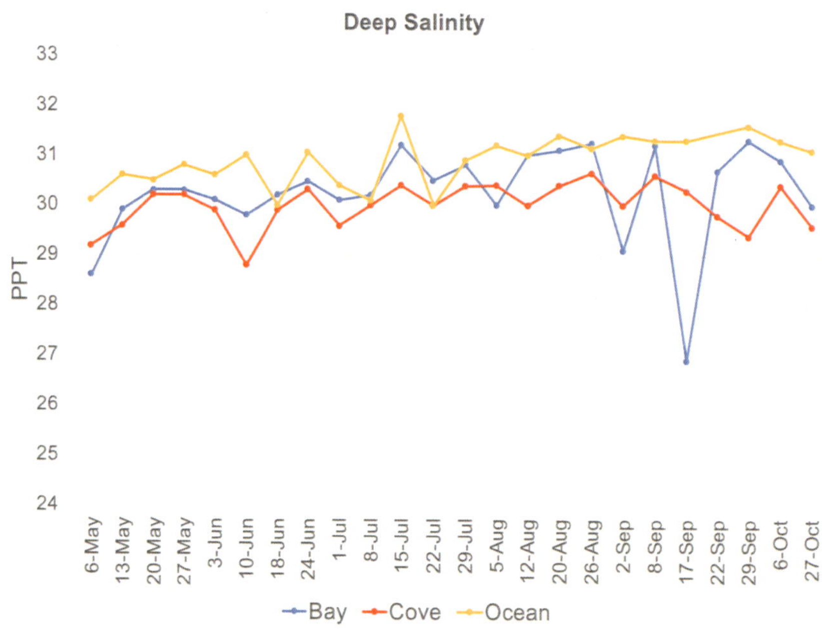


*Figure 6. Shallow and deep salinity concentrations in parts per thousand for all sites during the 2022 monitoring season.*

## Dissolved Oxygen

Dissolved oxygen (DO) levels generally remained at healthy concentrations at the shallow and deep depths for the majority of the monitoring season for all three locations (Figure 7). The cove, however, did experience potentially stressful levels of dissolved oxygen at deep levels on 12 August, 28 August, and 8 September. Aside from these three instances, dissolved oxygen remained at healthy levels averaging 8.5 mg/L one meter below the surface at the ocean site, 8.3 mg/L one meter below the surface at the bay site, and 8.0 mg/L one meter below the surface at the cove site.

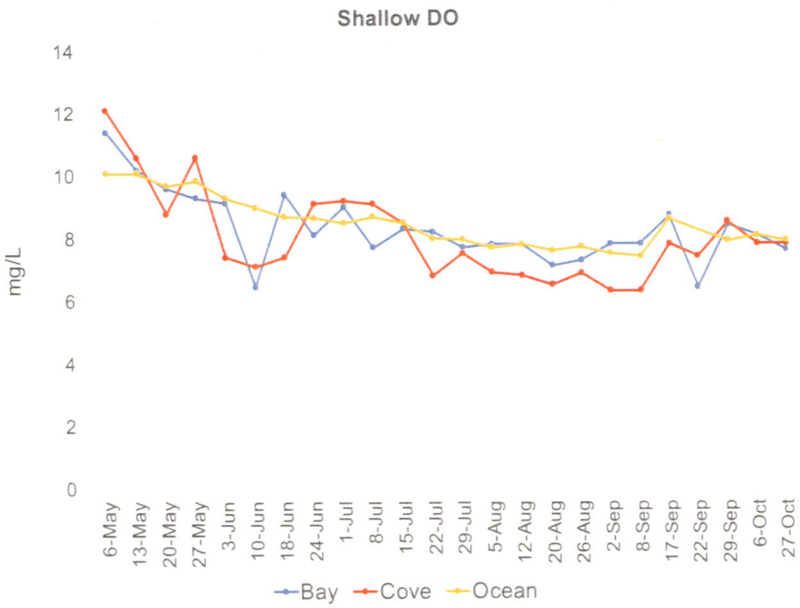


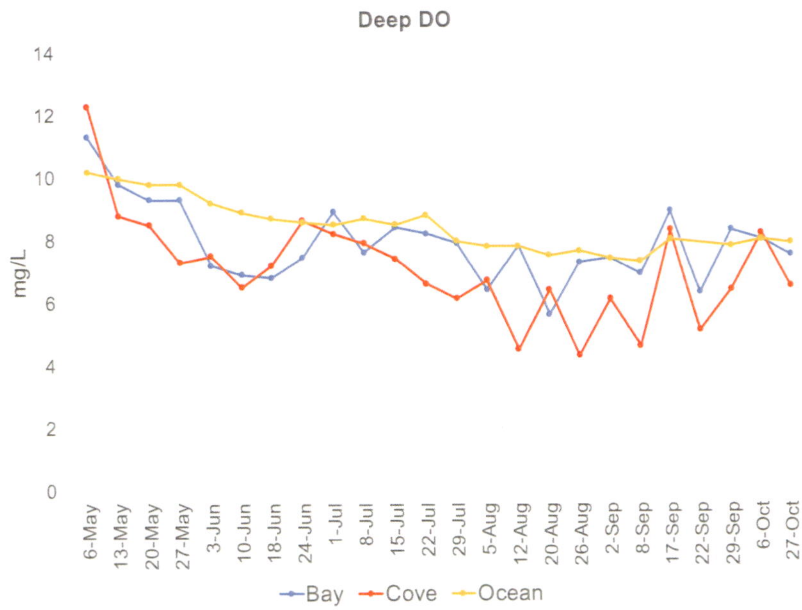


*Figure 7. Shallow and deep dissolved oxygen concentrations for all sites during the 2022 monitoring season.*

## Bacteria

Levels of enterococci remained below 60 per 100 mL throughout the sampling season, meeting the RI Department of Health standards for recreational use (Figure 8). The highest levels of enterococci were seen in October at the cove site, where they reached 42 per 100 mL. The highest levels of fecal coliform were recorded during August for all three sites. The bay site contained 84 fecal coliform per 100 mL while the cove site contained 132 fecal coliform per 100 mL. The ocean site contained 201 fecal coliform per 100 mL (Figure 9). These sites do not meet the RI Department of Environmental Management standards for shellfish waters of a geometric mean less than 14 fecal coliform per 100 mL. We are highly suspicious of the ocean site coliform counts in August. We cannot explain why the ocean site would be 2.5 times higher than the bay site and almost twice the coliform levels in Foster Cove. It is especially odd that in the months preceding and following August 2022 coliform counts were too low to measure.

| Watershed code | MONITORING LOCATION | MAY | JUNE | JULY | AUG. | SEPT. | OCT. | GEOMEAN |
|---|---|---|---|---|---|---|---|---|
| | | -- Most Probable Number of Enterococci per 100 mL -- | | | | | | |
| LN | Napatree Point - Cove | <10 | <10 | <10 | <10 | <10 | 42 | **<10** |
| LN | Napatree Point - Bayside | <10 | <10 | <10 | <10 | <10 | <10 | **<10** |
| CW | Napatree Point - Oceanside | <10 | 10 | <10 | <10 | <10 | <10 | **<10** |

*Figure 8. Enterococci concentrations for the three Napatree Point monitoring sites during the 2022 monitoring season. Source: URIWW: https://web.uri.edu/wp-content/uploads/sites/1667/22-WC-Bacteria.pdf.*

| Watershed code | MONITORING LOCATION | MAY | JUNE | JULY | AUG. | SEPT. | OCT. | GEOMEAN |
|---|---|---|---|---|---|---|---|---|
| | | ---- Most Probable Number of Fecal coliform per 100 mL -- | | | | | | |
| LNB | Napatree Point - Cove | <10 | <10 | <10 | **132** | <10 | <10 | **<10** |
| LNB | Napatree Point - Bayside | <10 | 20 | <10 | **84** | <10 | <10 | **<10** |
| CW | Napatree Point - Oceanside | <10 | <10 | <10 | **201** | <10 | <10 | **<10** |

*Figure 9. Fecal coliform concentrations for the three Napatree Point monitoring sites during the 2022 monitoring season. Source: URIWW, https://web.uri.edu/wp-content/uploads/sites/1667/22-WC-Bacteria.pdf.*

## CONCLUSIONS

The three water quality sampling sites at Napatree mostly remained at healthy, stable levels for the 2022 season. USGS maintains a record of freshwater discharge at a station on the Pawcatuck River (Figure 10). Data corresponding to the 2022 Napatree water quality season is shown in Figure 11. Typically, periods of high freshwater discharge from the Pawcatuck River are associated with lower salinity readings at the bay and cove sites. However, the summer of 2022 was an exceptionally dry year (see Cole *et al.*, 2022) and we did not see any surges in discharge as we have in previous years.

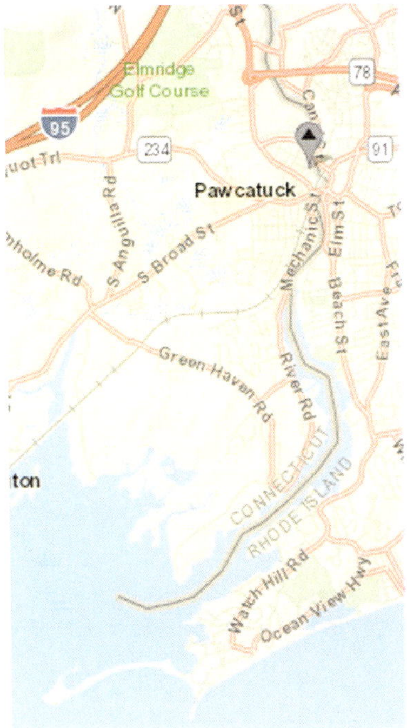


*Figure 10. Location of USGS water discharge station. Source: https://waterdata.usgs.gov/nwis/nwismap/?site_no=01118500&agency_cd=USGS.*

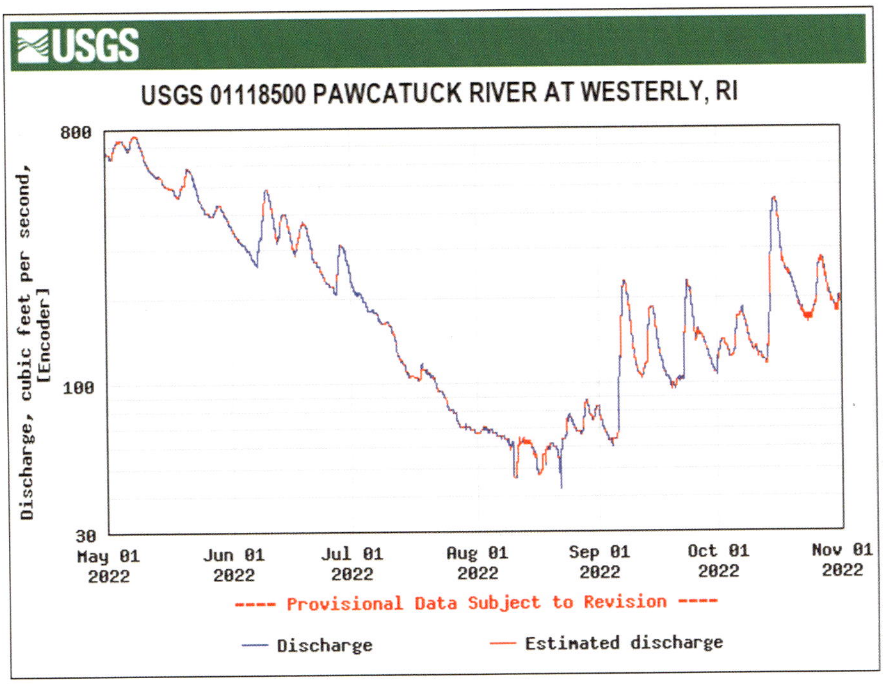


*Figure 11. Freshwater discharge from 2022 USGS Pawcatuck station. Source: https://waterdata.usgs.gov/usa/nwis/uv?01118500.*

## ACKNOWLEDGMENTS

The 2022 Napatree water quality monitoring would not have been possible without the help of Pete August and Maddison Smith who filled in when needed. In addition, Pete was kind enough to use his personal power boat to get us to the sampling sites when the Zodiak was not running.

## REFERENCES

Cole D., J. Sassi, E. Bodell, and T. Pappadia. 2022. "Visitor Activity on Napatree: 2022." In *The State of Napatree Report: 2022*, edited by Daniel Cole. The Watch Hill Conservancy. https://thewatchhillconservancy.org/napatree/napatree-resources/.

URI Watershed Watch. 2022. https://web.uri.edu/watershedwatch/.

USGS Pawcatuck River Station. 2022. https://waterdata.usgs.gov/usa/nwis/uv?01118500.

YSI. 2019. "ProSolo ODO/CT Specification Sheet." Xylem. https://www.ysi.com/File%20Library/Documents/Specification%20 Sheets/W145-ProSolo-ODOCT-Specification-Sheet.pdf.

*Dr. Nicole Rohr and Jim Turenne collect bathymetric data in the Napatree Lagoon.*
*Photo Credit: The Watch Hill Conservancy*

CHAPTER 8

# Updated Elevation (Bathymetry and Topography) Model and Subaqueous Soils of the Napatree Lagoon*

*Bryan A. Oakley,*[1] *Nicole Rohr,*[2] *James Turenne,*[3] *Michael Bradley,*[4] *Scott Rasmussen,*[4]
*Peter August*[4] *& Janice Sassi*[5]

[1] Department of Environmental Earth Science, Eastern Connecticut State University
[2] Coastal Institute, University of Rhode Island
[3] US Department of Agriculture, Natural Resources Conservation Service
[4] Environmental Data Center, University of Rhode Island
[5] Napatree Point Conservation Area, The Watch Hill Conservancy

## INTRODUCTION

The lagoon near the western end of Napatree Point is a defining feature of the Napatree Point Conservation Area (NTPCA) landscape. The lagoon marks the western end of the barrier dune ecosystem that extends to the east. It forms the eastern edge of the maritime shrubs and salt marsh of the western end of Napatree. The lagoon is a biodiversity hotspot for shorebirds and is important habitat for juvenile horseshoe crabs (*Limulus polyphemus*). The barrier spit enclosing the lagoon is one of the most dynamic areas of the NTPCA. The formation of the lagoon began following the Great New England Hurricane of 1938, when Sandy Point separated from the Napatree Barrier, then it was eroded during a hurricane in 1944 and subsequently reformed. The lagoon was fully enclosed by the migrating spit sometime between 1951 and 1962 (Figure 1). The configuration of the lagoon and its opening to Little Narragansett Bay changes with every major storm.

Despite its prominence on the Napatree landscape and its importance to biodiversity, surprisingly little data exists regarding the physical parameters, water quality, and biota of the lagoon. A long-term assessment of the lagoon began in 2014. The goals of the 2014 study were to (1) map the bathymetry, (2) survey the subaqueous soils, and (3) assess the water quality. The water quality data have been reported

* Originally published in 2018

in previous *State of Napatree Reports*. We updated the bathymetric and subaqueous soil surveys of the lagoon and report the results here.

The results of the bathymetric survey were originally reported by Rohr *et al.* (2015); however, these data have been reexamined in the context of our improved understanding of the tidal characteristics and tidal datums within the lagoon (Rasmussen, 2017). This reanalysis was conducted as part of an ongoing undergraduate research project at Eastern Connecticut State University examining the 2018 bathymetry of the lagoon.

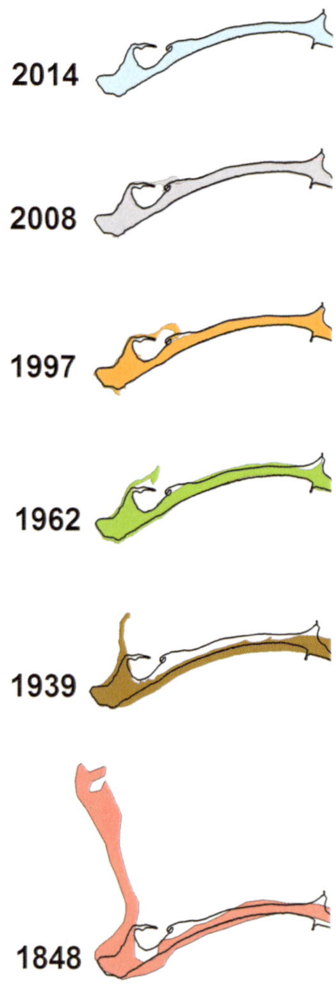


*Figure 1. Changes in the Napatree Point shoreline over time. Black outline on each map is the 2014 shoreline. Shorelines are taken from the 1848 NOAA nautical chart and RIGIS aerial photography.*

## METHODS

### Subaqueous Soil Survey

Using a 3.4 m (11-foot) Zodiac watercraft, subaqueous soil samples were taken at three locations (Figure 2) using a bucket auger lowered to a depth of approximately 50 cm (19.7 inches) into the soil. Soil properties were analyzed in the field by the State Soil Scientist of Rhode Island (James Turenne). An underwater GoPro video camera was deployed at each site to examine bottom conditions within the lagoon.

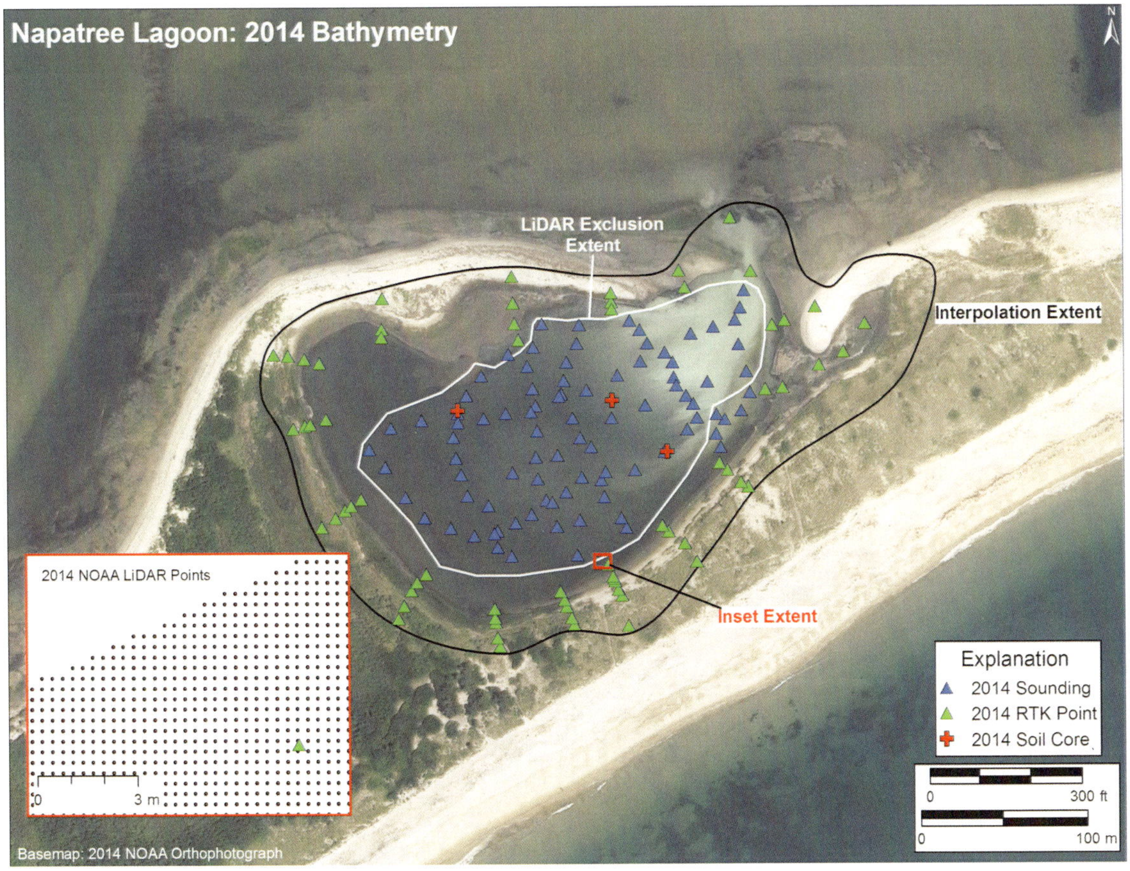


*Figure 2. Subaqueous soil, bathymetry, and land elevation sampling locations in and around the Napatree Lagoon. The black outline marks the extent of the interpolation; the area inside of the white line was removed from the LiDAR-derived points. We used the measured bathymetric points to create the interpolation inside the white zone. Inset box shows the density of the LiDAR-derived points.*

## Bathymetry

Depth measurements were taken from the Zodiac watercraft at ninety-one locations in the lagoon (Figure 2). A geographic coordinate was obtained at each site using a Garmin Oregon 650t GPS receiver capable of measuring horizontal positions to 3 m (9.8 feet). Depths were obtained with a measured survey rod that was submerged until the end hit the surface of the lagoon bottom. Collection of depth measurements occurred within a period of sixty minutes. The depth of a submerged cement block was taken with the measured rod at the beginning and end of the data collection period. An elevation measurement of the cement block was also obtained using a Trimble real-time kinematic global positioning system (RTK GPS) receiver. Depth measurements in the lagoon were measured relative to the North American Vertical Datum of 1988 (NAVD88), and elevations were adjusted to account for the change in tidal height during data collection (Figure 3). The final depth measurements are in feet relative to MLLW as determined by Rasmussen (2017).

Elevations of the area surrounding the lagoon were taken at sixty-five locations with a Trimble R8 Model 3 GPS receiver using a RTK survey protocol. The RTK base station was located at a surveyed control mark (Oakley *et al.*, 2018). The geographic coordinates for this control point were previously surveyed with RTK GPS, which measures elevation to within 6 cm (2.4 inches). All elevations were recorded in NAVD88 feet and referenced to RI State Plane NAD83 coordinates. Projection and horizontal datum conversions were done using ArcGIS software and the "NAD_1983 to WGS_1984_5 Transformation" parameter.

The elevation of the area surrounding the lagoon, including the intertidal and subtidal flats, were extracted from a digital elevation model (DEM) derived from aerial LiDAR data collected by NOAA in 2014 and downloaded from the Rhode Island Geographic Information System. The DEM was converted to a point file with a 0.3 m (1 foot) spacing (Figure 2, inset) using the ESRI ArcMap Raster to Point tool. The lagoon bathymetric depth measurements, the RTK elevations, and the LiDAR-derived points of the land adjacent to the lagoon were merged into a single mass point dataset using the ArcGIS MERGE utility. The merged point file was converted into a continuous surface using the Natural Neighbors interpolation within the ESRI ArcGIS Spatial Analyst extension. The resulting raster was clipped to an outline of the lagoon area derived from 2013 imagery. The portions of the interpolation >0.9 m (3 feet) MLLW were excluded using the Set Null function within ESRI ArcMap Raster Calculator tool. A simple hillshade model of the raster was created to better illustrate the bathymetry within the lagoon.

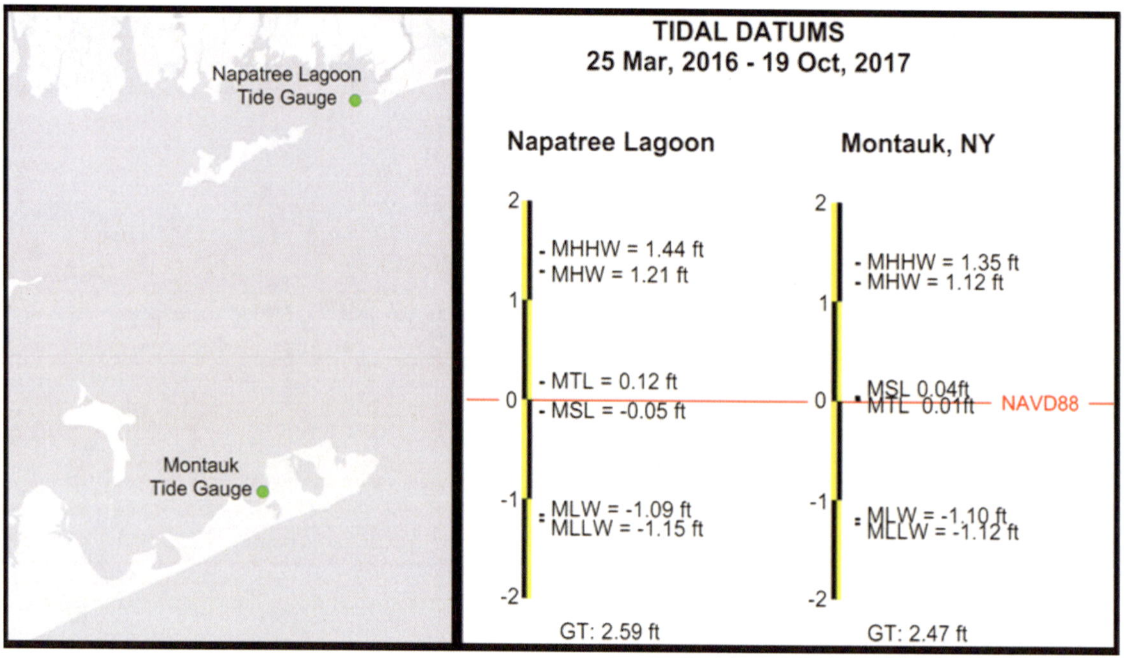


*Figure 3. Tidal datums for the Napatree Lagoon and Montauk, New York, tide gauges between 25 March 2016 and 19 October 2017 (from Rasmussen, 2017).*

## RESULTS

### Subaqueous Soil Survey

The subaqueous soils of the lagoon are Nagunt sand, a medium-to-fine sand with a thin fluid sand surface layer. Based on examination of the 1939 aerial photographs of Napatree, the whole lagoon is judged to be located on the surge platform of the Napatree Barrier. A detailed description of this particular soil type can be found at https://soilseries.sc.egov.usda.gov/OSD_Docs/N/NAGUNT.html. A mixture of green and red macroalgae was observed throughout the lagoon water column, and thick mats composed of filamentous algae and blade-forming *Ulva* were present on the lagoon bottom. Decomposing algal mats were prolific in the sandy areas of the exposed intertidal zone surrounding the lagoon.

## Bathymetry

A bathymetric map for the lagoon is shown in Figure 4. The deepest part of the lagoon is on the eastern portion near the 2014 inlet, reaching a maximum depth of -1 m (-3.3 feet) (MLLW). Overall, the lagoon is relatively flat, with a slight eastward and southern slope. Average depths within the lagoon basin are largely between 0.3–0.6 m (1–2 feet) below MLLW.

Cross sections of the bathymetry (Figures 5, 6) show relatively steep sides, particularly on the northern and southern shoreline, consistent with both a slip face of the washover fans deposited into the lagoon as both the Napatree Barrier and lagoon barrier spit migrate during storms as well as adjacent to inlet channels.

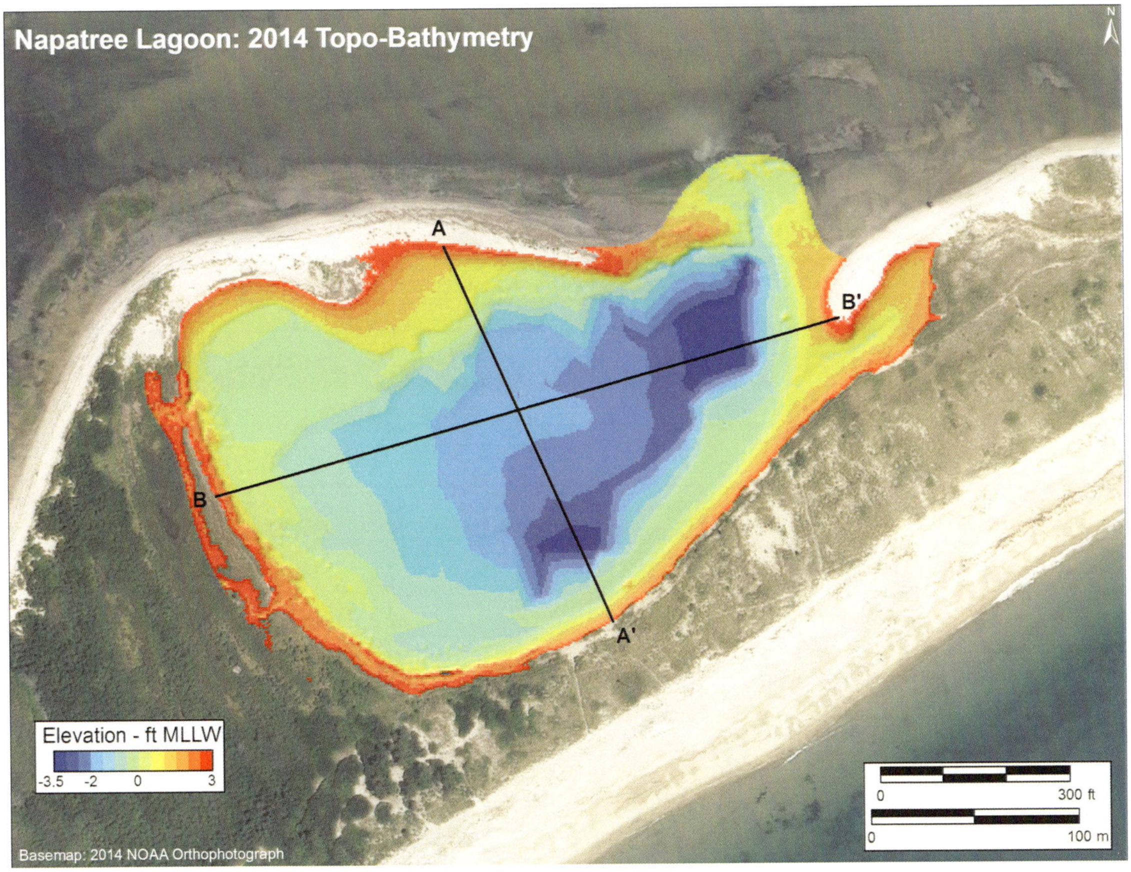


*Figure 4. Topography and bathymetry of the Napatree Lagoon and adjacent areas. Elevations are expressed in feet, relative to MLLW as calculated by Rasmussen (2017). Solid black lines represent cross sections A-A' (Figure 5) and B-B' (Figure 6).*

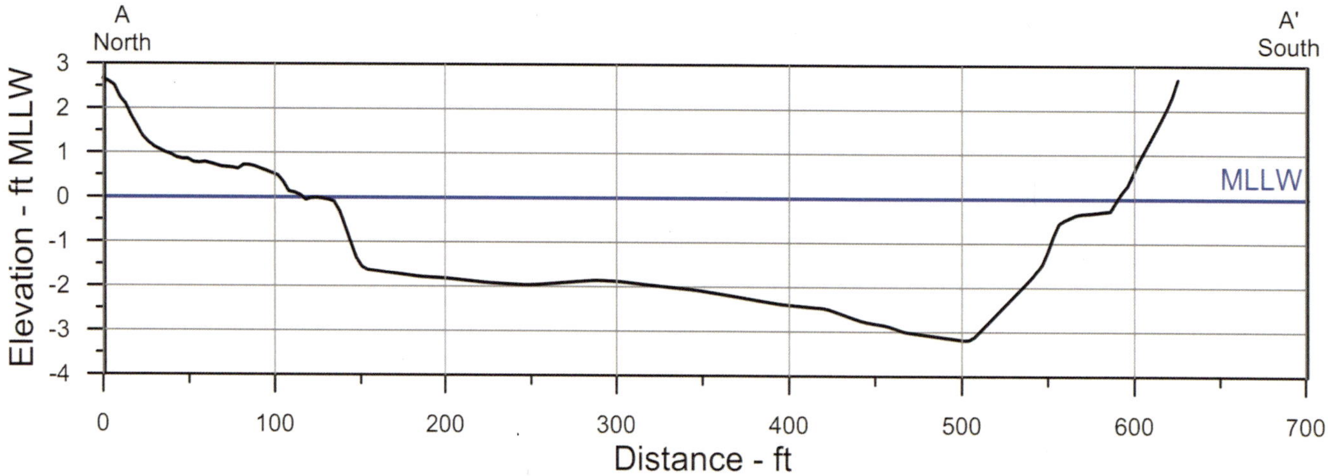


*Figure 5. North to south cross section of the lagoon bathymetry. See Figure 4 for the location of cross section. Vertical exaggeration approximately 25:1.*

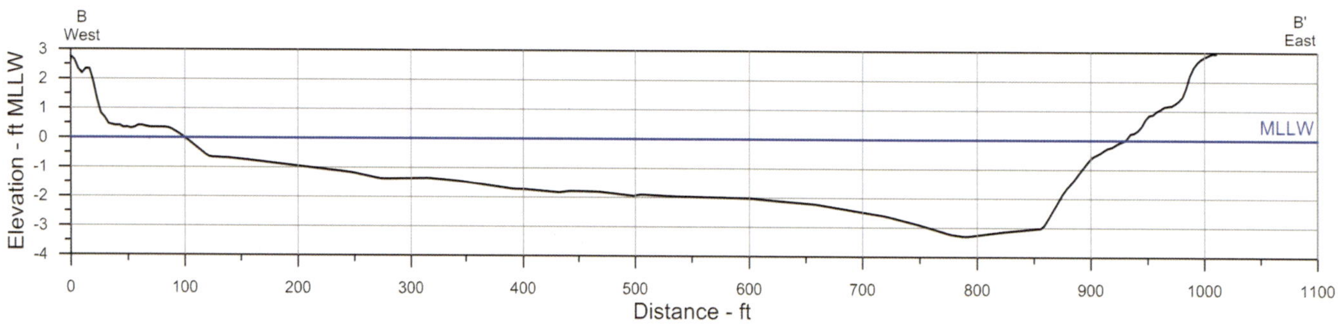


*Figure 6. North to south cross section of the lagoon bathymetry. See Figure 4 for the location of cross section.*

The configuration of the lagoon, with its current opening to Little Narragansett Bay and its depth profile, will likely change with the next large storm as the lagoon spit migrates south via overwash and washover fan deposition. The spit has continued to lengthen since 2014, and the inlet position has migrated east. New bathymetric data were collected in 2018 and will be compared to the updated 2014 bathymetry in a forthcoming study. The protocol we used here was relatively simple, and the merging of LiDAR-derived points with RTK and bathymetric data appears to be a valid approach to creating more detailed topobathymetric models of areas like the Napatree Lagoon if field surveys are conducted within close proximity of aerial LiDAR surveys.

## ACKNOWLEDGMENTS

We are grateful for the URI Coastal Institute, the URI EDC, and the USDA NRCS for allowing their staff to assist in this project. The technical advice we received from Drs. Candace Oviatt and Sandra Whitehouse are most appreciated. Scott Rasmussen assisted in converting rod depths to tidally adjusted NAVD88 depths in the original 2014 analysis.

## REFERENCES

Oakley, B.A. 2018. "Understanding the Short- and Long-Term Shoreline Change of Napatree Barrier Using RTK GPS Beach Profiles and Mapping of the Last High Tide Swash: 2018 Update." In *The State of Napatree Report: 2018*, edited by J. Sassi. The Watch Hill Conservancy. https://thewatchhillconservancy.org/napatree/napatree-resources/.

Rasmussen, S. 2017. "Tidal Characteristics of the Napatree Lagoon." In *The State of Napatree Report: 2017*, edited by J. Sassi. The Watch Hill Conservancy. https://thewatchhillconservancy.org/napatree/napatree-resources/.

Rohr, N., J. Turenne, M. Bradley, P. August, and J. Sassi. 2015. "An Ecological Reconnaissance of the Napatree Lagoon: Physical Properties." In *The State of Napatree Report: 2015*, edited by J. Sassi. The Watch Hill Conservancy. https://thewatchhillconservancy.org/napatree/napatree-resources/.

*Eastern Connecticut State University students Madeline Varney (left) and Alyson Augenstein (right) measuring the depth of the lagoon.*

*Photo Credit: The Watch Hill Conservancy*

CHAPTER 9

# Bathymetric Model of the Napatree Lagoon*

*Alyson Augenstein, Madeline Varney, Jennifer Croteau & Bryan A. Oakley*
Department of Environmental Earth Science, Eastern Connecticut State University

## INTRODUCTION

The Napatree Lagoon, located within the Napatree Point Conservation Area (NTPCA) in Watch Hill, Rhode Island, is an important habitat for shorebirds, including Piping Plovers *(Charadrius melodus)*, American Oystercatchers *(Haematopus palliatus)*, Least Terns *(Sternula antillarum)*, and various finfish and shellfish. A bathymetric survey of the lagoon was originally conducted in 2014 (Rohr *et al.*, 2015); this project updated the bathymetric model of the lagoon. Understanding the geomorphology of the lagoon is a crucial piece of information to aid conservation efforts and provide insight on how the lagoon may continue to change form in response to future storms. The lagoon was mapped over three days in May and June 2018 to get a detailed bathymetric model.

## METHODS

### Field Methods

Elevation measurements were recorded for the portion of the lagoon that were walkable (wadable) using a Trimble R10 real-time kinematic global positioning system (RTK GPS) mounted on a modified SECO all-terrain rover rod "Big Wheel" (Figure 1). Continuous topo mode within the R10 was used to continuously collect points every 1 m (3.3 feet). Lead line soundings deployed via kayak were used to measure deeper areas of the lagoon (cover photograph). The geographic position of each sounding was mapped with a Bad Elf GNSS Surveyor Differential GPS. The RTK GPS measurements collected in the field were recorded relative to NAVD88 and later corrected to report all depths relative to mean lower low water (MLLW) using previously established tidal datums (Rasmussen, 2017). The lead line soundings were corrected using a tide staff deployed during the survey. In addition to the bathymetric mapping, the cross section of the lagoon inlet was measured using the Trimble RTK GPS.

---

* Originally published in 2019

## Laboratory Methods

ESRI ArcMap Shapefiles of the RTK GPS measurements and the lead line soundings were merged using the ArcGIS MERGE utility to create a complete bathymetric dataset of the Napatree Lagoon. Adjacent points collected with the different techniques were compared and found to be in good agreement. The bathymetric points were interpolated with the ESRI ArcGIS Spatial Analysist extension into a continuous surface using the natural neighbor algorithm. The resulting raster of the lagoon was clipped to an extent determined by walking the last high tide swash of the lagoon using the ESRI ArcGIS Raster Processing tool. Volumes of water within the lagoon were calculated using the Functional Surface tool within the ESRI ArcGIS 3D Analyst extension. The lagoon water volume was calculated above both mean higher high water (MHHW) and MLLW.

*Figure 1. Eastern Connecticut State University students Alyson Augenstein (left), Jennifer Croteau (middle), and Madeline Varney (right) mapping the bathymetry of the flood-tidal delta near the lagoon entrance using the survey wheel and RTK GPS.*

# RESULTS

## Lagoon Bathymetry

A total of 8,425 points were measured with the Trimble R10 RTK GPS on 17–18 May 2018. These surveys spanned the majority of the Napatree Lagoon and the flood-tidal delta (Figure 2). An additional 94 points were collected with the lead line soundings (Figure 2). A bathymetric model of the lagoon is shown in Figure 3 interpolated with 0.25 m (0.82 feet) contours. The spatial extent of the lagoon was calculated at both MLLW (30,000 $m^2$ or 7.4 acres) and MHHW (40,800 $m^2$ or 10.1 acres). The volume of the Napatree Lagoon was determined to be 10,250 $m^3$ (13,406 cubic yards) at MLLW and 39,300 $m^3$ (51,402 cubic yards) at MHHW, resulting in a tidal prism (volume of water exiting/entering the lagoon between high tide and low tide) of 29,000 $m^3$ (37,931 cubic yards). The northern portion of the lagoon is the shallowest area with many portions intertidal. The eastern half of the lagoon is the deepest, with the maximum depth recorded at -0.66 m (2.2 feet) MLLW. The cross-sectional area of the inlet (measured relative to mean sea level) is 5 m$^2$ (53.8 square feet) (Figure 4).

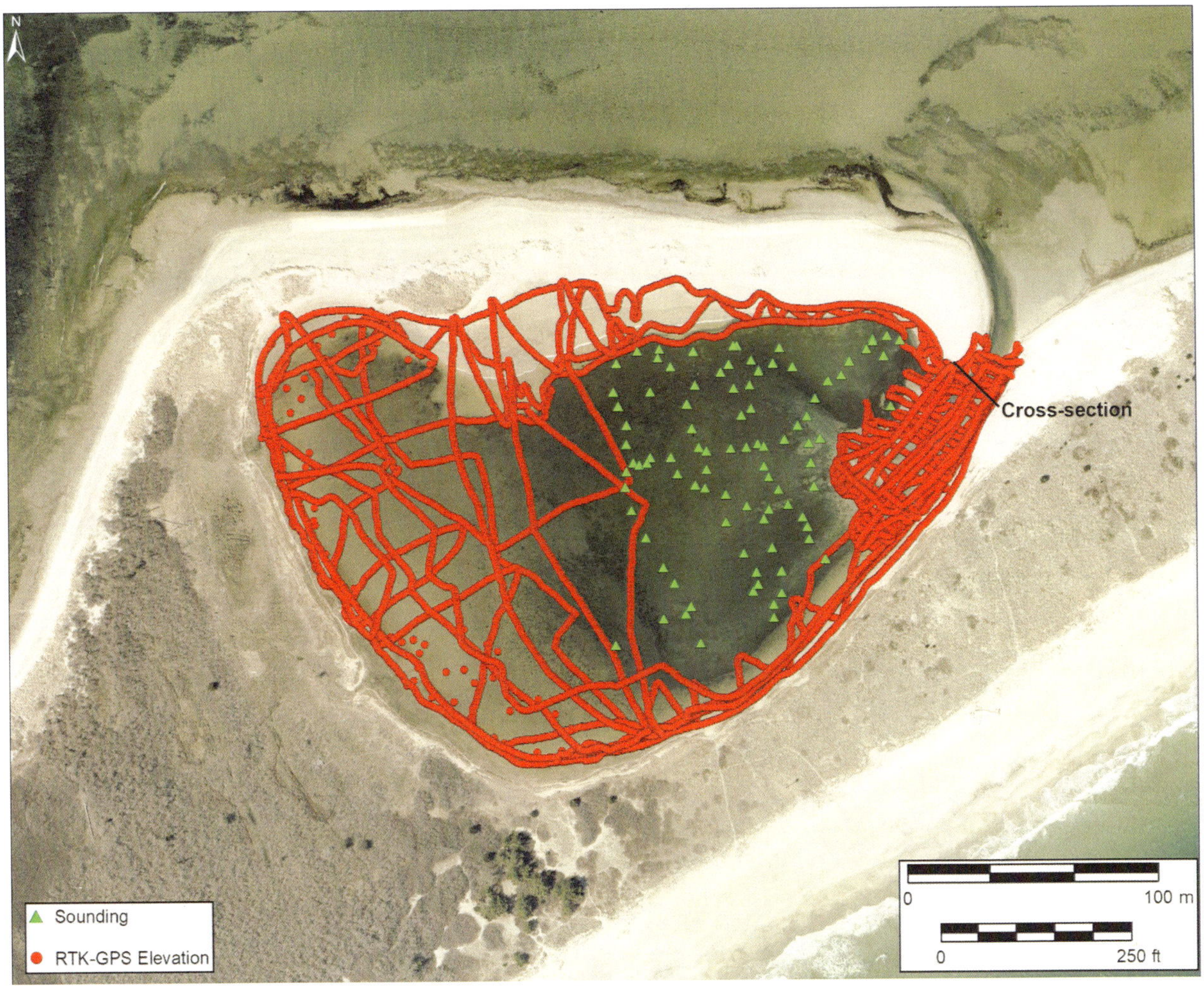


*Figure 2. Distribution of points collected using the Trimble RTK GPS and survey wheel (red circles) and lead line soundings (green triangles). The black line across the lagoon inlet is shown in Figure 4. Basemap is the spring 2018 RIDOT aerial photography.*

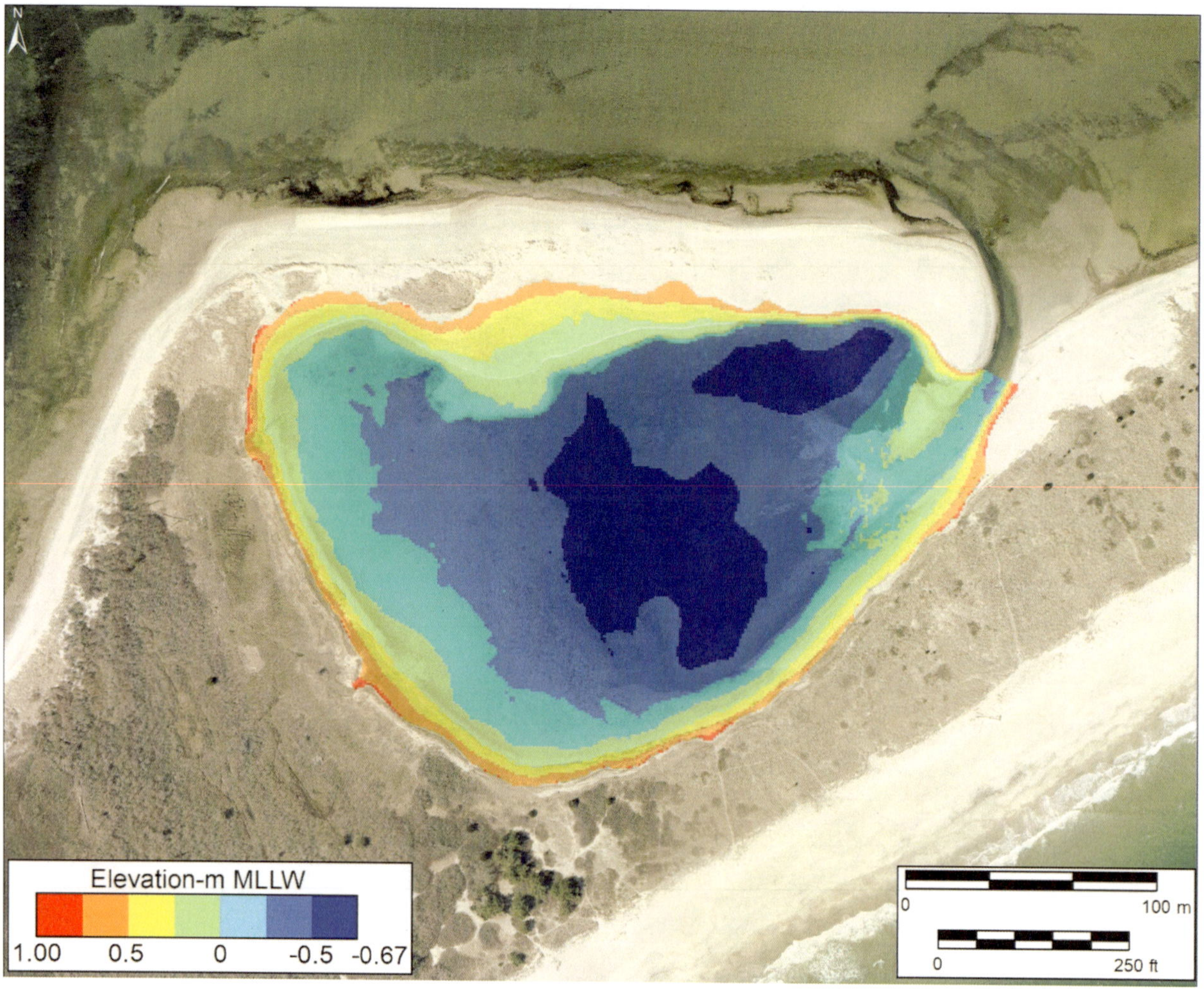


*Figure 3. Natural neighbor interpolated surface of the lagoon bathymetry based on the points shown in Figure 2. Contour interval is 0.25 m (0.82 feet). Basemap is the spring 2018 RIDOT aerial photography.*

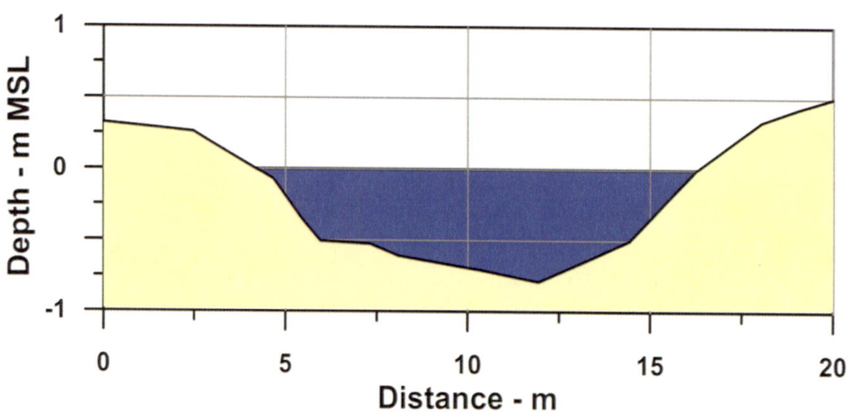


*Figure 4. Cross section across the Napatree Lagoon inlet. See Figure 2 for the location.*

## DISCUSSION

Overall, the Napatree Point Lagoon is shallow, matching the interpreted history that the lagoon is formed on the subtidal surge platform of Sandy Point (prior to breaching during the 1938 hurricane) and the Napatree Barriers. The central basin and cove northwest of the inlet are the deepest portions of the lagoon, with a maximum depth of 0.66 m (2.16 feet) below MLLW. Given the differences in data density, point distribution and vertical uncertainty of the 2014 survey (Oakley, 2018; Rohr *et al.*, 2015), quantitative differences between the surveys were not calculated. Qualitatively, the lagoon has shoaled and decreased in area since the last survey conducted in 2014 through a combination of tidal delta deposition and spit migration. The methods presented here can be replicated in the future and the lagoon should be remapped, particularly after a storm that overwashes the lagoon spit.

The size of a tidal inlet is related to the tidal prism. The O'Brien relationship quantifies this relationship and predicts the minimum cross-sectional area of an inlet below mean sea level (O'Brien, 1931; 1969) (Equation 1).

$$\text{Equation 1: } A = 7.49 \times 10^{-4} * P^{0.86}$$

Where:

$A$ = Inlet Area ($m^2$)

$P$ = Tidal Prism ($m^3$)

The minimum cross-sectional area of the inlet channel based on O'Brien (1931, 1969) was determined to be 5.3 $m^2$ (57.0 square feet), which is very similar to the cross section measured in 2018 with RTK GPS (5 $m^2$, 53.8 square feet). This suggests the inlet will remain open; however, changes in the sediment budget (i.e., an addition of sand) can alter inlet dynamics, potentially further reducing the inlet size. As the lagoon continues to shoal and decrease in area (and tidal prism) in the upcoming years, the inlet size and tidal prism are likely to decrease, and ultimately could close the lagoon off from Little Narragansett Bay.

## ACKNOWLEDGMENTS

Fieldwork has been assisted by Pete August and Janice Sassi (The Watch Hill Conservancy) and Jennifer Croteau and Madie Varney (Eastern Connecticut State University undergraduate students). This project has been supported by grants from The Watch Hill Conservancy and the Eastern Connecticut State University Exemplary Program Fund. Pete August provided constructive edits of this document. We thank Scott Rasmussen for establishing tidal characteristics of the lagoon in a previous project.

## REFERENCES

Oakley, B. A., N. Rohr, J. Turenne, M. Bradley, S. Rasmussen, P. August, and J. Sassi. 2018. "Updated Elevation (Bathymetry and Topography) Model and Subaqueous Soils of the Napatree Lagoon." In *The State of Napatree Report: 2018*, edited by J. Sassi. The Watch Hill Conservancy. https://thewatchhillconservancy.org/napatree/napatree-resources/.

O'Brien, M. O. 1931. "Estuary Tidal Prism Related to Entrance Areas." *Civil Engineering*, 1:738–739.

O'Brien, M.O. 1969. "Equilibrium Flow Areas of Inlets on Sandy Coasts." *Journal of Waterways*, Harbors Division, 95:43–52.

Rasmussen, S. 2017. "Tidal Characteristics of Napatree Lagoon." In *The State of Napatree Report: 2017*, edited by J. Sassi. The Watch Hill Conservancy. https://thewatchhillconservancy.org/napatree/napatree-resources/.

Rohr, N., J. Turenne, M. Bradley, P. August, and J. Sassi. 2014. "An Ecological Reconnaissance of the Napatree Lagoon: Physical Properties." In *The State of Napatree Report: 2014*, edited by J. Sassi. The Watch Hill Conservancy. https://thewatchhillconservancy.org/napatree/napatree-resources/.

*Launching the sUAS from Little Narragansett Bay.*
*Photo Credit: The Watch Hill Conservancy*

CHAPTER 10

# Mapping Napatree Point with Small Unmanned Aerial Systems*

*Christina Montello,*[1] *Gregory Bonynge,*[2] *Christopher Damon,*[2] *Peter August*[2] *& Janice Sassi*[3]
[1] Coastal Fellow, University of Rhode Island
[2] Department of Natural Resources Science, University of Rhode Island
[3] Napatree Point Conservation Area, The Watch Hill Conservancy

## INTRODUCTION

Detailed and timely aerial photographs support the Napatree scientific team and inform natural resource management decisions. The availability of up-to-date statewide orthophotographs is becoming increasingly scarce. Severe technical and legal limitations are placed upon the use of free publicly accessible imagery that is easily available from commercial sources such as Google Maps and Microsoft Bing. Small unmanned aerial systems (sUAS, a.k.a. drones) are quickly becoming a very practical means for safely collecting and producing timely aerial photographs of focused areas.

During 2017, an sUAS was used to collect aerial photographs and motion video of Napatree Point and surrounding waters in an inaugural partnership between The Watch Hill Conservancy, the Watch Hill Fire District, and the University of Rhode Island Environmental Data Center (EDC). The imagery developed in this project provided information for three primary scientific initiatives that were part of this year's project: (1) creating a seamless orthorectified image that serves as a basis for future habitat mapping and land cover change assessments on Napatree; (2) measuring the extent of a prominent patch of eelgrass off Napatree in Little Narragansett Bay; and (3) exploring methodologies for unobtrusively documenting boat density and anchoring locations.

## METHODS

Initial planning for this summer's fieldwork began in fall 2016. After a series of informal conversations with colleagues at URI and staff representing the Napatree Point Conservation Area, the EDC submitted a request to the Watch Hill Fire District Parks Commission to fly an sUAS for scientific research purposes

* Originally published in 2017

during summer 2017. This request was approved in May 2017. The EDC then sought approval to conduct sUAS operations at Napatree from URI risk management representatives which granted permission in June 2017. Since these operations were conducted under the supervision of a Federal Aviation Administration (FAA) certified remote pilot under FAA Part 107 rules, no FAA approvals were needed. With the necessary paperwork lined up, we were clear to begin planning and conducting sUAS flights over Napatree. Details of the equipment, software, and protocols we used in this study can be found in Damon (2016).

### Equipment

A DJI Phantom 4 Pro sUAS with a 20-megapixel camera was used for this project. Flight planning and control was accomplished with the Map Pilot App for iOS. Post-flight image processing was done using Pix4Dmapper software.

Camera and flight parameters were as follows:

| Camera | Mission |
|---|---|
| • Autofocus<br>• JPEG output format<br>• Sunny white balance | • Maximum altitude 122 m (400 feet)<br>• Maximum forward speed 9.6 m/s<br>• Forward/lateral overlap 70–75%<br>• Camera angle 90 degrees (straight down) |

### Regulatory Compliance

The aircraft was flown from a 5.5 m (18-foot) center console boat anchored off the northern shore of Napatree in Little Narragansett Bay (see cover photo). On board was an FAA certified remote pilot in command who oversaw all flight operations and monitored the area for potential hazards. We also had on board a dedicated spotter who kept the aircraft in visual contact. During all flights, air traffic radio communications at the Westerly Airport were monitored with a portable VHF aircraft radio.

### Baseline Aerial Photography

The first priority was flying the entire length of Napatree to collect baseline aerial photographs for the entire conservation area. Three missions were flown on the morning of 21 June 2017. A total of 651 photographs (3 GB) were collected. These images were subsequently processed at the EDC using Pix4Dmapper software. This step matches adjacent photos to create a single, seamless image that was geographically registered to real-time kinematic global positioning system-derived ground control points for the study area that were provided by Dr. Bryan Oakley.

### Osprey Nesting Site Inspection

While in the field on 21 June 2017, we also collected photographs of the nest platform surface of an Osprey nesting site. The nest, located south of the Napatree Lagoon, has been abandoned by Osprey for the last two summers. Prior to that, Osprey had successfully nested on the platform for many years. The purpose of the drone-based imaging was to see if there were any obvious defects in the nest surface. Such an inspection would have been otherwise impossible for Napatree staff to do without assistance from an sUAS.

### Submerged Aquatic Vegetation Mapping

Rhode Island's largest bed of eelgrass (*Zostera marina*, 33.1 ha, 82 acres) lies north of Napatree in Little

Narragansett Bay. We took advantage of a remarkable opportunity to map the eelgrass bed with the sUAS at the exact same time as the site was photographed by an airplane commissioned by US Fish and Wildlife Service (USFWS) to obtain imagery to support eelgrass mapping in eastern Long Island Sound. Keystone Aerial Survey and Quantum Spatial had an aircraft flying at 5,181 m (17,000 feet) on 28 June 2017. The true-color orthophotography this project produced had a pixel resolution of 25 cm (10 inches). At the time the USFWS aircraft flew the area, we were anchored on the south edge of the eelgrass patch obtaining imagery with the sUAS. The drone imagery had a spatial resolution of 3 cm (1.2 inches). The purpose of this experiment was to compare both sources of imagery for their use in eelgrass mapping.

### Visitor Boat Counts

Napatree Point is a very popular summer weekend destination for recreational boaters. A typical summer weekend will have up to five hundred boats anchored off Napatree's bayside shoreline. Napatree Naturalists typically conduct boat counts from shore. We explored the efficacy of using sUAS imagery to count boats at anchor and map their locations.

## RESULTS AND DISCUSSION

### Baseline Aerial Photography

A single seamless digital image of Napatree was successfully developed. The 1.5 GB image file has a spatial resolution of 3 cm (1.2 inches). It has already been put to use; Dr. Bryan Oakley has used the image for 2017 shoreline mapping (see chapter 6, Oakley, 2017). Examples of the imagery are shown in Figure 1. As an independent test of the accuracy of the georeferencing of the image, we compared seventeen clearly identifiable locations between the sUAS orthomosaic and the 2008 RIGIS aerial imagery. The average (± 1 standard deviation) error between the two datasets was 118.8 ± 51.1 cm (46.8 ± 20.1 inches).

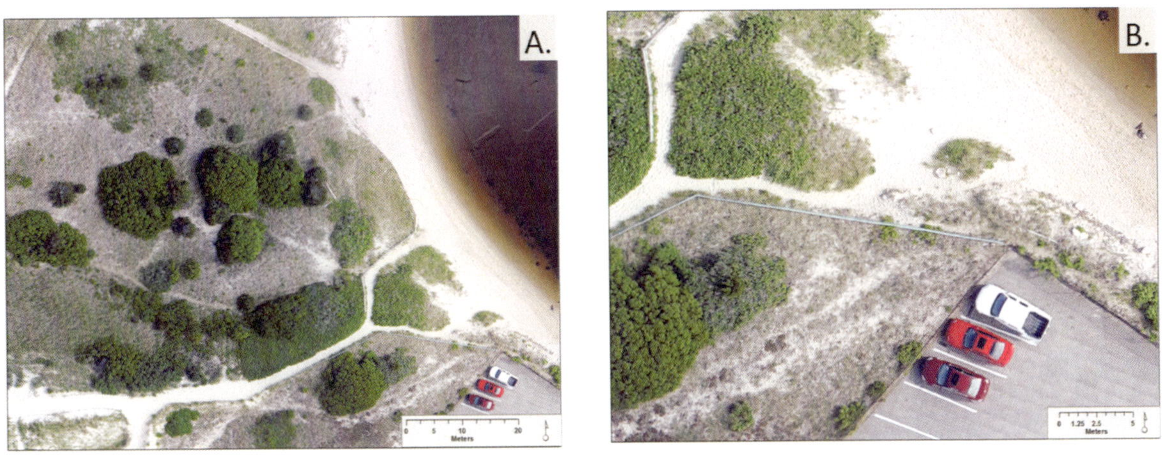


*Figure 1. Sample sUAS imagery of the entrance to the Napatree Point Conservation Area at two scales to show the resolution and detail of the data.*

### Osprey Nest Inspection

The sUAS photographs showed that the Osprey nesting surface was caked with mud that likely accumulated over the years from vegetation brought to the nest, debris from prey, etc. (Figure 2). An impermeable mud surface would render the nest unsuitable; pooling in the nest during rain events would be a stress on eggs and nestlings. This information was the basis for a decision by Napatree managers to replace the Osprey nest platform with a new one and to add other nesting platforms for Osprey to Napatree.

*Figure 2. Inspection of Osprey nest. A, The top of the Osprey nest platform as the sUAS approaches the site. B, Close-up of nest surface. Note what appears to be caked mud on the right (south) side of the nest surface.*

## Submerged Aquatic Vegetation

The sUAS imagery of the eelgrass bed was incredibly detailed, especially compared to the imagery obtained by the USFWS aircraft. The USFWS aircraft imagery was excellent; some of the best we have seen (Figure 3). Conditions for image acquisition were perfect in terms of sun glint and water clarity. At the fringes of the Napatree eelgrass patch were bright green regions which might represent macroalgae (Figure 4). Distinguishing eelgrass from macroalgae has proven to be very difficult using traditional aerial photography. The sUAS imagery might have great value in discerning these two habitat types in small areas. We anticipate furthering those observations next summer (2018). It is important to note that due to technical issues with our equipment, the relatively small area of eelgrass (~ 32 ha, 80 acres) surveyed required many hours in the field and many different sUAS flights to get full image coverage. The USFWS aircraft obtained complete near-shore imagery from Watch Hill, Rhode Island, to New London, Connecticut, in a matter of hours.

The NOAA Office for Coastal Management (formerly the Coastal Services Center) has developed a comprehensive document to assist with the planning of benthic habitat mapping (https://coast.noaa.gov/data/digitalcoast/pdf/bhm-guide.pdf). As compared to traditional imagery collection using airplanes, sUAS imagery can be especially sensitive to environmental parameters, such as turbidity, sea state, and sun glint, due to the low flight altitudes. To maximize image quality several guiding factors were considered at the time of collection:

- imagery was captured within two hours of low tide to maximize SAV visibility;
- the sun angle was between 20 and 40 degrees to maximize light penetration and minimize glint;
- water clarity must be high with little-to-no turbidity; and
- the sea state must be calm to minimize waves and ripples, which accentuate glint and degrade visibility.

Processing the sUAS imagery for the eelgrass bed was a challenge. Not surprisingly, Pix4D had a difficult time mosaicking adjacent images of plain sea surface. There are very few visual "tie points" for the imagery to find on pairs of images to establish a spatial relationship. Furthermore, the random texture of the eelgrass makes it very difficult for the software to mosaic images.

Finally, since there are essentially no high-accuracy ground control points in the eelgrass bed, it was challenging to georeference the sUAS imagery. The USFWS imagery was georeferenced (UTM Zone 18), so that was a very helpful resource.

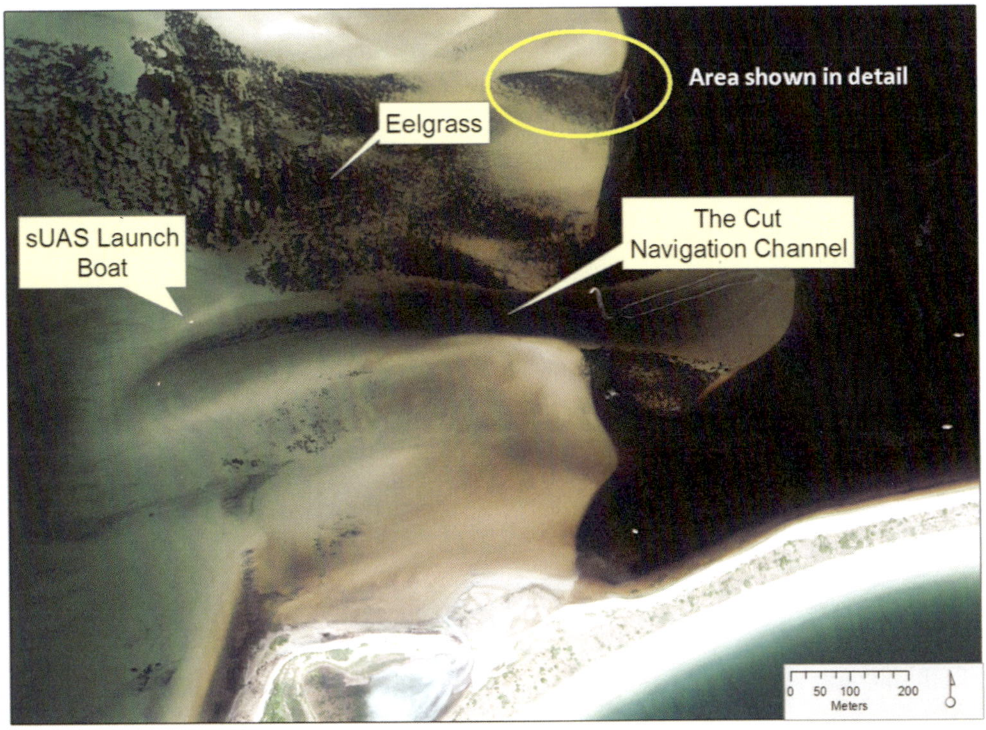


*Figure 3. Eelgrass bed off Napatree. Background image, USFWS, flight elevation 5,181 m (17,000 feet) on 28 June 2017. Small white dot is the anchored boat carrying the sUAS team flying the drone aircraft at the same time.*

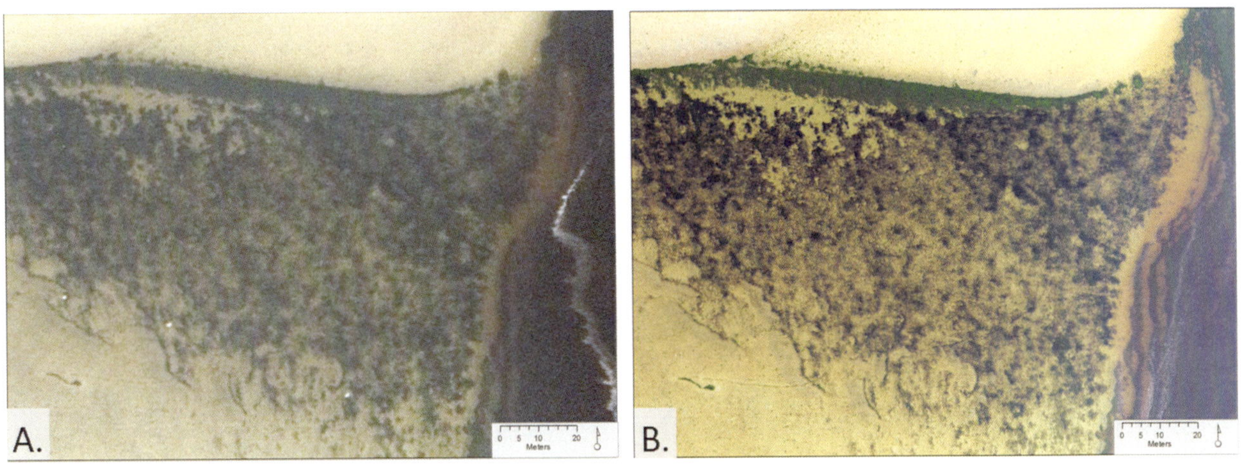


*Figure 4. Comparison of the resolution of airplane-based imagery. A, Pixel resolution of 25.4 cm (10 inches) with the sUAS imagery. B, Pixel resolution of 3 cm (1.2 inches) at a patch of eelgrass (see Figure 3 for location). The bright green band in B is possibly a swath of macroalgae.*

### Visitor Boat Count

Using the sUAS to map boat locations proved much harder than we thought. Since we were careful not to deliberately fly over people on boats (per FAA regulations), all of our imagery had to be obtained at oblique angles rather than straight down (planimetric). The oblique images are difficult, and sometimes impossible, to mosaic. Furthermore, images that contained only open water were unable to be mosaicked with adjacent photos.

The Esri software Full Motion Video was used to attempt to integrate the video imagery captured by the sUAS with a static basemap of the Napatree shoreline. Our initial results were not very successful. We must conduct additional research to develop a workflow for using sUAS technology over featureless areas such as open water.

## CONCLUSIONS

For small project areas, especially terrestrial sites, sUAS technology proved to be an effective and efficient way to collect large amounts of very detailed imagery. This was especially true when high quality ground control points were available.

Our mission planning software (Map Pilot App for iOS) performed exceptionally well for mission planning and flight control. Data collection for the Napatree Point Conservation Area went smoothly. For the 34.8 ha (86-acre) Napatree Point, four flights were required to gather complete coverage, as each flight was limited by battery reserves to approximately twenty minutes. Pix4Dmapper was very effective in creating a seamless image mosaic of land surfaces. Water surfaces were trickier to work with because of the paucity of tie points between adjacent images.

Launching and landing from a watercraft worked very well. We were relatively unobtrusive and could position the boat depending on where we were imaging. Monitoring Westerly Airport communications by VHF radio was very helpful to know when aircraft were approaching the area. Having a visual spotter on board being attentive to the position of the sUAS was essential. On two occasions the sUAS had to be temporarily dropped to a lower altitude to ensure we would be well-away from a US Coast Guard aircraft performing systemic search and rescue flights in the same area.

Mapping water surfaces and boats is a challenge and will require further experimentation to establish a reliable workflow.

## ACKNOWLEDGMENTS

This project was supported by the United States Geological Survey, AmericaView, and the University of Rhode Island Renewable Resources Extension Act Program. Napatree Point Conservation Area Park Commissioner Grant G. Simmons III enthusiastically supported our work and helped secure permission to work at Napatree. The DJI Phantom was purchased with funding from a grant-in-aid of research from the University of Rhode Island Coastal Institute.

## REFERENCES

Damon, C. 2016. *Small Unmanned Aircraft (SUA) Pilot Project. The USAID/Ghana Sustainable Fisheries Management Project (SFMP).* Narragansett, RI: Coastal Resources Center, Graduate School of Oceanography, University of Rhode Island and Environmental Data Center. GH2014_SCI019_EDC. https://www.crc.uri.edu/download/GH2014_SCI019_EDC_FIN508ok.pdf.

Oakley. 2017. "Understanding the Short and Long-Term Shoreline Change of Napatree Barrier Using RTK GPS Beach Profiles and Mapping of the Last High Tide Swash: 2017 Update." In *The State of Napatree Report: 2017*, edited by J. Sassi. The Watch Hill Conservancy. https://thewatchhillconservancy.org/napatree/napatree-resources/.

*Photo Credit: Janice Sassi, The Watch Hill Conservancy*

SECTION FOUR

# ANIMALS

*Rey Larsen conducting a bird survey on Napatree.*
*Photo Credit: The Watch Hill Conservancy*

CHAPTER 11

# Birds of the Napatree Point Conservation Area

*Reynold T. Larsen*
Napatree Point Conservation Area, The Watch Hill Conservancy

## INTRODUCTION

Napatree Point is a bird diversity hotspot. It has been designated a Globally Important Bird Area by the National Audubon Society due, in part, to its importance as a migratory stopover site (Wells *et al.*, 2005). Napatree is a breeding and foraging site for a number of endangered species, including Piping Plovers, Least Terns, Osprey, and American Oystercatchers.

I have been monitoring birds on Napatree Point since 1963.[*] With the advent of the Cornell University Lab of Ornithology data system—eBird—I have been logging my observations to this public database for approximately fifteen years. The species list presented here are the birds I have observed on Napatree.

## METHODS

With an assistant, I would begin a survey at the entrance to Napatree and slowly travel to the Napatree Lagoon in a utility terrain vehicle (UTV). A driver would operate the vehicle while I scanned the bayside shore and Little Narragansett Bay with binoculars. From the lagoon, I would walk the perimeter to the bayside shore, traverse the western tip of Napatree. At the Pines (Figure 1), I would examine the tree canopy for birds. I would continue the survey along the oceanside shoreline by UTV or on foot. A survey route was 4.8 km (3 miles) in length and would typically take 1.5 to 2 hours. All species seen or heard were noted, and I recorded the number of individuals of each species that were observed. Results were typically posted to eBird within twenty-four hours of a survey.

My records were supplemented by observations made by Christopher Raithel, the Rhode Island Department of Environmental Management non-game biologist, and Kevin Rogers, Napatree Naturalist.

The taxonomy follows the 2022 AOU checklist (Chesser *et al.*, 2022).

---

* *Note from the Editors:* For his more than sixty years of monitoring birds on Napatree, The Watch Hill Conservancy awarded Dr. Reynold Larsen the prestigious 2022 Whittemore Award for exceptional service. In 2021, the Rhode Island Natural History Survey designated Dr. Larsen the recipient of the Rhode Island Distinguished Naturalist Award.

*Figure 1. Bird survey route on Napatree.*

## RESULTS

I recorded 307 species of birds from fifty-five families on Napatree (Table 1). This list, as well as photographs of each species, can be found on The Watch Hill Conservancy website.

## BIRD SPECIES RECORDED ON NAPATREE POINT

*Bird species I observed in the Napatree Point Conservation Area up to August 2023. All records are documented in the eBird database (https://ebird.org/). The two exceptions are: species detected by other observers and recorded in eBird (#), and species detected by other observers, especially Christopher Raithel, and not recorded in eBird (*).*

## ANSERIFORMES:

### Anatidae—Ducks, Geese, and Swans

Greater White-fronted Goose*, *Anser albifrons*
Snow Goose, *Anser caerulescens*
Brant, *Branta bernicla*
Canada Goose, *Branta canadensis*
Mute Swan, *Cygnus olor*
Wood Duck, *Aix sponsa*
Gadwall, *Mareca strepera*
American Wigeon, *Mareca americana*
American Black Duck, *Anas rubripes*
Mallard, *Anas platyrhynchos*
Blue-winged Teal, *Spatula discors*
Northern Shoveler, *Spatula clypeata*
Northern Pintail, *Anas acuta*
Green-winged Teal, *Anas crecca*
Canvasback, *Aythya valisineria*
Redhead, *Aythya americana*
Ring-necked Duck, *Aythya collaris*
Greater Scaup, *Aythya marila*
Lesser Scaup, *Aythya affinis*
King Eider, *Somateria spectabilis*
Common Eider, *Somateria mollissima*
Harlequin Duck, *Histrionicus histrionicus*
Surf Scoter, *Melanitta perspicillata*
White-winged Scoter, *Melanitta deglandi*
Black Scoter, *Melanitta americana*
Long-tailed Duck, *Clangula hyemalis*
Bufflehead, *Bucephala albeola*
Common Goldeneye, *Bucephala clangula*
Barrow's Goldeneye*, *Bucephala islandica*
Hooded Merganser, *Lophodytes cucullatus*
Common Merganser, *Mergus merganser*
Red-breasted Merganser, *Mergus serrator*
Ruddy Duck, *Oxyura jamaicensis*

## PODICIPEDIFORMES:

### Podicipedidae—Grebes

Pied-billed Grebe, *Podilymbus podiceps*
Horned Grebe, *Podiceps auritus*
Red-necked Grebe, *Podiceps grisegena*
Eared Grebe, *Podiceps nigricollis*

## COLUMBIFORMES:

### Columbidae—Pigeons and Doves

Rock Pigeon, *Columba livia*
White-winged Dove, *Zenaida asiatica*
Mourning Dove, *Zenaida macroura*

## CUCULIFORMES:

### Cuculidae—Cuckoos, Roadrunners, and Anis

Yellow-billed Cuckoo, *Coccyzus americanus*
Black-billed Cuckoo, *Coccyzus erythropthalmus*

## CAPRIMULGIFORMES:

### Caprimulgidae—Goatsuckers

Common Nighthawk*, *Chordeiles minor*

## APODIFORMES:

### Apodidae—Swifts

Chimney Swift, *Chaetura pelagica*

### Trochilidae—Hummingbirds

Ruby-throated Hummingbird, *Archilochus colubris*

## GRUIFORMES:

### Rallidae—Rails, Gallinules, and Coots

Clapper Rail, *Rallus crepitans*
Virginia Rail, *Rallus limicola*
Sora, *Porzana carolina*
American Coot, *Fulica americana*

## CHARADRIIFORMES:

### Recurvirostridae—Stilts and Avocets

American Avocet, *Recurvirostra americana*

### Haematopodidae—Oystercatchers

American Oystercatcher, *Haematopus palliatus*

### Charadriidae—Lapwings and Plovers

Black-bellied Plover, *Pluvialis squatarola*
American Golden-Plover, *Pluvialis dominica*
Wilson's Plover, *Charadrius wilsonia*
Semipalmated Plover, *Charadrius semipalmatus*
Piping Plover, *Charadrius melodus*
Lesser Sand Plover[#], *Charadrius mongolus*
Killdeer, *Charadrius vociferus*

### Scolopacidae—Sandpipers, Phalaropes, and Allies

Spotted Sandpiper, *Actitis macularius*
Solitary Sandpiper, *Tringa solitaria*
Greater Yellowlegs, *Tringa melanoleuca*
Willet, *Tringa semipalmata*
Lesser Yellowlegs, *Tringa flavipes*
Upland Sandpiper, *Bartramia longicauda*
Whimbrel, *Numenius phaeopus*
Hudsonian Godwit, *Limosa haemastica*
Marbled Godwit, *Limosa fedoa*
Ruddy Turnstone, *Arenaria interpres*
Red Knot, *Calidris canutus*
Ruff, *Calidris pugnax*
Stilt Sandpiper, *Calidris himantopus*
Curlew Sandpiper, *Calidris ferruginea*
Red-necked Stint[#], *Calidris ruficollis*
Sanderling, *Calidris alba*
Dunlin, *Calidris alpina*
Purple Sandpiper, *Calidris maritima*
Baird's Sandpiper, *Calidris bairdii*
Least Sandpiper, *Calidris minutilla*
White-rumped Sandpiper, *Calidris fuscicollis*
Buff-breasted Sandpiper, *Calidris subruficollis*
Pectoral Sandpiper, *Calidris melanotos*
Semipalmated Sandpiper, *Calidris pusilla*
Western Sandpiper, *Calidris mauri*
Short-billed Dowitcher, *Limnodromus griseus*

Long-billed Dowitcher, *Limnodromus scolopaceus*
American Woodcock, *Scolopax minor*
Terek Sandpiper#, *Xenus cinereus*
Wilson's Phalarope, *Phalaropus tricolor*
Red-necked Phalarope, *Phalaropus lobatus*
Red Phalarope, *Phalaropus fulicarius*

## Stercorariidae—Skuas and Jaegers

Pomarine Jaeger*, *Stercorarius pomarinus*
Long-tailed Jaeger, *Stercorarius longicaudus*

## Alcidae—Auks, Murres, and Puffins

Dovekie*, *Alle alle*
Common Murre*, *Uria aalge*
Razorbill, *Alca torda*
Black Guillemot*, *Cepphus grylle*

## Laridae—Gulls, Terns, and Skimmers

Black-legged Kittiwake, *Rissa tridactyla*
Sabine's Gull, *Xema sabini*
Bonaparte's Gull, *Chroicocephalus philadelphia*
Laughing Gull, *Leucophaeus atricilla*
Franklin's Gull, *Leucophaeus pipixcan*
Ring-billed Gull, *Larus delawarensis*
Herring Gull, *Larus argentatus*
Iceland Gull, *Larus glaucoides*
Lesser Black-backed Gull, *Larus fuscus*
Glaucous Gull, *Larus hyperboreus*
Great Black-backed Gull, *Larus marinus*
Sooty Tern*, *Onychoprion fuscatus*
Bridled Tern, *Onychoppryon anaethetus*
Least Tern, *Sternula antillarum*
Caspian Tern, *Hydroprogne caspia*
Black Tern, *Chlidonias niger*
Roseate Tern, *Sterna dougallii*
Common Tern, *Sterna hirundo*
Arctic Tern, *Sterna paradisaea*
Forster's Tern, *Sterna forsteri*
Royal Tern, *Thalasseus maximus*
Sandwich Tern, *Thalasseus sandvicensis*
Black Skimmer, *Rynchops niger*

## GAVIIFORMES:

### Gaviidae—Loons

Red-throated Loon, *Gavia stellata*
Pacific Loon, *Gavia pacifica*
Common Loon, *Gavia immer*

## PROCELLARIIFORMES:

### Hydrobatidae—Storm-Petrels

Wilson's Storm-Petrel, *Oceanites oceanicus*
Leach's Storm-Petrel*, *Hydrobates leucorhous*

### Procellariidae—Shearwaters and Petrels

Sooty Shearwater, *Ardenna grisea*

## SULIFORMES:

### Sulidae—Boobies and Gannets

Northern Gannet, *Morus bassanus*

### Phalacrocoracidae—Cormorants

Double-crested Cormorant, *Nannopterum auritum*
Great Cormorant, *Phalacrocorax carbo*

## PELECANIFORMES:

### Pelecanidae—Pelicans

American White Pelican*, *Pelecanus erythrorhynchos*
Brown Pelican, *Pelecanus occidentalis*

### Ardeidae—Bitterns, Herons, and Allies

Least Bittern#, *Ixobrychus exilis*
American Bittern, *Botaurus lentiginosus*
Great Blue Heron, *Ardea herodias*
Great Egret, *Ardea alba*
Snowy Egret, *Egretta thula*
Little Egret, *Egretta garzetta*
Little Blue Heron, *Egretta caerulea*
Tricolored Heron, *Egretta tricolor*
Cattle Egret*, *Bubulcus ibis*
Black-crowned Night-Heron, *Nycticorax nycticorax*
Yellow-crowned Night-Heron, *Nyctanassa violacea*

### Threskiornithidae—Ibises and Spoonbills

Glossy Ibis, *Plegadis falcinellus*
White-faced Ibis*, *Plegadis chihi*

## ACCIPITRIFORMES:

### Cathartidae—New World Vultures

Turkey Vulture, *Cathartes aura*
Black Vulture#, *Coragyps astratus*

### Pandionidae—Ospreys

Osprey, *Pandion haliaetus*

### Accipitridae—Hawks, Kites, Eagles, and Allies

Northern Harrier, *Circus hudsonius*
Sharp-shinned Hawk, *Accipiter striatus*
Cooper's Hawk, *Accipiter cooperii*
Northern Goshawk, *Accipiter gentilis*
Bald Eagle, *Haliaeetus leucocephalus*
Red-shouldered Hawk, *Buteo lineatus*
Broad-winged Hawk, *Buteo platypterus*
Red-tailed Hawk, *Buteo jamaicensis*
Rough-legged Hawk, *Buteo lagopus*
Golden Eagle, *Aquila chrysaetos*

## STRIGIFORMES:

### Tytonidae—Barn Owls

Barn Owl, *Tyto alba*

### Strigidae—Typical Owls

Great Horned Owl*, *Bubo virginianus*
Snowy Owl, *Bubo scandiacus*
Barred Owl*, *Strix varia*
Short-eared Owl, *Asio flammeus*
Northern Saw-whet Owl*, *Aegolius acadicus*

## CORACIIFORMES:

### Alcedinidae—Kingfishers

Belted Kingfisher, *Megaceryle alcyon*

## PICIFORMES:

### Picidae—Woodpeckers and Allies

Red-headed Woodpecker*, *Melanerpes erythrocephalus*
Red-bellied Woodpecker, *Melanerpes carolinus*
Yellow-bellied Sapsucker*, *Sphyrapicus varius*
Downy Woodpecker, *Dryobates pubescens*
Hairy Woodpecker*, *Dryobates villosus*
Northern Flicker, *Colaptes auratus*

## FALCONIFORMES:

### Falconidae—Caracaras and Falcons

American Kestrel, *Falco sparverius*
Merlin, *Falco columbarius*
Gyrfalcon*, *Falco rusticolus*
Peregrine Falcon, *Falco peregrinus*

## PSITTACIFORMES:

### Psittacidae—Lories, Parakeets, Macaws, and Parrots

Monk Parakeet*, *Myiopsitta monachus*

## PASSERIFORMES:

### Tyrannidae—Tyrant Flycatchers

Eastern Wood-Pewee, *Contopus virens*
Yellow-bellied Flycatcher*, *Empidonax flaviventris*
Acadian Flycatcher, *Empidonax virescens*
Alder Flycatcher*, *Empidonax alnorum*
Willow Flycatcher, *Empidonax traillii*
Least Flycatcher*, *Empidonax minimus*
Eastern Phoebe, *Sayornis phoebe*
Great Crested Flycatcher, *Myiarchus crinitus*
Ash-throated Flycatcher#, *Myiarchus cinerascens*
Western Kingbird, *Tyrannus verticalis*
Eastern Kingbird, *Tyrannus tyrannus*

### Vireonidae—Vireos

White-eyed Vireo, *Vireo griseus*
Yellow-throated Vireo*, *Vireo flavifrons*
Blue-headed Vireo, *Vireo solitarius*
Warbling Vireo, *Vireo gilvus*
Red-eyed Vireo, *Vireo olivaceus*

### Laniidae—Shrikes

Northern Shrike, *Lanius borealis*

### Corvidae—Jays and Crows

Blue Jay, *Cyanocitta cristata*
American Crow, *Corvus brachyrhynchos*
Fish Crow, *Corvus ossifragus*
Common Raven, *Corvus corax*

### Paridae—Chickadees and Titmice

Black-capped Chickadee, *Poecile atricapillus*
Boreal Chickadee*, *Poecile hudsonicus*

### Alaudidae—Larks

Horned Lark, *Eremophila alpestris*

### Hirundinidae—Swallows

Northern Rough-winged Swallow, *Stelgidopteryx serripennis*
Purple Martin, *Progne subis*
Tree Swallow, *Tachycineta bicolor*
Bank Swallow, *Riparia riparia*
Barn Swallow, *Hirundo rustica*
Cliff Swallow, *Petrochelidon pyrrhonota*
Cave Swallow, *Petrochelidon fulva*

### Regulidae—Kinglets

Golden-crowned Kinglet, *Regulus satrapa*
Ruby-crowned Kinglet, *Corthylio calendula*

### Bombycillidae—Waxwings

Cedar Waxwing, *Bombycilla cedrorum*

### Sittidae—Nuthatches

Red-breasted Nuthatch, *Sitta canadensis*
White-breasted Nuthatch, *Sitta carolinensis*

### Certhiidae—Creepers

Brown Creeper, *Certhia americana*

### Polioptilidae—Gnatcatchers and Gnatwrens

Blue-gray Gnatcatcher, *Polioptila caerulea*

### Troglodytidae—Wrens

House Wren, *Troglodytes aedon*
Winter Wren, *Troglodytes hiemalis*
Sedge Wren[#], *Cistothorus stellaris*

Marsh Wren, *Cistothorus palustris*
Carolina Wren, *Thryothorus ludovicianus*

### Mimidae—Mockingbirds and Thrashers

Gray Catbird, *Dumetella carolinensis*
Brown Thrasher, *Toxostoma rufum*
Northern Mockingbird, *Mimus polyglottos*

### Sturnidae—Starlings

European Starling, *Sturnus vulgaris*

### Turdidae—Thrushes

Eastern Bluebird, *Sialia sialis*
Veery*, *Catharus fuscescens*
Gray-cheeked Thrush*, *Catharus minimus*
Swainson's Thrush, *Catharus ustulatus*
Hermit Thrush, *Catharus guttatus*
Wood Thrush*, *Hylocichla mustelina*
American Robin, *Turdus migratorius*

### Passeridae—Old World Sparrows

House Sparrow, *Passer domesticus*

### Motacillidae—Wagtails and Pipits

American Pipit, *Anthus rubescens*

### Fringillidae—Fringilline and Cardueline Finches and Allies

House Finch, *Haemorhous mexicanus*
Purple Finch, *Haemorhous purpureus*
Red Crossbill*, *Loxia curvirostra*
White-winged Crossbill, *Loxia leucoptera*
Common Redpoll, *Acanthis flammea*
Pine Siskin, *Spinus pinus*
American Goldfinch, *Spinus tristis*
Evening Grosbeak*, *Coccothraustes vespertinus*

### Calcariidae—Longspurs and Snow Buntings

Lapland Longspur, *Calcarius lapponicus*
Snow Bunting, *Plectrophenax nivalis*

### Passerellidae—New World Sparrows

Grasshopper Sparrow*, *Ammodramus savannarum*
LeConte's Sparrow*, *Ammospiza leconteii*
Nelson's Sparrow, *Ammospiza nelsoni*
Saltmarsh Sparrow, *Ammospiza caudacuta*

Seaside Sparrow, *Ammospiza maritima*
American Tree Sparrow, *Spizelloides arborea*
Chipping Sparrow, *Spizella passerina*
Clay-colored Sparrow, *Spizella pallida*
Field Sparrow, *Spizella pusilla*
Fox Sparrow, *Passerella iliaca*
Dark-eyed Junco, *Junco hyemalis*
White-crowned Sparrow, *Zonotrichia leucophrys*
White-throated Sparrow, *Zonotrichia albicollis*
Vesper Sparrow, *Pooecetes gramineus*
Savannah Sparrow, *Passerculus sandwichensis*
Savannah Sparrow (Ipswich), *Passerculus sandwichensis princeps*
Song Sparrow, *Melospiza melodia*
Lincoln's Sparrow, *Melospiza lincolnii*
Swamp Sparrow, *Melospiza georgiana*
Eastern Towhee, *Pipilo erythrophthalmus*

### Icteridae—Blackbirds

Yellow-breasted Chat*, *Icteria virens*
Bobolink, *Dolichonyx oryzivorus*
Red-winged Blackbird, *Agelaius phoeniceus*
Eastern Meadowlark, *Sturnella magna*
Yellow-headed Blackbird, *Xanthocephalus xanthocephalus*
Rusty Blackbird, *Euphagus carolinus*
Common Grackle, *Quiscalus quiscula*
Brown-headed Cowbird, *Molothrus ater*
Orchard Oriole, *Icterus spurius*
Baltimore Oriole, *Icterus galbula*

### Parulidae—Wood-Warblers

Ovenbird*, *Seiurus aurocapilla*
Louisiana Waterthrush, *Parkesia motacilla*
Northern Waterthrush, *Parkesia noveboracensis*
Blue-winged Warbler, *Vermivora cyanoptera*
Black-and-white Warbler, *Mniotilta varia*
Prothonotary Warbler*, *Protonotaria citrea*
Tennessee Warbler, *Leiothlypis peregrina*
Orange-crowned Warbler, *Leiothlypis celata*
Nashville Warbler, *Leiothlypis ruficapilla*
Connecticut Warbler*, *Oporornis agilis*
Mourning Warbler*, *Geothlypis philadelphia*
Kentucky Warbler*, *Geothlypis formosa*
Common Yellowthroat, *Geothlypis trichas*

Hooded Warbler*, *Setophaga citrina*
American Redstart, *Setophaga ruticilla*
Cape May Warbler, *Setophaga tigrina*
Northern Parula, *Setophaga americana*
Magnolia Warbler, *Setophaga magnolia*
Bay-breasted Warbler, *Setophaga castanea*
Blackburnian Warbler*, *Setophaga fusca*
Yellow Warbler, *Setophaga petechia*
Chestnut-sided Warbler, *Setophaga pensylvanica*
Blackpoll Warbler, *Setophaga striata*
Black-throated Blue Warbler, *Setophaga caerulescens*
Palm Warbler, *Setophaga palmarum*
Pine Warbler, *Setophaga pinus*
Yellow-rumped Warbler, *Setophaga coronata*
Prairie Warbler, *Setophaga discolor*
Black-throated Green Warbler, *Setophaga virens*
Canada Warbler*, *Cardellina canadensis*
Wilson's Warbler, *Cardellina pusilla*

### Cardinalidae—Cardinals, Piranga Tanagers, and Allies

Scarlet Tanager, *Piranga olivacea*
Northern Cardinal, *Cardinalis cardinalis*
Rose-breasted Grosbeak, *Pheucticus ludovicianus*
Blue Grosbeak*, *Passerina caerulea*
Indigo Bunting, *Passerina cyanea*
Dickcissel, *Spiza americana*

---


## ACKNOWLEDGMENTS

My surveys of Napatree were made possible by the help I received from Napatree staff, especially Janice Sassi, Daniel Cole, and Kevin Rogers, and fellow birders Barbara Gearhart and Christopher Raithel. Peter Paton, Janice Sassi, Daniel Cole, Avery MacLear, and Peter August assisted in presenting my bird list in the format required for this book.


## REFERENCES


Chesser, R. T., S. M. Billerman, K. J. Burns, C. Cicero, J. L. Dunn, B. E. Hernández-Baños, R. A. Jiménez, A. W. Kratter, N. A. Mason, P. C. Rasmussen, J. V. Remsen, Jr., D. F. Stotz, and K. Winker. 2022. "Check-list of North American Birds." American Ornithological Society (website). https://checklist.americanornithology.org/taxa/.

Wells, J. V., D. K. Niven, and J. Cecil. 2005. "The Important Bird Areas Program in the United States: Building a Network of Sites for Conservation, State by State." In *Bird Conservation Implementation and Integration in the Americas*. 191:1265–1269. Albany, California: USDA Forest Service, Pacific Southwest Research Station.

*Common Terns (Sterna hirundo)*
*Photo Credit: The Watch Hill Conservancy*

CHAPTER 12

# Responses of Birds to Humans at a Coastal Barrier Beach: Napatree Point, Rhode Island

*Thomas W. Mayo,*[1] *Peter W. C. Paton*[2] *& Peter V. August*[2]

[1] GIS Division, Water Department, City of Portsmouth, Rhode Island

[2] Department of Natural Resources Science, University of Rhode Island

**Citation**: Mayo, T., P. Paton, and P. August. 2015. "Responses of Birds to Humans at a Coastal Barrier Beach: Napatree Point, Rhode Island." *Northeastern Naturalist*, 22:501–512. Reprinted with permission.

2015 NORTHEASTERN NATURALIST 22(3):501–512

# Responses of Birds to Humans at a Coastal Barrier Beach: Napatree Point, Rhode Island

Thomas W. Mayo[1,*], Peter W.C. Paton[2], and Peter V. August[2]

**Abstract** - Human activity is one of the most important factors affecting disturbance to birds that use coastal barrier beaches in southern New England. The barrier beach at Napatree Point, RI, provides important breeding habitat for several bird species, key stopover habitat for thousands of migrating shorebirds, and is also a popular destination for people. Anecdotal evidence suggested that walkers, joggers, dogs, and watercraft were disrupting birds that foraged and roosted at this site. Our objectives were to characterize the frequency and sources of disturbance to birds and use this information to develop management recommendations to minimize the frequency of human disturbances to birds at Napatree Point. We conducted 106 hours of observation from May through August in 2013. Of 211 flight responses, the most common sources of disturbance to birds were beach walkers (57.8%), motorboats (8.5%), kayaks (8.5%), bird watchers (7.6%), and anglers (6.2%). Birds typically flushed when pedestrians (e.g., walkers, bird watchers, and anglers) were within 39 ± 24 m (median ± inter-quartile range) and watercraft (e.g., motorboats and kayaks) were within 38 ± 33 m. Flight responses were positively correlated with the number of people on the beach and the number of boats in the area. Disturbances to birds peaked in July when human visitation was highest. Using a spatially explicit density map of flight-initiation distance vectors, we identified the most important area to set as a buffer zone for human access if managers seek to reduce the frequency of human disturbances to birds at Napatree Point.

## Introduction

Coastal barrier beaches in southern New England provide important habitat for breeding and migrating bird species. Avian use of critical habitats such as these can be impacted by a variety of anthropogenic factors (Glover et al. 2011, Madsen 1995, Weston et al. 2012a). Birds typically exhibit a flight response when approached by humans. The distance at which the response occurs varies among species and various stimuli including pedestrians (Burger 1981, McLeod et al. 2013), watercraft activities (Peters and Otis 2006, Rodgers and Schwikert 2002), and aircraft (Smit and Visser 1993). Previous research has shown that human activities can effectively degrade habitat quality by directly disturbing breeding, foraging, and roosting waterbirds (e.g., waterfowl, wading birds, and shorebirds), with shorebirds demonstrated to be particularly vulnerable to human disturbance (Brown et al. 2001; Burger 1981, 1986; Pfister et al. 1992). Detrimental effects include reduced foraging rates and decreased use of foraging and roosting sites

[1]Masters of Environmental Science and Management Program, University of Rhode Island, Kingston, RI 02881. [2]Department of Natural Resources Science, University of Rhode Island, Kingston, RI 02881. [*]Corresponding author - tmayo@portsmouthwater.org.

Manuscript Editor: Greg Roberston

(Fitzpatrick and Bouchez 1998, Thomas et al. 2003) and the disruption of essential breeding behaviors (Weston and Elgar 2005, 2007).

In New Jersey, Burger (1986) found that 40% of the disturbances to shorebirds were from people walking on the beach, with anglers accounting for 10–20% of disturbances. Indeed, understanding which stimuli are present in an environment and at what density they occur is important because birds respond to different anthropogenic stimuli (e.g., walkers, cars, boats) at different rates or distances (Glover et al. 2011, McLeod et al. 2013). There can also be a strong seasonal component to disturbances (Stillman and Goss-Custard 2002), with Burger (1986) documenting the highest rates of disturbance in May and August, when most migrating shorebirds are present and human use of beaches also tends to peak in the northeastern US. There is also interspecific variation in response to disturbance, with larger shorebirds (and birds in general) flushing at greater distances than smaller shorebirds in response to similar disturbances (Koch and Paton 2014, Weston et al. 2012a).

Napatree Point in southwestern Rhode Island is a local hotspot for breeding and migratory birds, with 1 federally listed species, *Charadrius melodus* (Piping Plover), and 3 state-listed species—*Haematopus palliatus* (American Oystercatcher), *Sternula antillarum* (Least Tern), and *Pandion haliaetus* (Osprey)—regularly nesting on the barrier beach. In addition, numerous shorebirds (e.g., sandpipers, *Sterna* spp. [terns], *Larus* spp. [gulls]) and waterbirds (e.g., *Phalacrocorax* spp. [cormorants], egrets, waterfowl) congregate at Napatree Point during spring and fall migration. Scientists from the US Fish and Wildlife Service make regular trips throughout the year to conduct bird surveys of the area, and Watch Hill Conservancy staff conduct weekly surveys. Because of the relatively large numbers of birds that forage and roost in this area, naturalists often visit the lagoon to search for rare birds. On warm summer days, over 1000 people and 300 boats have been recorded at Napatree Point (Sassi 2013). The beach on the southern side of Napatree Point is a popular destination for beach enthusiasts, while several hundred boats often moor overnight to the northeast of the lagoon in southern Little Narragansett Bay, particularly on weekends in July and August (Sassi 2013). Thus, there is high potential for humans to have a negative impact on avian use of Napatree Point.

Our objectives were to (1) quantify the various sources of human disturbance to birds (e.g., walker, angler, or bird watcher) using the lagoon at Napatree Point, (2) assess interspecific variation in flight responses among various sources of human stimuli by measuring the distance between the disturbance stimuli and the disturbed bird (Blumstein et al. 2003), and (3) use these data to develop recommendations to inform resource managers at Napatree Point on how best to minimize disturbance to resident and migratory shorebirds and waterbirds.

## Methods

### Study site

We conducted fieldwork in the Napatree Point Conservation Area (NPCA, hereafter Napatree Point), a 2.4-km-long x 100–300-m-wide (27.7 ha) barrier beach located in the village of Watch Hill, Westerly, RI (Fig. 1). This barrier beach is bounded to the north by Little Narragansett Bay and to the south by the Atlantic Ocean. Napatree Point is a biological preserve that is owned and managed by the Watch Hill Fire District and the Watch Hill Conservancy (Westerly, RI). Terrestrial habitats present include maritime shrubland, salt marsh, maritime beach strand, and maritime herbaceous dune (Enser et al. 2011). A 5.2-ha tidal lagoon and adjacent sandy beaches provide breeding habitat for *Limulus polyphemus* L. (Horseshoe Crab), whose eggs are a vital food resource for migratory shorebirds (Gillings et al. 2007). The lagoon and adjacent areas also have extensive *Mytilus edulis* L. (Blue Mussel) and *Zostera marina* L. (Marine Eelgrass) beds that offer key foraging habitat for shorebirds and waterbirds; therefore, this area is one of the most important biodiversity hotspots in the region.

**Avian surveys**

One observer (T.W. Mayo) conducted all surveys from a fixed point, surrounded by 0.5-m-tall *Ammophila breviligulata* Fern. (American Beachgrass), keeping well hidden in the dunes 115 m from the eastern edge of the lagoon (Fig. 1). This fixed

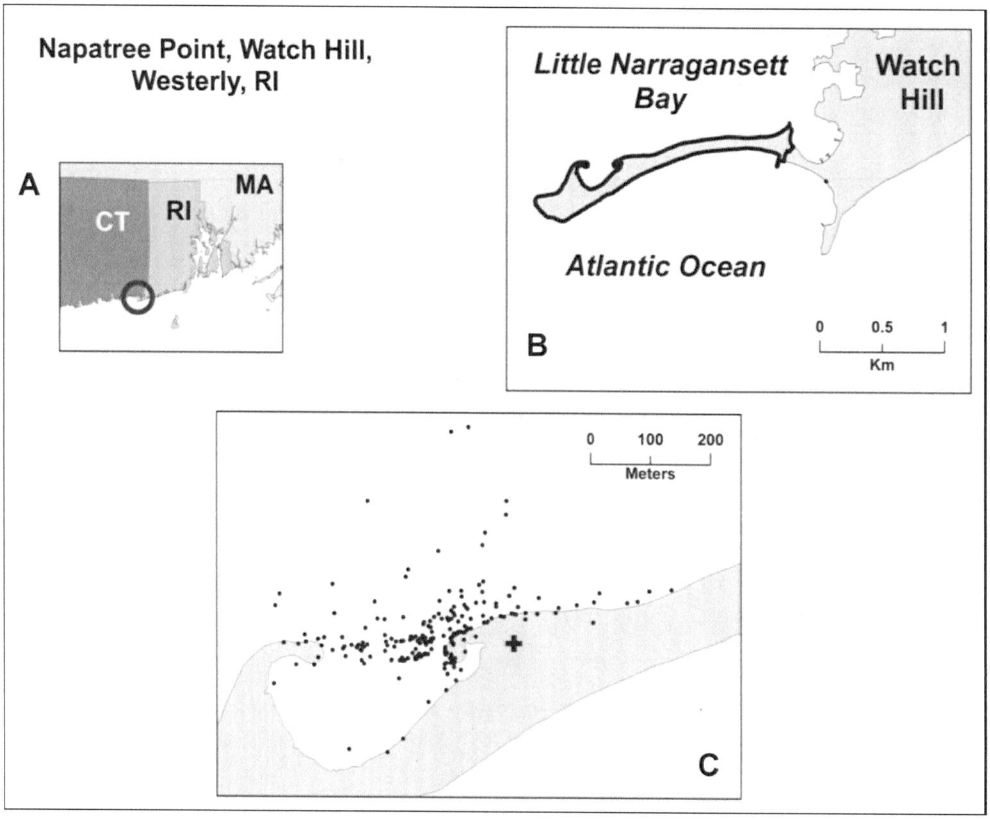


Figure 1. Map of the Napatree Point Conservation Area, the peninsula extending out from Watch Hill, RI. (A) Regional map showing the location of the study area (indicated by a circle). (B) Map of Napatree Point. The boundary of the Napatree Point Conservation Area is shown in the bold outline. (C) Observation zone for this study. The location of the observer (+) and locations of flocks (•) that were disturbed during observations are shown.

point was far enough from foraging and roosting birds to prevent disturbance while simultaneously allowing the observer to quantify when birds were flushed by anthropogenic sources within the study-area boundaries. These boundaries extended from the western edge of the lagoon to the first abandoned Osprey nest pole east of the vantage point. All disturbances witnessed were within the observation zone shown in Figure 1C; disturbances witnessed outside of the observation zone were not recorded. T.W. Mayo used a 60x spotting scope and 10 x 42 binoculars to make observations from 09:00 to 16:00 on weekends and holidays, when human visitation rates were highest. He completed observations under all weather conditions except during inclement weather (heavy rains and thunderstorms) that shortened some of the observation periods.

The observer counted potential sources of anthropogenic disturbance (i.e., pedestrians, watercraft, or aircraft) and weather information (i.e., air temperature, wind speed, and percent cloud-cover in 25% increments) every 30 min within the observation zone. He categorized watercraft as a motorboat, sailboat, canoe/kayak, paddleboard, or jet ski; and pedestrians as a walker (walking only), jogger (obviously running), birder (actively using binoculars or a spotting scope), angler (carrying a fishing rod or actively fishing), or ATV (driving in an all-terrain vehicle, which was restricted to management staff because no other vehicles are allowed on Napatree Point). We categorized aircraft as a helicopter, airplane (always single-engine fixed-wing craft), or a kite on a string.

Each time an individual bird or flock of birds took flight within the observation zone, the observer recorded the location of the anthropogenic-disturbance stimulus and the species and location of the bird(s) that flushed. These locations were recorded in the field on an analog map of the study zone. We employed ArcGIS (Environmental Systems Research Institute, v10.2 software, Redlands, CA) to create analog maps developed from high-resolution (10-cm pixel size) digital orthophotography. We identified from the orthophotography or with GPS coordinates (<3 m horizontal accuracy) permanent beach landmarks, coastline, and natural features as reference points on the analog map. There were enough reference features on the analog map to accurately locate disturbance events on the ground to within ~5–10 m. We could not accurately map the exact location of airborne sources of disturbance because of the lack of spatial reference in the sky and the speed with which aircraft moved.

We recorded whether the disturbance source caused flight responses by individual or multiple birds and, when possible, recorded the species that were flushed. This identification was relatively easy when individual birds flushed but was difficult when a large, multispecies flock flushed at once. Because our study design was based on observations of naturally occurring human disturbances and the concomitant flight responses in birds, we did not experimentally create stimuli to disturb birds.

To determine flight-initiation distances (Blumstein et al. 2003), we scanned the analog maps recorded in the field into digital image format, georegistered the digital images, and converted bird locations when initially flushed and their corresponding disturbance source to point-features in a GIS. In order to display the spatial pattern of disturbance stimuli and where birds were when they responded, we performed a number of GIS-based analyses. For each disturbance event, we created a line vector that extended from the first bird flushed to the source of disturbance. The length of the vector represented the flight-initiation distance (FID) between the flushed bird and the disturbance source. We mapped these vectors to provide a visual representation of the spatial distribution of disturbances in and around the lagoon.

We used the default kernel-density algorithm in ArcGIS v 10.2 software to create a density map (meters of disturbance vector per km$^2$) of disturbance vectors. To develop a spatially-explicit management zone to protect nesting, foraging, and roosting shorebirds and waterbirds at Napatree Point, we extracted the top 50% (i.e., greater than the median disturbance density) of areas in the disturbance-density map to determine where most disturbance events took place. By buffering this area by 40 m, the recommended additional buffer by Rodgers and Smith (1995), we established a management zone where pedestrians and watercraft were likely to intrude in the highest disturbance-density area (Fig. 2). We excluded disturbances caused by aircraft from the density analysis because they were not ground-based events and we could not map their locations as accurately as disturbances that occurred on the ground.

We conducted all statistical analyses using Microsoft Excel and R software (R Core Team 2013). Results of a Shapiro-Wilk test showed that our data were not

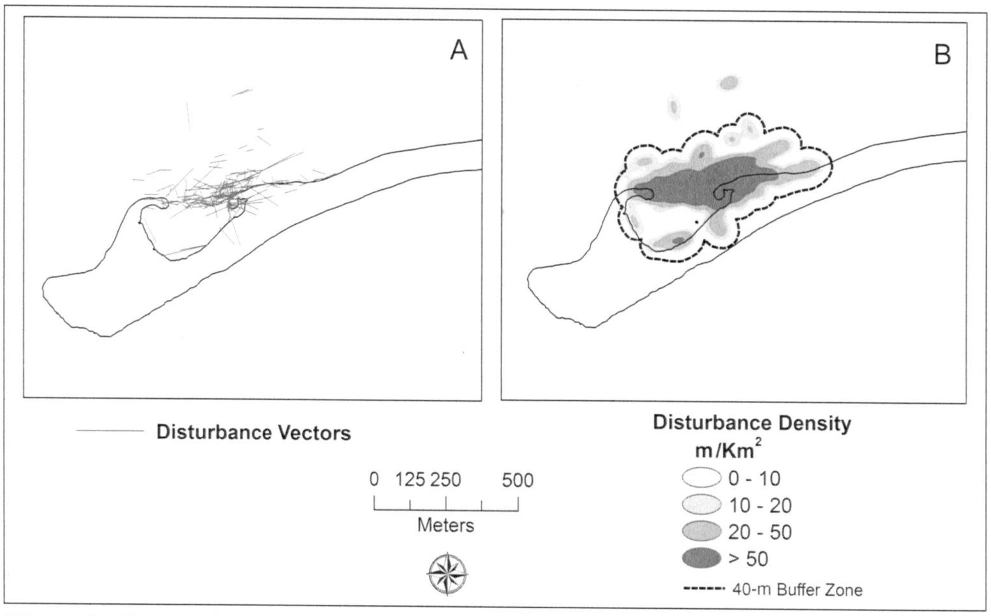


Figure 2. The spatial distribution of disturbance events around the lagoon in the Napatree Point Conservation Area during 2013. Left panel shows vectors depicting locations of anthropogenic stimuli and the closest disturbed avian flock. Right panel is a density map of the disturbance vectors. We classified disturbance vectors into approximate quartiles of disturbance density that are expressed as meters of disturbance vector per square kilometer (m/km$^2$). The broken line is a 40-m buffer away from the top 50% of disturbance densities.

normally distributed; therefore, we used non-parametric Spearman rank-correlation tests to measure dependence among variables.

## Results

We detected a total of 211 disturbance events during 106 hours of observation (average = 1.99 disturbances per hour) over 21 days of field observations from mid-May to late-August 2013. We detected an average of 11 ± 9 (median ± interquartile range, absolute range = 3–19) disturbance events per observation day, peaking from early to mid-August. The frequency of disturbances (disturbances per observation hour/day) tended to increase as weather became warmer (mean temperature; Spearman $r = 0.43$, $P < 0.06$), and when pedestrian use of the observation zone increased (number visitors/day; Spearman $r = 0.76$, $P < 0.001$, $n = 19$, Fig. 3). We counted a total of 2779 pedestrians during the 30-min observation windows on 19 of the 21 days of sampling; no pedestrian or watercraft counts were done in May. On average, we observed 28.9 pedestrians per hour per day, with the highest human-activity rates between 12:00 and 14:30; indeed, 97% of all disturbance events occurred between the hours of 10:00 and 15:00. We counted a total of 2936 watercraft (primarily motorboats) during the study. We found a significant positive correlation between the number of watercraft and pedestrian counts (Spearman $r = 0.82$, $P < 0.001$, $n = 19$, Fig. 3).

Disturbances to birds were caused most often by people walking along the beach (57.8%), motorboats (8.5%), kayakers (8.5%), birders (7.6%), aircraft (6.6%), and anglers (6.2%) (Table 1). Single-disturbance events caused multiple flocks to flush slightly more often than they caused a single flock to flush (114 and 97 disturbances, respectively; Table 1). Multiple disturbances typically occurred when a jogger, birder, or walker continued to travel along the shore or lagoon edge, whereas single disturbances typically occurred when a walker approached the lagoon and then decided to turn back. Stationary disturbances (e.g., anglers) generally caused fewer multiple flushes (Table 1).

We documented disturbances to 25 avian species, with *Larus argentatus* (Herring Gull), *Sterna hirundo* (Common Tern), Least Tern, and American Oystercatcher the most frequently disturbed species (Appendix 1). We documented 16 occasions when Piping Plovers (federally listed as threatened), and 15 occasions when *Sterna dougallii* (Roseate Tern; federally listed as endangered) were flushed (Appendix 1). The FID was greatest for aircraft (median distance = 117.5 m) and lowest for pedestrians and watercraft (38.7 m and 38.4 m, respectively; Table 1). Disturbances were concentrated at the entrance to the lagoon, where large numbers of birds and disturbance stimuli (walkers, birders, and boats) tended to congregate (Fig. 2).

## Discussion

Previous research has shown that human disturbance can have negative impacts on migratory-bird use of a coastal stopover site (Pfister et al. 1992, Thomas et al.

2003); therefore, we were interested in relationships between anthropogenic stimuli and bird responses at Napatree Point. We found that pedestrians walking along the

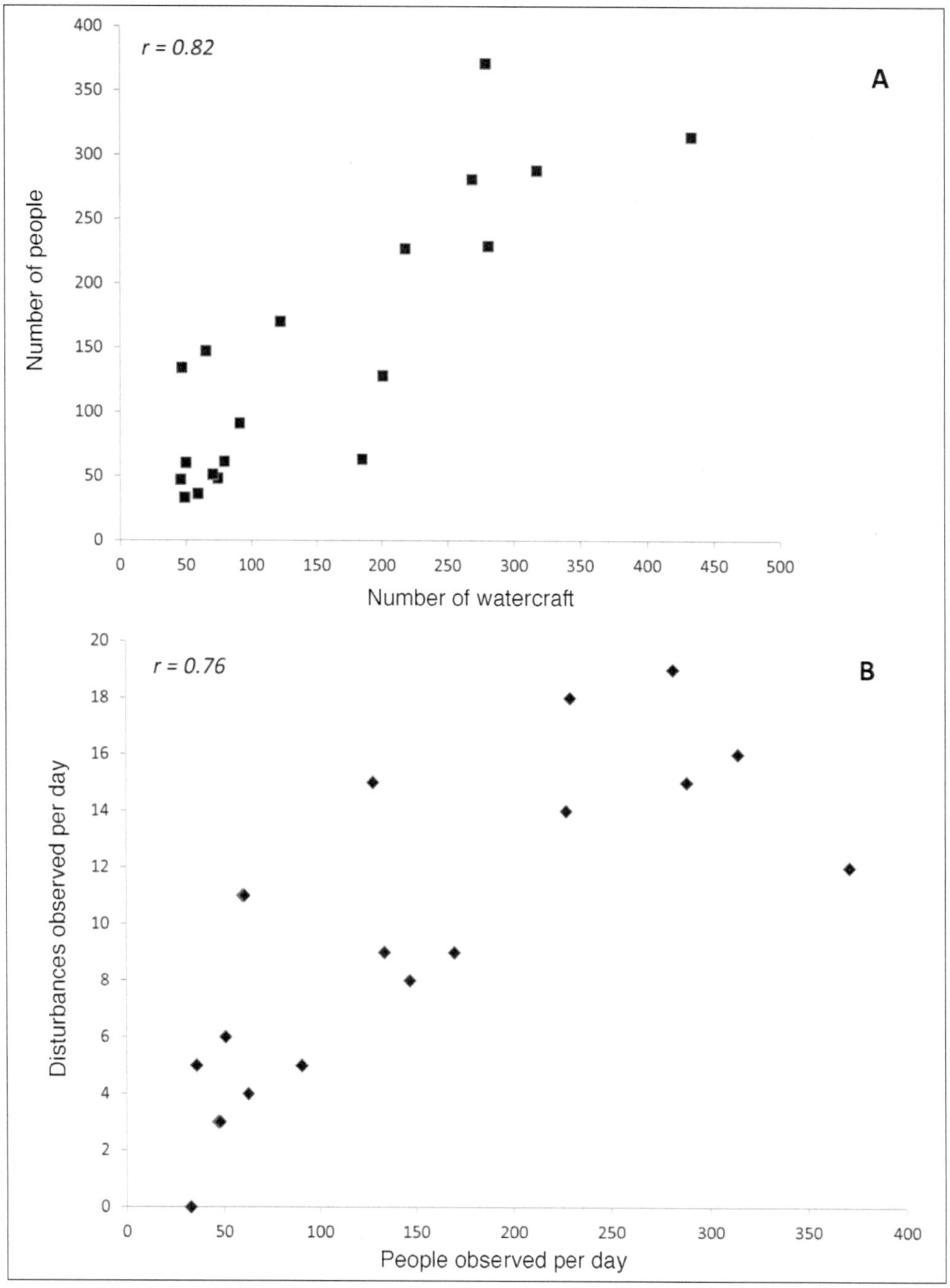


Figure 3. Relationship between the daily total number of people observed at Napatree Point Conservation Area and the number of boats anchored or moving through the site (A) and between disturbance rates (disturbances per day) and visitors per day (B). Spearman rank correlations were used to calculate *r* values ($n$ = 19 days).

beach and the eastern edge of the lagoon accounted for approximately 75% of all bird-disturbance events, which was similar to previous research on coastal barrier beaches (Burger 1986, Lafferty 2001) and beaches worldwide (Weston and Elgar 2005, 2007), except where vehicles dominated (Schlacher et al. 2013). In general, we found that birds were less likely to flush when approached by a slow-moving pedestrian, such as a person actively fishing (i.e., anglers), compared to an active jogger or walker, a result which concurred with past research (Burger 1981, Glover et al. 2011). Other studies have shown that, compared to people walking alone, people walking dogs can have a greater impact on bird responses at a beach (Thomas et al. 2003) and longer FIDs (Glover et al. 2011), but dogs were only observed near the lagoon on 2 occasions during this study. In both cases, it was our subjective assessment that the people accompanying the dogs appeared to be the source of disturbance and not the pets. In one instance, the dog was leashed, while in the other, the dog was free-running. An active campaign by NPCA staff to alert beach enthusiasts that dogs are prohibited on Napatree Point by town ordinance from 8:00 AM to 17:00 PM during the summer apparently has been quite effective (unlike many other efforts to manage dogs on beaches; Williams et al. 2011). However, boaters and local beach enthusiasts often walk their dogs in the conservation area in the early morning hours when many

Table 1. Sources and relative frequency of disturbances to birds at the Napatree Point Conservation Area Lagoon. Multiple flushes occur when 1 source of disturbance causes multiple flocks of birds to flush their foraging or resting grounds at different times or cause a single flock to flush more than once. Each distinct disturbance event per source was counted separately. Number - number of distrubance events; % total = percent total disturbances; % multiple = percent of disturbance type causing multiple flushes

| Type of disturbance | Number | % total | % multiple | Median (25th percentile, 75th percentile) of distance (m) between disturbance and flushed birds |
|---|---|---|---|---|
| Pedestrians | | | | |
| Total | 157 | 74.4 | 59.2 | 38.7 (27.7, 51.7) |
| Bird watchers | 16 | 7.6 | 62.5 | 32.8 (26.6, 48.8) |
| Anglers | 13 | 6.2 | 46.2 | 41.1 (39.1, 54.4) |
| Jogger | 5 | 2.4 | 80.0 | 32.6 (28.4, 61.5) |
| Walker | 122 | 57.8 | 59.0 | 37.7 (27.8, 51.2) |
| ATV | 1 | 0.5 | 100.0 | |
| Watercraft | | | | |
| Total | 40 | 19.0 | 40.0 | 38.4 (29.7, 62.6) |
| Motorboat | 18 | 8.5 | 38.9 | 38.4 (29.6, 52.9) |
| Sail boat | 1 | 0.5 | 100.0 | |
| Kayak/canoe | 18 | 8.5 | 38.9 | 45.5 (29.8, 61.8) |
| Jet ski | 1 | 0.5 | 0.0 | |
| Paddleboard | 2 | 0.9 | 50.0 | |
| Aircraft | | | | |
| Total | 14 | 6.6 | 35.7 | 117.6 (86.6, 141.4) |
| Helicopter | 4 | 1.9 | 75.0 | 95.0 (82.0, 112.6) |
| Kite | 1 | 0.5 | 0.0 | |
| Airplane | 9 | 4.3 | 22.2 | 135.0 (89.8, 157.3) |

birds are actively foraging; thus, we are concerned that dogs are not banned on the beach 24 hours per day to minimize disturbance to birds using this key habitat (Burger 1986, Burger et al. 2004).

Estimates of FID have significant ramifications for management of shorebird and waterbird breeding and stopover habitat (Blumstein et al. 2003, Koch and Paton 2014, Rodgers and Smith 1995, Weston et al. 2012a). The median FID we documented when birds were disturbed by walkers or boats (~40 m) is similar to distances reported in previous studies (e.g., Blumstein 2003, Burger and Gochfeld 1991, Thomas et al. 2003), but less than values reported for some larger species of shorebirds documented in other areas (e.g., Koch and Paton 2014), perhaps due to different methods of observation and analysis between studies. The disturbance-density map we developed provides guidance for Napatree Point resource managers on possible regions from which pedestrians and boats should be excluded during migration periods and breeding season in order to reduce disturbance to birds (Ikuta and Blumstein 2003, Rodgers and Smith 1995).

An area to the east of the lagoon is generally fenced off most years with symbolic fencing and signage to prevent people from entering Piping Plover nesting areas, which is a standard operating procedure for Rhode Island beaches where this threatened species nests (Melvin et al. 1991). Such fencing appears associated with high compliance in some international beaches (Weston et al. 2012b). We propose placement of symbolic fencing and signage around our buffered protection zone (Fig. 2) so that foraging, roosting, and nesting shorebirds and waterbirds would be protected from anthropogenic disturbance sources. We have no easy way to reduce aircraft disturbance to the site, but based on our surveys, aircraft do not appear to have a major impact on birds at Napatree Point. Elsewhere throughout the region, kite-boarders are becoming a disturbance source at migratory stopover locations, but this recreational activity is currently not prevalent at Napatree Point. Furthermore, understanding the hourly pattern of disturbances to birds is helpful to managers. Stationing beach patrols and naturalists near the lagoon between 12:00 PM and 14:30 on warm weekend days to educate pedestrians and boaters about this significant resource could be the most effective opportunity to teach beach patrons of the importance of minimizing disturbance to wildlife.

## Acknowledgments

We would like to thank the NPCA naturalist team for general support in the implementation of this study. Janice Sassi, the NPCA Manager, provided logistic support, moral support, and generously shared her knowledge of the Napatree ecosystem. Jessica Cressman, Kevin Rogers, and Jesse Malo provided field assistance and conservation knowledge. Steve Brown provided various supplies and many thoughtful brainstorming sessions with us. The Watch Hill Fire District and Watch Hill Conservancy allowed the use of their parking resources and ATV for transiting out to the study site. Grant Simmons (Watch Hill Fire District) and the late Chaplin B. Barnes (Watch Hill Conservancy) provided inspiration, support, and encouragement throughout the study. This research was funded, in part, by a Dean's Grant from the URI MESM program. This publication was supported by the Rhode Island Agricultural Experiment Station through funding from the National Institute of Food and Agriculture, United States Department of Agriculture, Hatch activities. The thoughtful comments by two anonymous reviewers significantly improved the clarity of the paper.

## Literature Cited

Blumstein, D.T. 2003. Flight-initiation distance in birds is dependent on intruder-starting distance. Journal of Wildlife Management 67:852–857.

Blumstein, D.T., L.L. Anthony, R. Harcourt, and G. Ross. 2003. Testing a key assumption of wildlife buffer zones: Is flight-initiation distance a species-specific trait? Biological Conservation 110:97–100.

Brown, S.C., C. Hickey, B. Harrington, and R. Gill (Eds.). 2001. The US Shorebird Conservation Plan. 2nd Edition. Manomet Center for Conservation Sciences, Manomet, MA. 50 pp.

Burger, J. 1981. The effect of human activity on birds at a coastal bay. Biological Conservation 21:231–241.

Burger, J. 1986. The effect of human activity on shorebirds in two coastal bays in northeastern United States. Environmental Conservation 13:123–130.

Burger, J., and M. Gochfeld. 1991. Human distance and birds: Tolerance and response distance of resident and migrant species in India. Environmental Conservation 18:158–165.

Burger, J., C. Jeitner, K. Clark, and L.J. Niles. 2004. The effect of human activities on migrant shorebirds: Successful adaptive management. Environmental Conservation 4:283–288.

Enser, R., D. Gregg, C. Sparks, P. August, P. Jordan, J. Coit, C. Raithel, B. Tefft, B. Payton, C. Brown, C. LaBash, S. Comings, and K. Ruddock. 2011. Rhode Island ecological communities classification. Technical Report. Rhode Island Natural History Survey, Kingston, RI. 33 pp.

Fitzpatrick, S., and B. Bouchez. 1998. Effects of recreational disturbance on the foraging behavior of waders on a rocky beach. Bird Study 45:157–171.

Gillings, S., P.W. Atkinson, S.L. Bardsley, N.A. Clark, S.E. Love, R.A. Robinson, R.A. Stillman, and R.G. Weber. 2007. Shorebird predation of Horseshoe Crab eggs in Delaware Bay: Species contrasts and availability constraints. Journal of Animal Ecology 76:53–414.

Glover, H.K., M.A. Weston, G.S. Maguire, K.K. Miller, and B.A. Christie. 2011. Towards ecologically meaningful and socially acceptable buffers: Response distances of shorebirds in Victoria, Australia, to human disturbance. Landscape and Urban Planning 103:326–334.

Ikuta, L.A., and D.T. Blumstein. 2003. Do fences protect birds from human disturbance? Biological Conservation 112:447–452.

Koch, S.L., and P.W.C. Paton. 2014. Assessing anthropogenic disturbances to develop buffer zones for shorebirds using a stopover site. Journal of Wildlife Management 78:58–67.

Lafferty, K.D. 2001. Birds at a southern California beach: Seasonality, habitat use, and disturbance by human activity. Biodiversity and Conservation 10:1949–1962.

Madsen, J. 1995. Impacts of disturbance on migratory waterfowl. Ibis 137:S67–S74.

McLeod, E.M., P.-J. Guay, A.J. Taysom, R.W. Robinson, and M.A. Weston. 2013. Buses, cars, bicycles, and walkers: The influence of the type of human transport on the flight responses of waterbirds. PLoS ONE 8(12): e82008. doi:10.1371/journal.pone.0082008.

Melvin, S.M., C.R. Griffin, and L.H. MacIvor. 1991. Recovery strategies for Piping Plovers in managed coastal landscapes. Coastal Management 19:21–34.

Peters, K.A., and D.L. Otis. 2006. Wading-bird response to recreational boat traffic: Does flushing translate into avoidance? Wildlife Society Bulletin 34:1383–1291.

Pfister, C., B.A. Harrington, and M. Levine. 1992. The impact of human disturbance on shorebirds at a migration-staging area. Biological Conservation 60:115–126.

R Core Team. 2013. R: A language and environment for statistical computing. R. Foundation for Statistical Computing, Vienna, Austria. Available online at http://www.R-project.org.

Rodgers, J.A., and S.T. Schwikert. 2002. Buffer-zone distances to protect foraging and loafing waterbirds from disturbance by personal watercraft and outboard-powered boats. Conservation Biology 16:216–24.

Rodgers, J.A., and H.T. Smith. 1995. Set-back distances to protect nesting-bird colonies from human disturbance in Florida. Conservation Biology 9:89–99.

Sassi, J. 2013. The state of Napatree report. Watch Hill Conservancy, Watch Hill, RI. 51 pp.

Schlacher, T.A., T. Nielsen, and M.A. Weston. 2013. Human recreation alters behavior profiles of non-breeding birds on open-coast sandy shores. Estuarine, Coastal, and Shelf Science 118:31–42.

Smit, C.J. and G.J.M. Visser. 1993. Effect of disturbance on shorebirds: A summary of existing knowledge from the Dutch Wadden Sea and Delta area. Wader Group Study Bulletin 68:6–19.

Stillman, R.A., and J.D. Goss-Custard. 2002. Seasonal changes in the response of Oystercatchers (*Haematopus ostralegus*) to human disturbance. Journal of Avian Biology 33:358–65.

Thomas, K., R.G. Kvitek, and C. Bretz. 2003. Effects of human activity on the foraging behavior of Sanderlings, *Calidris alba*. Conservation Biology 109:67–71.

Weston M.A., and M.A. Elgar. 2005. Disturbance to brood-rearing Hooded Plover (*Thinornis rubricollis*): Responses and consequences. Bird Conservation International 15:193–209.

Weston M.A., and M.A. Elgar. 2007. Responses of incubating Hooded Plovers (*Thinornis rubricollis*) to disturbance. Journal of Coastal Research 23:569–576.

Weston, M.A., E.M. McLeod, D.T. Blumstein, and P.-J. Guay. 2012a. A review of flight-initiation distances and their application to managing disturbances to Australian birds. Emu 112:269–286.

Weston, M.A., F. Dodge, A. Bunce, D.G. Nimmo, and K.K. Miller. 2012b. Do temporary beach closures assist in the conservation of breeding shorebirds on recreational beaches? Pacific Conservation Biology 18:47–55.

Williams, K.J.H., M.A. Weston, S. Henry, and G.S. Maguire. 2009. Birds and beaches, dogs, and leashes: Dog owners' sense of obligation to leash dogs on beaches in Victoria, Australia. Human Dimensions of Wildlife 14:89–101.

**Appendix 1**. Interspecific variation in the total number of disturbance events by various anthropogenic stimuli at Napatree Point Conservation Area during 2013.

| | Aircraft | | | Pedestrians | | | | | Watercraft | | | | |
|---|---|---|---|---|---|---|---|---|---|---|---|---|---|
| Species | Heli-copter | Kite | Plane | ATV | Birder | Angler | Runner | Walker | Motor boat | Kayak/ canoe | Paddle-board | Sail boat | *n* |
| *Haematopus palliatus* Temminck (American Oystercatcher) | 1 | 1 | 5 | | 6 | 4 | 2 | 41 | 4 | 3 | | | 67 |
| *Gavia imer* (Brünnich) (Common Loon) | 1 | | | | 1 | 1 | | 2 | 1 | | | | 6 |
| *Phalacrocorax auritus* (Lesson) (Double-crested Cormorant) | 2 | | 5 | | 2 | 2 | | 10 | 5 | 2 | 1 | 1 | 30 |
| *Egretta thula* (Molina) (Snowy Egret) | | 1 | | 1 | 2 | 1 | | 12 | | 3 | | | 20 |
| *Ardea alba* L.(Great Egret) | | | | | 1 | | | 4 | | 1 | | | 6 |
| *Pandion haliaetus* L. (Osprey) | | | | | 2 | 1 | 1 | 3 | 1 | 1 | | | 9 |
| *Pluvialis squatarola* L. (Black-bellied Plover) | 1 | 1 | 2 | | 1 | 3 | 2 | 9 | 1 | 2 | | | 22 |
| *Charadrius semipalmatus* Bonaparte (Semipalmated Plover) | 1 | 1 | 3 | | 2 | 3 | | 6 | 1 | | | | 17 |
| *Charadrius melodus* Ord (Piping Plover) | 1 | 1 | 2 | | 2 | 1 | | 9 | | | | | 16 |
| *Actitis macularius* L. (Spotted Sandpiper) | | | | | | | | 3 | | | | | 3 |
| *Tringa semipalmata* (Gmelin) (Willet) | | | 1 | | 3 | 1 | | 10 | | | | | 15 |
| *Arenaria interpres* (L.) (Ruddy Turnstone) | 1 | 1 | 2 | | 2 | 4 | 1 | 13 | 2 | | | | 26 |
| *Calidris alba* (Pallas) (Sanderling) | | 1 | | | 1 | 2 | | 12 | 2 | 2 | | | 20 |
| *Calidris canutus* (L.) (Red Knot) | | | | | 1 | | 1 | 4 | | | | | 6 |
| *Calidris alpina* (L.) (Dunlin) | | | | | | | | | | 1 | | | 1 |
| *Calidris pusilla* (L.) (Semipalmated Sandpiper) | | | | | | | | 3 | | | | | 3 |
| *Limnodromus griseus* (Gmelin) (Short-billed Dowitcher) | | 1 | | | | | | 3 | | | | | 4 |
| *Larus smithsonianus* Coues (American Herring Gull) | 1 | 1 | 6 | | 5 | 7 | 2 | 56 | 9 | 7 | 1 | 1 | 96 |
| *Larus marinus* L. (Great Black-backed Gull) | 1 | | 2 | | 2 | 1 | | 16 | 1 | 1 | 1 | 1 | 26 |
| *Sterna hirundo* L. (Common Tern) | 2 | | 6 | 1 | 7 | 4 | 3 | 52 | 7 | 8 | 1 | 1 | 92 |
| *Sterna dougallii* Montagu (Roseate Tern) | | 1 | | | 1 | 1 | | 9 | | 1 | 1 | 1 | 15 |
| *Sternula antillarum* Lesson (Least Tern) | 2 | 1 | 6 | 1 | 9 | 4 | 3 | 53 | 4 | 7 | 1 | 1 | 92 |
| *Rhynchops niger* L. (Black Skimmer) | | | 1 | 1 | 1 | 2 | | 4 | | | | | 9 |
| *Hirundo rustica* L. (Barn Swallow) | | | 1 | | | | | 2 | | | | | 3 |
| *Aegelaius phoeniceus* (L.) (Red-winged Blackbird) | | | | | | 1 | | 2 | | | | | 3 |
| % of all events | 2.3 | 1.8 | 6.9 | 0.7 | 8.4 | 7.1 | 2.5 | 55.7 | 6.3 | 6.4 | 1.0 | 1.0 | 607 |

*Piping Plover*
*Photo Credit: Finn Harty*

CHAPTER 13

# Piping Plover Monitoring at Napatree Point Conservation Area: 2022

*Noah Barreto & Daniel Cole*
Napatree Point Conservation Area, The Watch Hill Conservancy

## INTRODUCTION

Piping Plovers (*Charadrius melodus*) are small shorebirds that nest in open, sandy areas along the Atlantic coast and Great Lakes region. They return to breeding grounds, such as Napatree, in late March or early April. Nesting pairs form small depressions in the sand, often marked with small pebbles or pieces of broken shells, somewhere on the high beach close to the dunes. If a pair successfully produces young, the hatchlings may leave the nest only a few hours after being born, quickly learning to feed themselves. By mid-September, both adult and young plovers usually have begun their winter migration to the southeastern United States, the Gulf Coast, and occasionally even northern Mexico.

Piping Plovers are federally protected under the Endangered Species Act, and the Atlantic population is currently listed as "threatened." These shorebirds were common in the nineteenth century, but they became a victim of the millinery trade and rapid development of coastal areas. Until the Migratory Bird Treaty Act of 1918, Piping Plovers' recovery peaked in the mid-twentieth century, but then began a second period of decline. Land development had continued to encroach on their potential nesting areas, and visitation to beaches increased following World War II. Many nesting habitats are accessible to people and dogs, both of which can disturb plovers enough to abandon their nests. Because plovers, as well as their chicks, eggs, and nests, are so well camouflaged into the sand, nesting habitats are now often marked with signs and rope. Beyond human-based threats, the exposed nature of Piping Plover nests makes them highly susceptible to nest predators, as well as high surf and extreme tides (Center for Biological Diversity, 2023; USFWS, 1994).

Napatree Point Conservation Area (NTPCA) provides a large area of nesting habitat for Piping Plovers, making it an important site for the recovery of the Atlantic population. The US Fish and Wildlife Service (USFWS) has monitored Piping Plovers on Napatree since 2001. The Watch Hill Conservancy, the Watch Hill Fire District, and the USFWS work together to manage land use, as recommended by the USFWS Piping Plover Recovery Plan. According to the USFWS, a threshold goal of 1.5 fledglings per pair equates to an increasing Piping Plover population (Durkin, 2021). Napatree has not met this target since 2010 (Durkin, 2022).

## METHODS

In early April, staff from the NTPCA and USFWS erected symbolic fencing around predicted nesting areas, which were based on past nesting locations and the expertise of USFWS biologists (Figure 1). The fencing consists of 2.4 m (8-foot) galvanized U-poles driven into the sand with a strand of rope running between each pole. A sign requesting that beachgoers stay out of closed areas was attached to every other pole.

*Figure 1. USFWS symbolic fencing and poles, 2022.*

The monitoring methods used at Napatree followed USFWS protocols. Once Piping Plovers started arriving from their wintering grounds, Napatree was surveyed three times a week by USFWS or NTPCA staff. A survey is a standardized search for breeding activity. All areas with plover tracks and scrapes are noted to designate territories, and all Piping Plovers encountered are tallied. Once eggs are laid, the geographic coordinates of each nest were obtained by GPS, and the nest was monitored from a distance. If no birds were found to be incubating a known nest, it was checked for failure. If the nest had failed, monitors attempted to determine the cause. Once the eggs in a nest had hatched, the chicks were monitored every other day until they had fledged.

Although USFWS often uses exclosures to protect nests at other monitoring sites, they did not use them on Napatree (Durkin, 2021, 2022). Exclosures will often signal to predators that a nest is at that location, increasing the likelihood of predation once plovers leave the exclosure. At Napatree, the predominant animals that might target an exclosure were Peregrine Falcons (*Falco peregrinus*), which were frequently seen hunting shorebirds along the beach this year. Fish Crows (*Corvus ossifragus*) are also a common predator of Piping Plover chicks on Napatree.

## RESULTS

Nesting data reported here are taken from the annual Piping Plover monitoring and management report published by USFWS (Durkin, 2022). In 2022, three pairs of Piping Plovers established territories on Napatree. The number of nesting pairs was well below the twenty-two-year average of 5.3 nesting pairs per year, a number unsurpassed since 2016. However, because three nesting pairs were also found in 2021, the plover population abundance is considered stable (Figure 2).

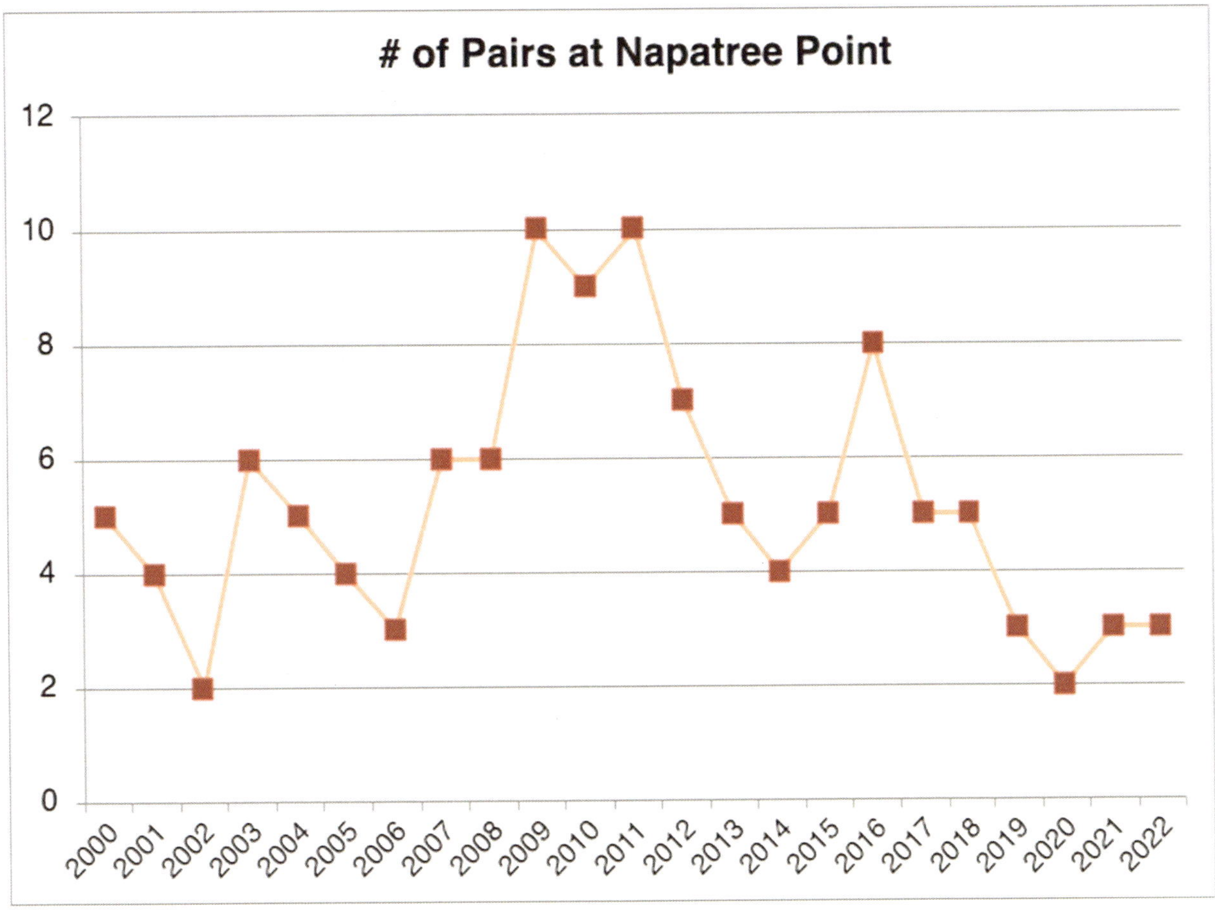


*Figure 2. Piping Plover nesting history at Napatree Point (Durkin, 2022).*

Among the three pairs, two nested on the oceanside of Napatree, while one nested on the bayside. One pair abandoned a viable nest on Napatree, still containing one egg, before renesting in a separate location. There were five nest attempts between the pairs, with only one being successful. All three hatchlings from this nest fledged successfully. With three total fledglings being produced from the three pairs attempting to nest, Napatree's 2022 average productivity was one chick per pair (Figure 3). This is a decrease from the 2021 average, but well above the overall average productivity since 2000, which is 0.63 chicks per pair (Figure 3).

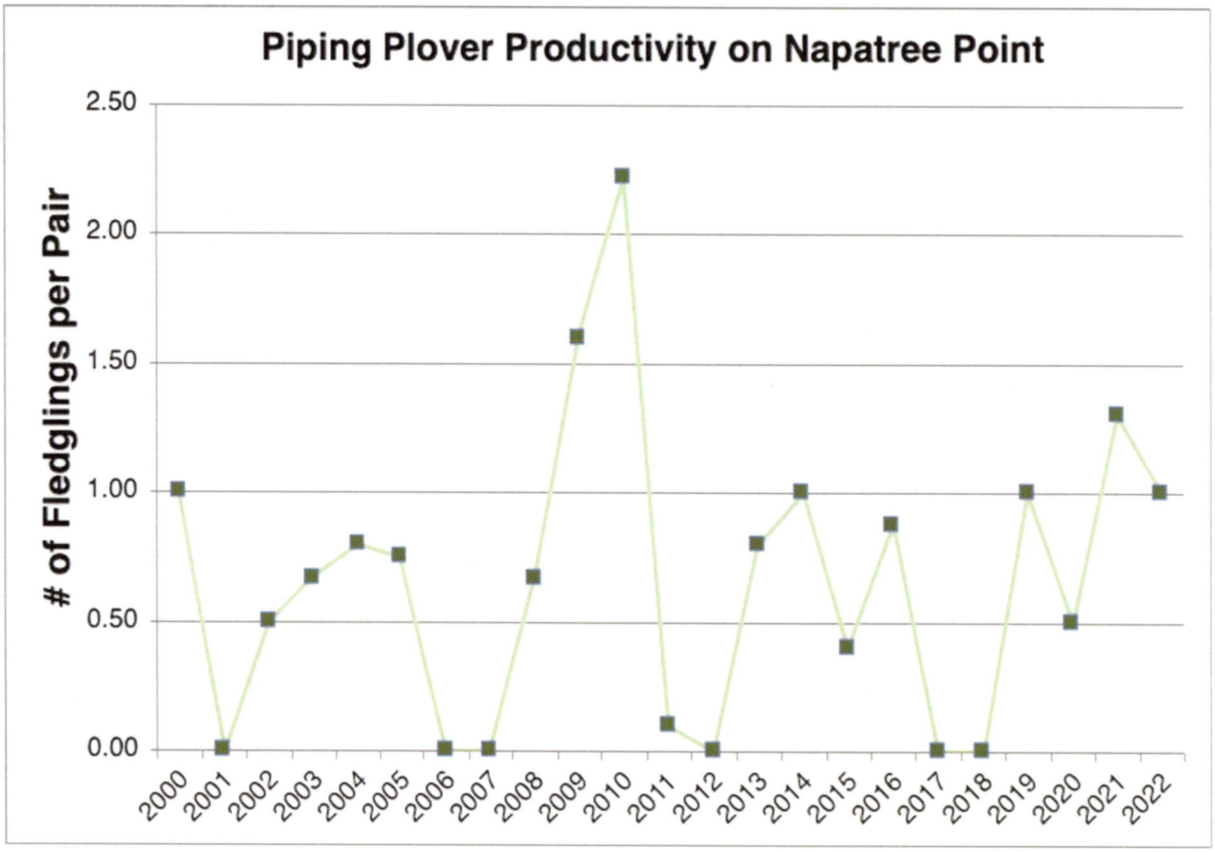


*Figure 3. Piping Plover nesting productivity at Napatree Point (Durkin, 2022).*

In addition to Peregrine Falcons, other historically common threats to plovers, such as human disturbance, disturbance by dogs, and overwash, continued throughout this season (Durkin, 2022). With nesting plover abundance increasing at neighboring locations, despite similar threats being present, it is not fully clear why Napatree's plover nest abundance has been low in recent years. However, Napatree is one of Rhode Island's most significant sites for plovers; the bayside continued to be an essential staging and stopover site with up to forty-five individuals appearing at the lagoon in late summer (Durkin, 2022).

## CONCLUSIONS

The number of nesting pairs (Figure 2) continued to be low, and productivity (Figure 3) dropped slightly from the previous year. Predation, notably by Peregrine Falcons, hindered the effectiveness of exclosures on Napatree (Durkin, 2022). The plovers nesting on the oceanside were successful, even without the help of exclosures. The bayside produced no successful nests. Disturbance to nesting birds, especially on the bayside, by pedestrians and dogs was a prime cause for the lack of nesting success (Durkin, 2022). However, despite the overall low numbers, Napatree's Piping Plover fledge rate was 100 percent, as all three hatchlings successfully fledged.

The Watch Hill Conservancy has little ability to control predator numbers on Napatree, but we can be more vigilant in ensuring that visitors observe municipal leash laws for dogs. As in past years, this should be a high priority management goal for 2023. The Conservancy's Naturalists will continue to remind pedestrians to always leash their dogs during the hours they are permitted on the beach. Naturalists will also insist that all visitors, including dogs, stay on denoted trails in the dunes to further protect areas where Piping Plovers might hide when feeling threatened.

The number of nesting Piping Plovers in USFWS-managed beaches in Rhode Island has been steadily increasing since 1992. The statewide results reflected stable numbers in 2022, with a record count of ninety-nine total pairs across Rhode Island (Durkin, 2022). East Beach, Napatree's neighboring beach, is still considered by USFWS to be "the largest contributor to productivity in Rhode Island," and reflects the importance of Watch Hill to the Piping Plover's statewide recovery (Durkin, 2022).

## ACKNOWLEDGMENTS

We thank everyone who aided in the Piping Plover conservation efforts at Napatree in 2022, especially USFWS Piping Plover Coordinator Maureen Durkin and her staff and volunteers.

## REFERENCES

Center for Biological Diversity. 2023. "Natural History of the Piping Plover." https://www.biologicaldiversity.org/species/birds/piping_plover/natural_history.html.

Durkin, Maureen. 2021. *Piping Plover Breeding Season Monitoring and Management Report, 2021*. Rhode Island National Wildlife Refuge Complex.

Durkin, Maureen. 2022. *Piping Plover Breeding Season Monitoring and Management Report, 2022*. Rhode Island National Wildlife Refuge Complex.

US Fish and Wildlife Service. 1994. "Guidelines for Managing Recreational Activities in Piping Plover Breeding Habitat on the US Atlantic Coast to Avoid Take Under Section 9 of the Endangered Species Act." US Fish and Wildlife Service, Northeastern Region.

Peregrine Falcon (*Falco peregrinus*) refueling during fall migration on Napatree (November 2017).
Photo Credit: Janice M. Sassi

CHAPTER 14

# Diurnal Raptor Trends during Fall Migration at Napatree: 1994–2020

*Joe Zbyrowski,[1] Clara Cooper-Mullin[2] & Peter Paton[2]*
[1] Master Bird Bander, Westerly, Rhode Island
[2] Department of Natural Resources Science, University of Rhode Island

## INTRODUCTION

Starting in the fall of 1994, Joe Zbyrowski initiated a diurnal raptor migration monitoring program at Napatree Point. Following HawkWatch protocols, he and colleagues stood on the crest of the eastern dune and recorded hawks migrating past Napatree, including both raptors migrating on the southern beachside of the crest and birds north of the Watch Hill cove (Figure 1).

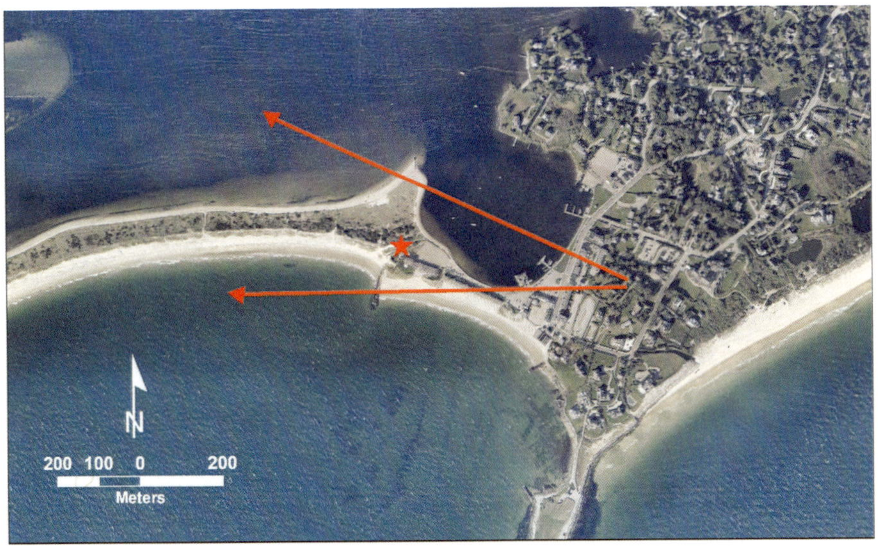


*Figure 1. Location of raptor survey site (red star) at Napatree conducted by Joe Zbyrowski and colleagues from 1994 to 2020 and primary flight directions of raptors at Napatree Point (red arrows) during fall migration.*

*Aerial Photograph: RIGIS 2022.*

## METHODS

From 1994 to2020, Joe Zbyrowski and fellow raptor observers spent a total of 402 days (1,181 total hours) conducting hawk surveys at Napatree (Tables 1, 2). They followed HawkWatch survey protocols to assess the relative abundance of diurnal raptors migrating past Napatree (Vekasy and Smith, 2002).

| Year | Number of Hours | Number of Days |
|---|---|---|
| 1994 | 58 | 17 |
| 1995 | 54 | 15 |
| 1996 | 42 | 13 |
| 1997 | 49 | 17 |
| 1998 | 89 | 28 |
| 1999 | 121 | 31 |
| 2000 | 123 | 39 |
| 2001 | 99 | 31 |
| 2002 | 130 | 40 |
| 2003 | 77 | 26 |
| 2004 | 28 | 16 |
| 2005 | 24 | 9 |
| 2006 | 8 | 3 |
| 2007 | 9 | 6 |
| 2008 | 22 | 8 |
| 2009 | 10 | 3 |
| 2010 | 48 | 14 |
| 2011 | 16 | 6 |
| 2012 | 15 | 7 |
| 2013 | 23 | 9 |
| 2014 | 19 | 8 |
| 2015 | 6 | 3 |
| 2016 | 11 | 5 |
| 2017 | 12 | 6 |
| 2018 | 35 | 19 |
| 2019 | 32 | 13 |
| 2020 | 21 | 10 |
| **Total** | **1,181** | **402** |

*Table 1. Annual variation in total diurnal raptor survey effort (number of hours and number of days) at Napatree from 1994 to 2020*

| Month | 1994–2003 Survey Days/Year (Mean ± SD) | 2004–2020 Survey Days/Year (Mean ± SD) |
|---|---|---|
| September | 13.3 ± 0.92 | 3.0 ± 0.41 |
| October | 11.5 ± 2.19 | 4.1 ± 0.54 |
| November | 2.6 ± 0.51 | 1.9 ± 0.71 |

*Table 2. Summary of monthly variation in raptor survey effort at Napatree for September, October, and November over two time periods from 1994 to 2020.*

## RESULTS

### Survey Effort

There were more surveys conducted in September and October than November (Table 2, Figure 2). Since November was not consistently surveyed each year, we exclude those data going forward for species-specific analyses of population trends. In addition, we only were able to calculate trends for species that were observed consistently in larger numbers (>0.05 individuals/day/year) across at least twenty years of the survey.

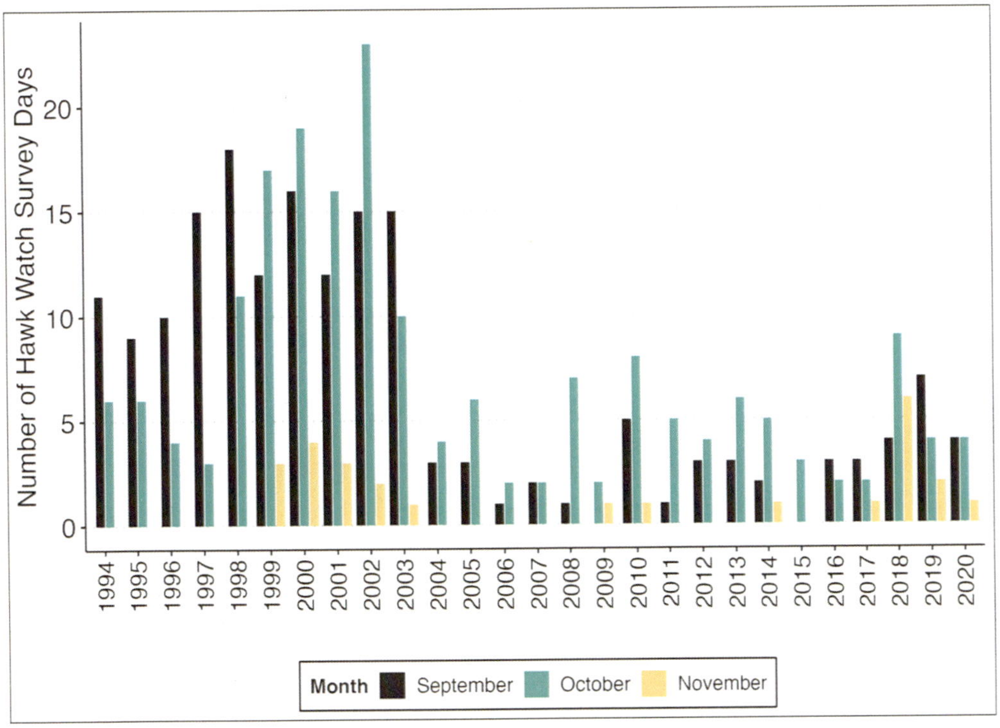


*Figure 2. Number of days raptor surveys were conducted at Napatree Point, Rhode Island, annually during September (black bars), October (turquoise bars), and November (yellow bars).*

Because annual survey effort declined after 2003 (Table 1, Figure 2), we analyzed trends from 1994 to 2003 separately from trends between 2004 and 2020 for each species. We aggregated hourly data to daily totals and calculated trends using generalized additive models (GAM) to determine non-linear population trends for each species and time period using the R package *poptrend* (Knape, 2016; Hornik and R Core

Team, 2022). Unless otherwise noted the frequency of observations are expressed as the mean (± standard deviation (SD)) number of raptors counted per sampling day.

## Abundance

There were sixteen species detected by the HawkWatch surveys during fall migration from 1994 to 2020 (Table 3). The most abundant and common species detected were American Kestrel, Osprey, Sharp-shinned Hawk, Cooper's Hawk, Merlin, and Northern Harrier (Table 3, Figure 3). Although Peregrine Falcons and Red-tailed Hawks were detected in at least twenty-one of the twenty-seven years of the survey, individuals of these species were more abundant between 1994 and 2003 (Peregrine Falcon: 0.25 ± 0.04; Red-tailed Hawk: 0.23 ± 0.04) but were infrequently counted after 2003 (Peregrine Falcon: 0.05 ± 0.02; Red-tailed Hawk: 0.05 ± 0.01). All other species were less common as they were detected in lower numbers and inconsistently across the twenty-seven years of the survey (Table 3, Figure 2).

| Species | Total Number of Detections | Mean Daily Count | SD Daily Count | Maximum Daily Count | Number of Years Detected 1994-2020 |
|---|---|---|---|---|---|
| Black Vulture, *Coragyps atratus* | 1 | NA | NA | 1 | 1 |
| Turkey Vulture, *Cathartes aura* | 54 | 0.13 | 0.89 | 10 | 8 |
| Osprey, *Pandion haliaetus* | 2,963 | 7.37 | 10.37 | 93 | 27 |
| Golden Eagle, *Aquila chrysaetos* | 2 | NA | NA | 1 | 2 |
| Northern Harrier, *Circus hudsonius* | 818 | 2.03 | 2.73 | 16 | 27 |
| Sharp-shinned Hawk, *Accipiter striatus* | 11,099 | 27.61 | 34.80 | 188 | 27 |
| Cooper's Hawk, *Accipiter cooperii* | 997 | 2.48 | 3.17 | 22 | 27 |
| Northern Goshawk, *Accipiter gentilis* | 18 | 0.04 | 0.43 | 7 | 8 |
| Unidentified Accipiter | 5 | 0.01 | 0.11 | 1 | 3 |
| Red-shouldered Hawk, *Buteo lineatus* | 21 | 0.05 | 0.30 | 3 | 9 |
| Broad-winged Hawk, *Buteo platypterus* | 89 | 0.22 | 1.18 | 17 | 11 |
| Red-tailed Hawk, *Buteo jamaicensis* | 225 | 0.56 | 1.48 | 13 | 22 |
| Rough-legged Hawk, *Buteo lagopus* | 1 | NA | NA | 1 | 1 |
| Unidentified *Buteo* | 10 | 0.02 | 0.20 | 2 | 5 |
| American Kestrel, *Falco sparverius* | 3,347 | 8.33 | 25.68 | 394 | 27 |
| Merlin, *Falco columbarius* | 997 | 2.67 | 3.74 | 23 | 27 |
| Peregrine Falcon, *Falco peregrinus* | 165 | 0.41 | 1.06 | 11 | 21 |
| Bald Eagle, *Haliaeetus leucocephalus* | 60 | 0.15 | 0.56 | 6 | 15 |
| Unidentified falcon | 11 | 0.03 | 0.20 | 3 | 6 |
| Unidentified raptor | 2 | NA | NA | 2 | 2 |

*Table 3. Estimated abundance (total number of detections, mean and SD of daily counts, and the number of years each species was detected) for all species of diurnal raptors identified at Napatree Point between 1994 and 2020. Since only one Black Vulture, one Rough-legged Hawk, two Golden Eagles, and two unidentified raptors were detected across all surveys in twenty-seven years, we did not calculate a mean count per survey for those species.*

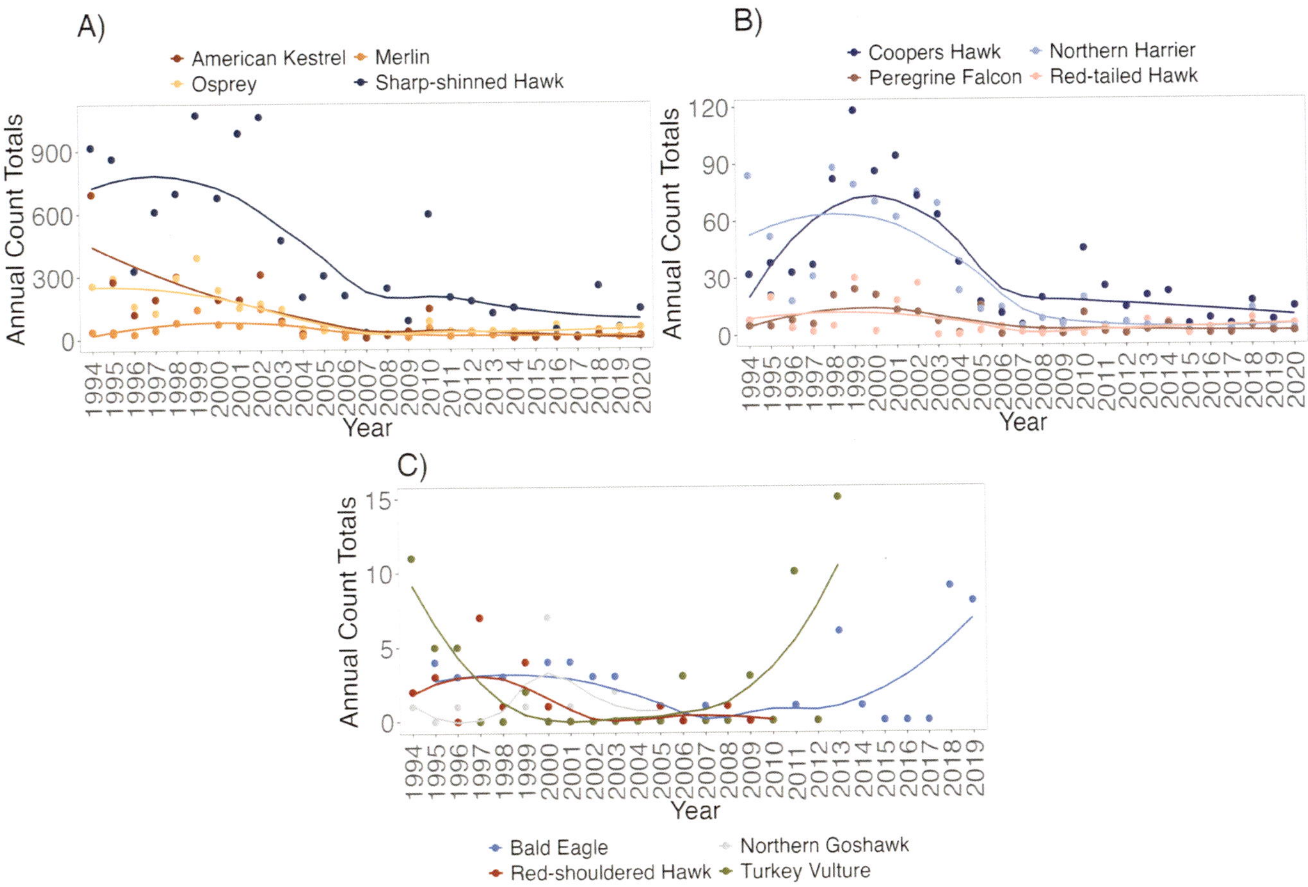


*Figure 3. A, Variation in annual counts of American Kestrel, Merlin, Osprey, and Sharp-shinned Hawk individuals at Napatree Point, Rhode Island, during fall migrations from 1994 to 2020. B, Total annual counts of Cooper's Hawk, Northern Harrier, Peregrine Falcon, and Red-tailed Hawk individuals at Napatree Point, Rhode Island, during fall migrations from 1994 to 2020. C, Total annual counts of Bald Eagle, Northern Goshawk, Red-shouldered Hawk, and Turkey Vulture individuals at Napatree Point, Rhode Island, during fall migrations from 1994 to 2020. Unknown and rare species are not included. Trend lines depict a smoothed (loess) visualization of change over time, but do not indicate significance.*

## Count Trends

Between 1994 and 2003, we detected significant declines in migrating American Kestrels, Osprey, and Sharp-shinned Hawks (Table 4, Figure 4*A–C*). There were no significant trends (declining or increasing) in migrating Cooper's Hawks, Merlin, Northern Harriers (Figure 4*D–F*), Northern Goshawks, Peregrine Falcons, and Red-tailed Hawks between 1994 and 2003 (Table 4). We also found no significant trends for any species for counts conducted between 2004 and 2020 (Table 4, Figure 4). There were enough Northern Goshawks, Peregrine Falcons, and Red-tailed Hawks detected between 1994 and 2003 to calculate species trends for that time period. However, after 2003 there were not enough individuals detected each year to provide trend information.

| Species | Trend (95% CI) 1994–2003 | Trend (95% CI) 2004–2020 |
|---|---|---|
| American Kestrel | ***-82% (-95%, -31%)*** | -66% (-98%, 947%) |
| Cooper's Hawk | 74% (-19%, 285%) | -47% (-95%, 507%) |
| Merlin | -50% (-83%, 34%) | -54% (-96%, 453%) |
| Northern Harrier | -17% (-58%, 59%) | -80 (-97%, 34%) |
| Osprey | ***-43% (-77%, -18%)*** | 59% (-91%, 2316%) |
| Sharp-shinned Hawk | ***-29% (-73%, -67%)*** | -46% (-99%, 5114%) |
| Peregrine Falcon | 49% (-66%, 503%) | NA |
| Red-tailed Hawk | -98% (-99%, 2658%) | NA |

*Table 4. Annual trends for each common raptor species counted at Napatree Point for two time periods: 1994 to 2003 and 2004 to 2020. Estimated percentage of population changes (trend) were calculated by generalized additive mixed models. Significant trends are bolded and italicized.*

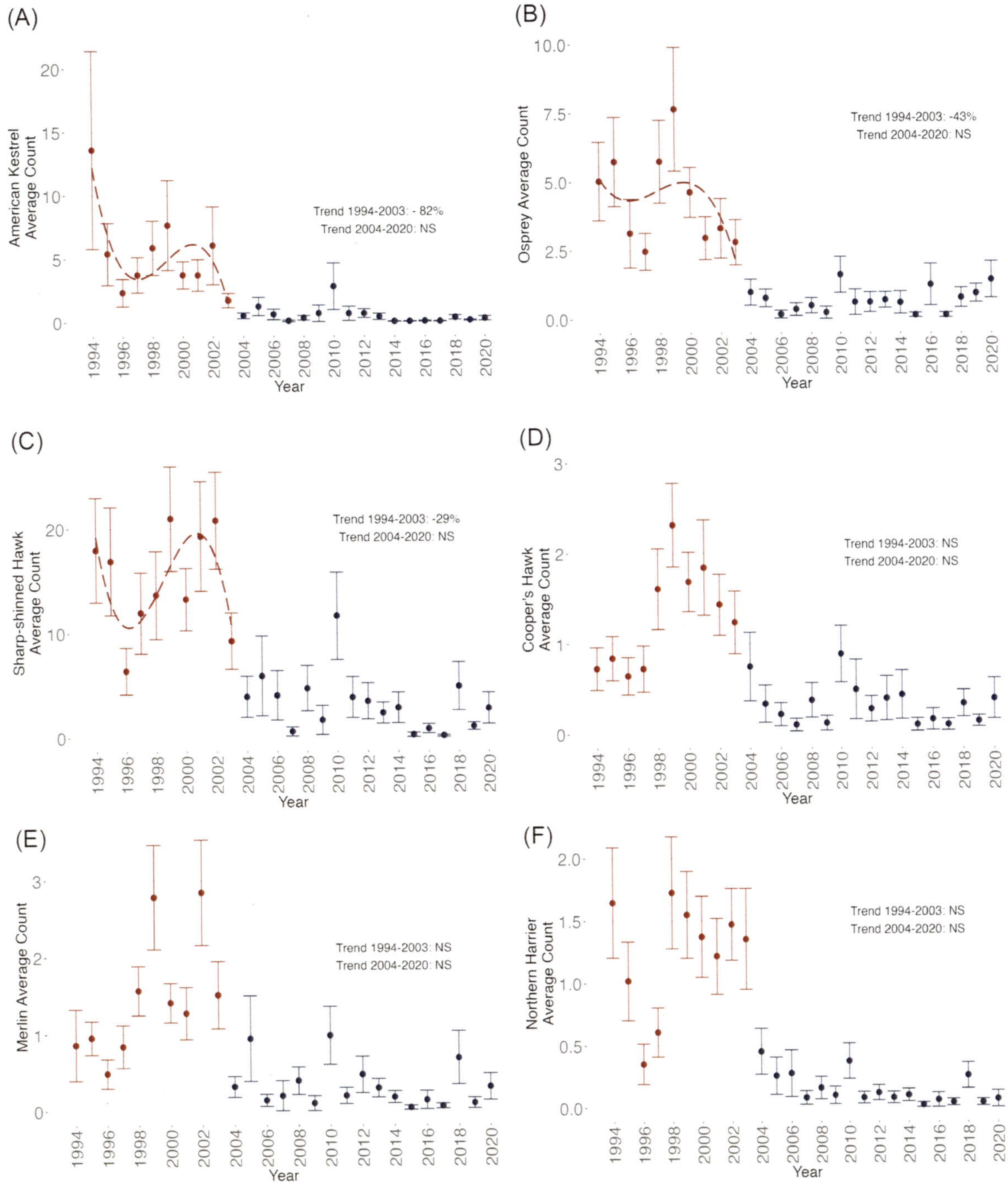


*Figure 4. Mean yearly counts (± standard error) of diurnal raptors at Napatree Point from 1994 to 2020 of (A) American Kestrel, (B) Osprey, (C) Sharp-shinned Hawk, (D) Cooper's Hawk, (E) Merlin, and (F) Northern Harrier individuals. Red dots represent counts conducted between 1994 and 2003, and blue dots represent counts conducted between 2004 and 2020. Trends for each species and time period are indicated by percent change or not significant (NS). A dashed line indicates a significant trend for a given species and/or time period.*

## DISCUSSION

Declines of the more abundant species at Napatree are consistent with regional and national trends in some cases but not others.

American Kestrel: The decline of this small falcon at Napatree is not surprising given our current understanding of population trends across North America. Available evidence suggests kestrel have been declining since the 1960s along the Atlantic coast and the Appalachian Mountains, although reasons for these declines are uncertain and the focus of recent research efforts (Bednarz and Therrien, 2023).

Osprey: The decline of Osprey at Napatree might seem surprising given what we know about the species' recovery from cessation of DDT use in southern New England. Bierregaard *et al.* (2014) provide an excellent summary of the decline of the species from New York City to Boston from approximately 1,000 pairs in 1940 to about 110 pairs in 1970. After DDT use ceased, the population rebounded, although high density colonies in Narragansett Bay and Mount Hope Bay never re-established. However, there is a new concentration of Ospreys at Westport, Massachusetts, and the species has expanded northward (Bierregaard *et al.*, 2014). Regional numbers of ospreys counted at inland HawkWatch sites in the northeastern United States also suggest a decline in Ospreys (Brandes *et al.*, 2016; Crewe *et al.*, 2016). Thus, the decline of Osprey detections at Napatree is perplexing.

Sharp-shinned Hawk: The Breeding Bird Survey suggests no evidence of a decline or increase in the Sharp-shinned Hawk population for New England or the Mid-Atlantic from 1966 to 2019 (www.mbr-pwrc.usgs.gov/bbs/reglist19v3.html). However, counts of migrating Sharp-shinned Hawks have been steadily declining at Hawk Mountain, Pennsylvania, since 1983 (Dumandan *et al.*, 2021, see supplemental material Figure 1).

Cooper's Hawk: In contrast to Sharp-shinned Hawk, Cooper's Hawk breeding numbers have been increasing dramatically in southern New England, with an estimated 3.4 percent annual increase in abundance from 1966 to 2019 based on the Breeding Bird Survey (www.mbr-pwrc.usgs.gov/bbs/reglist19v3.html). A similar trend has been documented at Hawk Mountain Sanctuary in Pennsylvania, with counts there steadily increasing since 1970 (Dumandan *et al.*, 2021, see supplemental material Figure 1). Thus, it is somewhat surprising that counts of this forest-dwelling species have declined at Napatree since the 1990s.

Merlin: Although counts of Merlin have declined over time at Napatree, available evidence suggests the number of migrants is actually increasing in the region. For example, at Hawk Mountain, Pennsylvania, Merlin counts have steadily increased since 1973 (Dumandan *et al.*, 2021, see supplemental material Figure 1).

Northern Harrier: This species declined at Napatree starting about 2004. At Hawk Mountain, Pennsylvania, counts of migrant harriers have been declining since about 1990 (Dumandan *et al.*, 2021, see supplemental material Figure 1). This might be related to increases in woodlands in the region and declines in grassland habitat where the species breeds.

## ACKNOWLEDGMENTS

Joe Zbyrowski was assisted in the field by Arnold Moorhouse, Arthur Payte, Moe Parenteau, and Helen Jankoski. This analysis was supported by funding from the University of Rhode Island Coastal Institute, The Watch Hill Conservancy, and the A. M. Roberts Charitable Foundation. Daniel Cole, Noah Barreto, and Peter August assisted in data entry and reviewing the manuscript.

## REFERENCES


Bednarz, J. C., and J. F. Therrien. 2023. “The Full Annual Cycle of the American Kestrel: State of the Knowledge, Information Gaps, and Conservation Needs.” *Journal of Raptor Research*, 57:125–130.

Bierregaard Jr, R. O., A. B. David, L. Gibson, R. S. Kennedy, A. F. Poole, M.S. Scheibel, and J. Victoria. 2014. “Post-DDT Recovery of Osprey (*Pandion haliaetus*) Populations in Southern New England and Long Island, New York, 1970–2013.” *Journal of Raptor Research*, 48:361–374.

Brandes, D., D. Oleyar, S. Hoffman, and L. Goodrich. 2016. “The Raptor Population Index, 2016 Regional Trend Summaries and Conservation Assessments.” Retrieved from https://www.rpi-project.org/2019-analysis/#2019-map.

Crewe, T., P. Taylor, D. Lepage, L. Goodrich, J. Brown, and J. Sodergren. 2016. “The Raptor Population Index, 2016 Analysis Methods and Trend Results.” Retrieved fromhttps://www.rpi-project.org/data-collection-and-analysis-methods/.

Dumandan, P. K. T., K. Bildstein, L. Goodrich, A. Zaiats, T. Caughlin, and T. Katzner. 2021. “Shared Functional Traits Explain Synchronous Changes in Long-Term Count Trends of Migratory Raptors.” *Global Ecology and Biogeography*, 30:640–650.

Hornik and R Core Team. 2022. “The Official FAQ.” https://cran.r-project.org/doc/FAQ/.

Knape, J. 2016. “Decomposing Trends in Swedish Bird Populations Using Generalized Additive Mixed Models.” *Journal of Applied Ecology*, 53:1852–1861.

Vekasy, M., and J. Smith. 2002. *HawkWatch International Raptor Migration Observer Procedures Manual*. Salt Lake City, UT: HawkWatch International.

*Ginsberg and Rothwell sampling bees.*
*Photo Credit: The Watch Hill Conservancy*

CHAPTER 15

# Bee Fauna of Coastal Napatree Point and Two Inland Sites in Southern Rhode Island

*Aya Rothwell*[1] *& Howard Ginsberg*[2]

[1] Department of Natural Resources Science, University of Rhode Island

[2] United States Geological Survey Patuxent Wildlife Research Center, Rhode Island Field Station, University of Rhode Island

**Citation:** Rothwell, A., and H. S. Ginsberg. 2019. "The Bee Fauna of Coastal Napatree Point and Two Inland Sites in Southern Rhode Island." *Northeastern Naturalist*, 26:446–464. Reprinted with permission.

# The Bee Fauna of Coastal Napatree Point and Two Inland Sites in Southern Rhode Island

Aya Rothwell[1] and Howard S. Ginsberg[2,*]

**Abstract** - We surveyed the bee fauna at Napatree Point, a coastal barrier beach in southwestern Rhode Island, using bee-bowl and netting samples, and compared results to bee-bowl samples at 2 inland sites. We collected a total of 53 species and morphospecies at Napatree Point, including 5 likely Rhode Island state records and several coastal dune and sand-nesting species that were not found inland. The comparative bee-bowl samples (colored bowls with soapy water placed at the sites to collect visiting bees) captured 35 species at Napatree Point and 66 at the inland sites (which included 6 likely state records, 2 shared with Napatree). The Napatree fauna shared numerous species with the inland sites, but overall species composition differed substantially. Both Napatree and inland sites showed greatest bee activity and species richness in spring. During spring, the most common bees at Napatree were twig- and cavity-nesting species such as *Ceratina dupla* and *Osmia simillima*, and the wood-nesting *Lasioglossum oblongum*, while the most abundant bees inland were the soil-nesting *Andrena nasonii* and *Augochlorella aurata*. Netting samples differed from bee-bowl samples in that they captured larger species and species foraging at flowers distant from the bee-bowl transects, but they missed several diminutive species that were captured by bee bowls. Use of 2 sampling methods, therefore, provided a broader view of the bee fauna than would have been possible with a single collection method.

## Introduction

Declines of pollinator species and the lack of monitoring programs to track their status have engendered increasing concern (Allen-Wardell et al. 1998, National Research Council 2007, Tepedino and Ginsberg 2000, Winfree 2010). Coastal bees and their habitats, in particular, are not well studied. Coastal areas are considered to be particularly vulnerable by the National Park Service and other organizations because of the potential effects of storms and sea-level rise associated with climate change (Rykken et al. 2014).

Coastal dune habitats have distinctive floras and faunas (Ehrenfeld 1990), with many dune specialists including rare and endemic species (Howe et al. 2010, Rykken et al. 2014). Recently, Ascher et al. (2014) surveyed the bees at Gardiners Island and surrounding islands near Long Island, NY, and Goldstein and Ascher (2016) surveyed bees at Martha's Vineyard, MA. Bees have also been surveyed at inland dune sites in Maryland (Selfridge et al. 2017) and at coastal sand dunes in Wales (Howe et al. 2010).

---

[1]Department of Natural Resources Science, University of Rhode Island, Kingston, RI 02881.
[2]US Geological Survey Patuxent Wildlife Research Center, Rhode Island Field Station, University of Rhode Island, Kingston, RI 02881. [*]Corresponding author - hginsberg@usgs.gov.

Manuscript Editor: David Orwig

Napatree Point, a barrier beach in Westerly at the southwestern corner of Rhode Island, is managed as a conservation area, but the bee fauna of this site has never been surveyed. We studied the bee fauna of Napatree Point using standardized sampling protocols to create a baseline for future monitoring efforts there, and we collected comparative inland samples at 2 sites in southern Rhode Island. We conducted 6 bee-bowl samplings over the spring and summer of 2017, where we set up bee bowls along transects at 2 sites in Napatree and 2 sites inland (Francis C. Carter Preserve and Great Swamp Management Area). We also netted at flowering patches at Napatree Point throughout the field season to provide a more complete survey of the Napatree bee fauna.

## Methods

### Study sites

We conducted the fieldwork for this study at 3 sites in Rhode Island: Napatree Point Conservation Area, Francis C. Carter Memorial Preserve, and Great Swamp Management Area (Fig. 1).

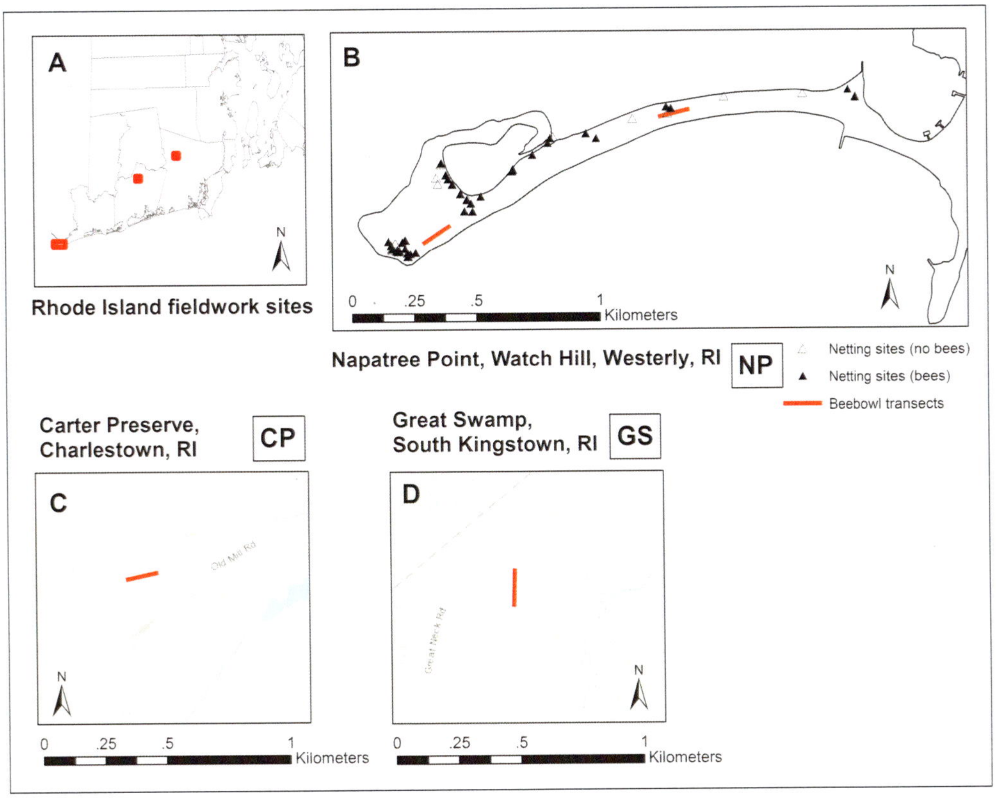


Figure 1. Map of Napatree Point field sites and inland field sites. (A) Map of lower half of Rhode Island showing sampling locations (as squares). (B) Map of Napatree Point showing bee-bowl transects and netting sites. (C) Bee-bowl transect location in Carter Preserve. (D) Bee-bowl transect location in Great Swamp. Maps created with ArcGIS using Esri World Topo Map.

Napatree Point Conservation Area, located in Westerly, RI, is a barrier beach habitat (sandy ocean beach, primary dune, secondary dune area, bay) on a moraine, with an abandoned fort and surrounding forested habitat at the west end. The area is bounded to the north by Little Narragansett Bay and to the south by the Atlantic Ocean (Mayo et al. 2015). Napatree Point is managed by the Watch Hill Conservancy and the Watch Hill Fire District. One bee-bowl transect was located in central Napatree Point (41°18'39.1"N, 71°52'19.5"W to 41°18'40.1"N, 71°52'14.0"W) in an area of secondary dunes dominated by *Ammophila breviligulata* Fernald (Beach Grass) meadows with patches of barrier-island shrubs (Ehrenfeld 1990); high winds are frequent. The other bee-bowl transect was located on the western end of Napatree (41°18'22.9"N, 71°53'00.6"W to 41°18'25.3"N, 71°52'55.7"W), closer to trees (including coastal shrub thickets and planted non-native species such as *Pinus thunbergii* (Parlatore) (Japanese Black Pine), a small lagoon, and the abandoned military fort (Fig. 1B). Additional information on Napatree Point is available from the Napatree Point Conservation Area (http://portal.napatreepoint.info/).

We collected comparative samples at 2 inland sites. One was at the Francis C. Carter Memorial Preserve in Charlestown, RI (Fig. 1C). The field site was located in a power line right-of-way (41°25'54.5"N, 71°40'16.1"W to 41°25'55.4"N, 71°40'10.4"W). The site is an open, dry area with trees on either side; the dominant vegetation consists of *Kalmia angustifolia* L. (Sheep Laurel), *Gaylussacia baccata* (Wangenh.) K. Koch (Black Huckleberry), and *Solidago* spp. (goldenrods). Carter Preserve is managed by the Nature Conservancy. Carter Preserve is 2.7 km inland and 21 km from Napatree Point.

The second inland site was in the Great Swamp Management Area in South Kingston, RI (Fig. 1D). The field was located on a walking trail in an old field surrounded by trees (41°28'23.2"N, 71°34'19.1"W to 41°28'28.1"N, 71°34'18.9"W). This site is a dense meadow of numerous herbaceous species, including *Potentilla* spp. (cinquefoils), *Trifolium repens* L. (White Clover), *Euthamia graminifolia* L. (Nuttall) (Grass-leaved Goldenrod), and *S. rugosa* Miller (Rough-stemmed Goldenrod). Great Swamp is managed by the RI Department of Environmental Management. Great Swamp is 3.4 km inland and 31 km from Napatree Point.

**Sampling methods**

Variations of pan or bowl traps have been widely used to sample bees (Cane et al. 2000, Droege et al. 2010, Westphal et al. 2008, Williams et al 2001). Bowl traps have some advantages over traditional capture by aerial netting in that they avoid investigator bias and can catch small bees that may be missed by netting (Droege et al. 2010, Selfridge et al. 2017, Westphal et al. 2008). However, the technique does not provide information on floral associations, and large bees may escape from bowls more easily (Westphal et al. 2008; A. Rothwell, pers. observ.). Though collecting bees with aerial nets can be influenced by collector bias, it readily captures larger bees, and so pairing the 2 methods can sample a broader range of bees than either alone (Cane et al. 2000).

We used the 2010 National Park Service Native Bee/Climate Change Study sampling protocol for bee-bowl sampling (Rykken et al. 2014). This sampling procedure compares bees found in vulnerable habitats to bees found in common inland habitats sampled using standardized bee-bowl transects. We set and collected bee bowls 6 times over the summer in 2017 (Fig. 2) on 16 May, 7 June, 27 June, 19 July, 23 August, and 10 September. (Great Swamp was re-sampled on 22 July, after a truck drove through the site and destroyed the 19 July samples). We placed bee bowls at coastal and inland sites on the same day.

The sampling transects were 150 m long, with 30 bee bowls spaced 5 m apart. The bee bowls consisted of plastic cups (7.5 cm diameter, 3.5 cm height), alternating white (unpainted) or painted blue or yellow (fluorescent colors, Guerra Paint and Pigment, New York, NY), in a pattern of blue, white, yellow. We filled the bowls 7/8 full with a solution of store-bought spring or distilled water mixed with blue Dawn dish detergent and left them out in the field for 24 h. While setting up and collecting the bee bowls, we wore lab gloves to avoid any effects of odor contamination (e.g., sweat bees lick human sweat). Insects collected from the bowls were placed into whirlpack bags with 70% ethanol, processed, and pinned prior to identification by S. Droege (Patuxent Wildlife Research Center, Laurel, MD), who also identified many of the netted specimens. We based bee sizes, for comparisons of bees captured in bee bowl vs. netting samples, on length measurements in Mitchell (1960, 1962) and Discover Life (www.discoverlife.org).

We surveyed flowering vegetation within 1 m on either side of each bee-bowl transect and tallied the number of 5-m transect sections (between bowls) with flowering vegetation. We recorded the occurrence of each flowering species within 1 m on either side of each transect.

For netting samples (taken only at Napatree Point), we monitored flowering plants and netted bees at flowering patches through the summer as flower species of interest (mostly common species in the barrier beach environment) entered peak anthesis. During the summer of 2017, we conducted netting on 10 days: 2 June, 14 June, 29 June, 5 July, 31 July, 10 August, 22 August, 5 September, 25 September, and 13 October (Fig. 2). A. Rothwell obtained all netting samples by collecting bees at each flowering patch for 30 min, and limited collections to 15 individuals per sample to avoid over-collecting. We transferred netted bees to labeled containers, and the specimens were held in a freezer at least overnight before being pinned and labeled. We identified the netted bees to species under

Figure 2. Dates for bee-bowl sampling and netting sampling. The bowl graphic symbolizes bee-bowl events and the net graphic symbolizes netting-sampling events. Netting only took place at Napatree Point. Bee bowls were placed at coastal and inland sites on the same day.

a dissecting microscope, using standard keys (Discoverlife.com; Mitchell 1960, 1962), with additional identifications and confirmations by S. Droege and J. Gibbs (University of Manitoba, Winnepeg, MN, Canada). We assessed apparent new records for Rhode Island by comparing our records to those listed as occurring in Rhode Island as per John Ascher's list of species records for the state on the Discover Life web site. We placed voucher specimens of most species in the University of Rhode Island Insect Collection, with additional specimens placed in the US National Collection (Smithsonian Institution, Washington, DC). We identified most flowering plants at the sample sites; difficult identifications were later verified by botanists (see Acknowledgments).

**Data analysis**

We compared the bee-bowl data from the 2 Napatree coastal sites and 2 inland sites for species diversity, species richness, and evenness. We collected bee-bowl samples in exactly the same way at each site; thus, we quantified species richness simply by totaling the number of species collected at each site. We also estimated the total numbers of species present at each bee-bowl transect using the online estimation program SPECRICH (www.mbr-pwrc.usgs.gov/software/specrich.html, accessed 20 July 2018), written by J.E. Hines (Patuxent Wildlife Research Center, Laurel, MD) based on the method of Burnham and Overton (1979). We employed the Shannon–Weiner diversity index (H') to quantify diversity (Peet 1974, Shannon and Weaver 1949, Southwood and Henderson 2000). We calculated evenness as 1 minus the Berger–Parker dominance index, or the proportion of the sample that consisted of species other than the dominant species (Southwood and Henderson 2000).

We compared species compositions of the 4 sample sites using canonical correspondence analysis (CCA) in the PAST package (paleontological statistics software package; Hammer et al. 2001). The samples were from the bee bowls at 4 transects, each sampled 6 times over the season (Fig. 2) for a total of 24 samples. The environmental variables included soil type (proportion of sand in soil types), forest cover, and distance from the coast. To assess these variables, we used ArcGIS to create a 200-m circular buffer from each bee-bowl transect. We analyzed soils data and the Rhode Island ecological communities classification data from the Rhode Island GIS database system (RIGIS, http://www.rigis.org/).

We compared taxonomic composition between the Napatree and inland sites by chi-square (SAS, version 9.3, FREQ procedure; SAS Institute, Inc., Cary, NC) using 10 taxonomic categories (Colletidae, Andrenidae, green Halictidae, *Halictus*, *Lasioglossum*, *Sphecodes*, Megachilidae, *Ceratina*, corbiculate Apidae, other Apidae). We combined the genera *Halictus*, *Lasioglossum*, and *Sphecodes* into 1 category (because of small sample sizes of some groups) for the comparison of netting vs. bee-bowl samples at Napatree Point.

To examine the phenology at each site, we plotted both bee abundance and numbers of species in each sample over the season. We tested these patterns for differences by comparing numbers at Napatree vs. inland sites during each of the 6 samples with chi-square tests, using SAS, version 9.3, FREQ procedure. We also compared differences in captures of bees in different size categories using chi-square tests.

## Results

At Napatree, we caught 654 individual bees at the combined bee-bowl sites and 176 individual bees through netted samples, for a total of 830 individuals. These collections represented 35 species and morphospecies at the bee-bowl sites and 33 species and morphospecies in the netted samples, totaling 53 species (Table 1). In the bee-bowl samples at the inland sites, we caught 448 individual bees at the Great Swamp and 262 individual bees at the Carter Preserve, for a total of 710 individuals. There were 46 species and morphospecies in the Great Swamp samples and 45 species and morphospecies at the Carter Preserve, totaling 66 species at the inland sites (Table 2). There is no official list of the bees of Rhode Island, but based on currently available compilations, our samples collected 9 species that are likely new records for the state (S. Droege, pers. comm., based on current records in Discover

Table 1. Species and number of individuals from netting sampling and bee-bowl sampling at Napatree. An asterisk (*) denotes possible Rhode Island State record. † denotes native pollen-specialist bees (oligolectic). Plant species: AB = *Ampelopsis brevipedunculata* (Porcelainberry), CE = *Cakile edentula* (Bigelow) Hook (Sea Rocket), CV = *Cirsium vulgare* (Bull Thistle), DC = *Daucus carota* L. (Queen Anne's Lace), HM = *Heracleum maximum* Bartram (Cow-parsnip), LaJ = *Lathyrus japonicus* (Beach Pea), LC = *Limonium carolinianum* (Sea-lavender), LoJ = *Lonicera japonica* Thunberg (Japanese Honeysuckle), RaR = *Raphanus raphanistrum* (Wild Radish), RoR = *Rosa rugosa* (Rugosa Rose), SC = *Solidago canadensis/altissima* (Canada Goldenrod/Tall Goldenrod), SS = *Solidago sempirvirens* (Seaside Goldenrod), TC = *Teucrium canadense* L. (American Germander). ‡ = Bee was near plant: not on flower. **14 males from netting samples were identified by J. Gibbs as possibly *L. atwoodi* Gibbs, *L. viridatum* Lovell or *L. oblongum* Lovell. These were included in with *L. oblongum*. ***4 males, tentatively identified by J. Gibbs as *L. hitchensii/subviridatum* (Cockerell), were included in *L.* sp. **** Possible *B. sandersoni* Franklin? *Augochlora pura* Say was collected at Napatree Point in additional samples in 2018. [Table continued on following page.]

| Family/species | Napatree bee bowls | Napatree netted bees | Plant species on which netted bees were collected |
|---|---|---|---|
| Colletidae | | | |
| *Colletes compactus* Cresson | 0 | 1 | SS |
| *Colletes kincaidii* Cockerell* | 0 | 3 | SS |
| *Colletes simulans Cresson*† | 0 | 3 | SS |
| *Hylaeus affinis* (Smith)/modestus Say | 2 | 3 | RaR, SC |
| *Hylaeus mesillae* Cockerell | 1 | 0 | |
| *Hylaeus schwarzii* (Cockerell) | 1 | 0 | |
| Andrenidae | | | |
| *Andrena alleghaniensis* Viereck* | 1 | 0 | |
| *Andrena asteris* Robertson† | 0 | 5 | SS |
| *Andrena commoda* Smith | 3 | 0 | |
| *Andrena hirticincta* Provancher† | 0 | 3 | SS |
| *Andrena nasonii* Robertson | 2 | 0 | |
| *Andrena perplexa* Smith | 1 | 0 | |
| *Andrena pruni* Robertson | 4 | 0 | |
| *Andrena thaspii* Graenicher | 0 | 1 | HM‡ |
| *Perdita octomaculata* Say† | 0 | 24 | SS |

Life: www.discoverlife.org; Scott et al. 2016). According to a recent compilation of northeastern native pollen-specialist or oligolectic bees (Fowler 2016), there were 4 native specialist bees in our Napatree Point samples (Table 1) and 1 at the Great Swamp (Table 2). We collected the specialist species during the expected time of the season, based on the flower taxa upon which they specialize.

Table 1, continued.

| Family/species | Napatree bee bowls | Napatree netted bees | Plant species on which netted bees were collected |
|---|---|---|---|
| Halictidae | | | |
| *Agapostemon sericeus* Forster | 1 | 1 | SS |
| *Agapostemon virescens* Fabricius | 5 | 1 | CV |
| *Augochlorella aurata* Smith | 6 | 1 | SC |
| *Halictus confusus* Smith | 2 | 1 | RaR |
| *Halictus ligatus* Say | 1 | 1 | AB |
| *Lasioglossum coriaceum* Smith | 4 | 0 | |
| *Lasioglossum ephialtum* Gibbs* | 0 | 9 | LC, SC |
| *Lasioglossum georgeickworti* Gibbs* | 27 | 0 | |
| *Lasioglossum leucozonium* Schrank | 1 | 0 | |
| *Lasioglossum marinum* Crawford | 53 | 19 | CE, LaJ, LC, RaR |
| *Lasioglossum oblongum* Lovell | 97 | 14** | LC |
| *Lasioglossum tegulare* Robertson | 5 | 1 | DC |
| *Lasioglossum versatum* Robertson | 4 | 0 | |
| *Lasioglossum zephyrum* Smith | 0 | 1 | |
| *Lasioglossum* sp. | 2 | 14*** | LC |
| *Sphecodes* sp. | 0 | 1 | AB |
| Megachilidae | | | |
| *Hoplitis pilosifrons* Cresson | 3 | 0 | |
| *Hoplitis producta* Cresson | 0 | 1 | |
| *Megachile melanophaea* Smith | 0 | 1 | LaJ |
| *Osmia atriventris* Cresson | 1 | 0 | |
| *Osmia bucephala* Cresson* | 1 | 0 | |
| *Osmia pumila* Cresson | 1 | 0 | |
| *Osmia simillima* Smith | 104 | 7 | LaJ, RaR |
| Apidae | | | |
| *Apis mellifera* L. | 0 | 6 | RaR, SS |
| *Bombus griseocollis* De Geer | 0 | 1 | RaR |
| *Bombus impatiens* Cresson | 0 | 24 | AB, CV, LaJ, RaR, RoR, SC, SS |
| *Bombus vagans* Smith**** | 1 | 11 | AB, LaJ, LC, LoJ, RaR, SC, TC |
| *Ceratina calcarata* Robertson | 48 | 2 | CV, SS |
| *Ceratina dupla* Say | 243 | 7 | LC, RaR, SC |
| *Ceratina mikmaqi* Rehan and Sheffield | 5 | 2 | RaR, CV |
| *Ceratina* sp. | 16 | 2 | LC, SC |
| *Melissodes druriellus* Latreille | 0 | 3 | SS |
| *Nomada articulata* Smith | 1 | 0 | |
| *Nomada* sp. (bidentate group) | 2 | 0 | |
| *Nomada luteolodies* Robertson | 1 | 0 | |
| *Nomada maculata* Cresson | 4 | 0 | |
| *Xylocopa virginica* L. | 0 | 2 | SS |
| Total individuals | 654 | 176 | |
| Total species | 35 | 33 | |

Table 2. Species and number of individuals from bee bowls at inland sites. An asterisk (*) denotes possible Rhode Island State record. † denotes native pollen-specialist bees (oligolectic), and ** indicates that some specimens may be *H. poeyi* Lepeletier. [Table continued on following page.]

| Family/species | Great Swamp bee bowls | Carter Preserve bee bowls |
|---|---|---|
| Colletidae | | |
| *Hylaeus affinis* (Smith)/*modestus* Say | 14 | 2 |
| Andrenidae | | |
| *Andrena bradleyi* Viereck† | 1 | 0 |
| *Andrena carlini* Cockerell | 4 | 9 |
| *Andrena nasonii* Robertson | 67 | 87 |
| *Andrena perplexa* Smith | 1 | 0 |
| *Calliopsis andreniformis* Smith | 0 | 2 |
| Halictidae | | |
| *Agapostemon sericeus* Forster | 0 | 3 |
| *Agapostemon texanus* Cresson | 1 | 6 |
| *Agapostemon virescens* Fabricius | 28 | 5 |
| *Augochlora pura* Say | 0 | 2 |
| *Augochlorella aurata* Smith | 92 | 32 |
| *Augochlorella persimilis* (Viereck)* | 1 | 2 |
| *Augochloropsis metallica* (Fabricius) | 1 | 0 |
| *Halictus confusus* Smith | 3 | 1 |
| *Halictus ligatus* Say | 13** | 8** |
| *Halictus parallelus* Say | 1 | 2 |
| *Halictus rubicundus* (Christ) | 0 | 3 |
| *Lasioglossum abanci* (Crawford) | 1 | 0 |
| *Lasioglossum acuminatum* McGinley | 0 | 3 |
| *Lasioglossum bruneri* (Crawford) | 0 | 1 |
| *Lasioglossum coeruleum* (Robertson)* | 1 | 0 |
| *Lasioglossum coreopsis* (Robertson)* | 0 | 1 |
| *Lasioglossum coriaceum* Smith | 11 | 3 |
| *Lasioglossum cressonii* (Robertson) | 28 | 0 |
| *Lasioglossum ephialtum* Gibbs* | 1 | 0 |
| *Lasioglossum leucocomum* (Lovell) | 0 | 3 |
| *Lasioglossum leucozonium* Schrank | 6 | 2 |
| *Lasioglossum oblongum* Lovell | 1 | 0 |
| *Lasioglossum oceanicum* (Cockerell) | 2 | 3 |
| *Lasioglossum pectorale* (Smith) | 0 | 6 |
| *Lasioglossum smilacinae* (Robertson) | 1 | 0 |
| *Lasioglossum tegulare* Robertson | 6 | 14 |
| *Lasioglossum timothyi* Gibbs | 0 | 1 |
| *Lasioglossum versatum* Robertson | 38 | 10 |
| *Lasioglossum* sp. | 0 | 1 |
| *Sphecodes coronus* Mitchell | 1 | 0 |
| *Sphecodes mandibularis* Cresson | 1 | 0 |
| *Sphecodes ranunculi* Robertson | 0 | 1 |
| *Sphecodes* sp. | 0 | 1 |
| Megachilidae | | |
| *Hoplitis producta* Cresson | 1 | 0 |
| *Hoplitis spoliata* (Provancher) | 0 | 1 |
| *Megachile brevis* Say | 0 | 1 |
| *Osmia atriventris* Cresson | 5 | 4 |
| *Osmia bucephala* Cresson* | 0 | 1 |

The most common flowering species near each transect are listed below. Napatree 1: *Lathyrus japonicus* Willdenow (Beach Pea), *Rosa rugosa* Thunberg (Rugosa Rose), *Raphanus raphanistrum* L. (Wild Radish), *Oenothera biennis* L. (Common Evening Primrose), and *Erigeron canadensis* L. (Horseweed). Napatree 2: Rugosa Rose, *Toxicodendron radicans* (L.) Kuntze (Eastern Poison Ivy), and *Lepidium virginicum* L. (Virginia Pepperweed); Great Swamp: *Taraxacum offiniale* Weber ex Wiggers (Dandelion), *Potentilla* spp. (cinquefoils), *Rosa multiflora* Thunberg (Multiflora Rose), *Rubus* spp. (brambles), *Stellaria graminea* L. (Lesser Stitchwort), White Clover, *Achillea millefolium* L. (Yarrow), *Vicia* sp. (a vetch), *Hypericum perforatum* L. (St. Johnswort), *Spiranthes vernalis* Engelmann & Gray (Spring Lady's Tresses), abundant *Solidago* (especially *S. juncea* Aiton [Early Goldenrod] and Rough-stemmed Goldenrod), Grass-leaved Goldenrod, and *Symphyotrichum racemosum* (Elliott) Nesom (Small White Aster). Carter Preserve: *Kalmia* spp. (laurels), Black Huckleberry, *Lysimachia quadrifolia* L. (Whorled Loosestrife), *Rubus hispidus* L. (Swamp Dewberry), goldenrods (especially *S. odora* Aiton [Sweet Goldenrod]), and Small White Aster.

We collected a total of 56 netting samples from 24 flowering species, with most bees collected at *Solidago sempervirens* L. (Seaside Goldenrod) and other golden-

Table 2, continued.

| Family/species | Great Swamp bee bowls | Carter Preserve bee bowls |
|---|---|---|
| *Osmia collinsiae* Robertson | 1 | 0 |
| *Osmia inspergens* Lovell and Cockerell | 0 | 1 |
| *Osmia pumila* Cresson | 11 | 2 |
| Apidae | | |
| *Apis mellifera* L. | 2 | 0 |
| *Bombus fervidus* Fabricius | 0 | 1 |
| *Bombus griseocollis* De Geer | 3 | 0 |
| *Bombus impatiens* Cresson | 5 | 2 |
| *Bombus vagans* Smith | 2 | 0 |
| *Ceratina calcarata* Robertson | 43 | 4 |
| *Ceratina dupla* Say | 19 | 6 |
| *Ceratina mikmaqi* Rehan and Sheffield | 9 | 6 |
| *Ceratina* sp. | 7 | 3 |
| *Melissodes bimaculatus* (Lepeletier) | 0 | 1 |
| *Nomada articulata* Smith | 1 | 5 |
| *Nomada* sp. (bidentate group) | 1 | 0 |
| *Nomada cressonii* Robertson | 3 | 0 |
| *Nomada imbricata* Scopoli* | 1 | 0 |
| *Nomada maculata* Cresson | 2 | 0 |
| *Nomada pygmaea* Cresson | 3 | 1 |
| *Nomada sayi/illinoensis* Robertson | 3 | 0 |
| *Nomada* sp. | 0 | 7 |
| *Peponapis pruinosa* (Say) | 1 | 2 |
| Total individuals | 448 | 262 |
| Total species | 46 | 45 |

rods, *Limonium carolinianum* (Walter) Britton (Sea-lavender), Wild Radish, Beach Pea, *Ampelopsis brevipedunculata* (Maximovich) Trautvetter (Porcelain Vine), and *Cirsium vulgare* (Savi) Tenore (Bull Thistle).

The bee fauna collected in bee bowls at Napatree differed substantially in taxonomic composition from that at the inland sites ($\chi^2 = 530.3$, df = 9, $P < 0.0001$). Estimates of species diversity, richness, and evenness tended to be higher at inland than at Napatree transects (Table 3). A CCA shows Napatree samples distinctly clumped toward the left along the first (horizontal) axis, with inland samples to the right (Fig. 3). Inland samples were scattered far more broadly along the second

Table 3. Species diversity, richness, and evenness at the bee-bowl transects. Total number of species at each bee-bowl transect estimated using SPECRICH (https://www.mbr-pwrc.usgs.gov/software/specrich.html).

| | Napatree Site 1 | Napatree Site 2 | Great Swamp | Carter Preserve |
|---|---|---|---|---|
| Diversity (Shannon–Weiner index) | 0.82 | 0.75 | 1.23 | 1.23 |
| Species richness (total species collected) | 22 | 26 | 46 | 45 |
| Estimate of total number of species (± SE) | 49.8 (± 10.37) | 34.0 (± 4.00) | 111.4 (± 22.15) | 59.0 (± 5.29) |
| Evenness (1 - Berger–Parker index) | 0.62 | 0.51 | 0.79 | 0.67 |

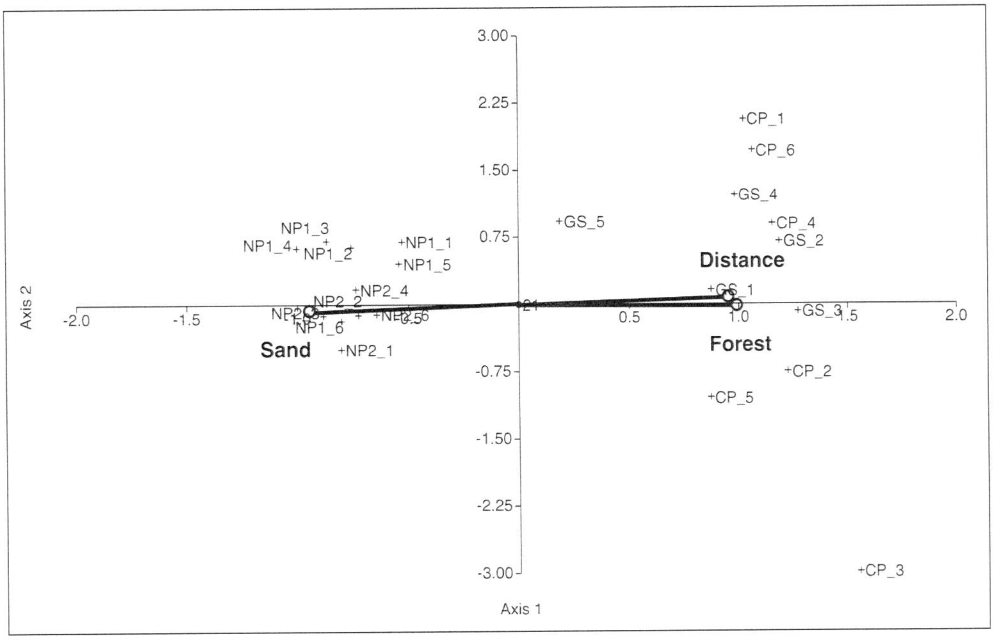


Figure 3. Canonical correspondence analysis (CCA) ordination of bee-bowl samples at Napatree sites and inland sites. "Sand" indicates the percent of sandy soil composition, "Distance" indicates the distance to the coast, "Forest" indicates the percent of forest cover. NP1 indicates Napatree Site 1, NP2 indicates Napatree Site 2, GS indicates Great Swamp and CP indicates Carter Preserve. A number indicating the sampling date session follows each site.

(vertical) axis than the Napatree samples. Each data point in Figure 3 represents a sample for a single site on a single day (except that no bees were caught at Napatree Site 2 on 23 August). The vectors for all of the environmental variables (soil type, forest cover, and distance from the coast) fell along the horizontal axis, suggesting that these environmental factors were associated with the difference in bee communities between coastal and inland sites.

Bee phenologies at Napatree Point and at inland sites displayed the greatest numbers of individuals and species in the spring (Fig. 4), but they differed in several details (individuals: $\chi^2 = 187.9$, df = 5, $P < 0.0001$; species: $\chi^2 = 16.5$, df = 5, $P = 0.0056$). Bee numbers apparently declined through the season, but by early June, individual numbers decreased sharply at Napatree but more gradually at the inland

Figure 4. Phenology of bee-bowl samples at Napatree sites compared to inland sites.

sites. Late in the season, individual numbers dropped inland but rose at Napatree Point. Species numbers displayed similar patterns at Napatree Point and inland sites (with inland species numbers being higher). However, the number of species at Napatree increased in the September samples.

Common species (those with greater than 45 individuals collected) were most often collected early in the season at Napatree Point (Fig. 5). *Ceratina dupla*, a twig-nesting species, was the most abundant species at Napatree early in the season, and numbers again increased slightly at the end of the season. The most

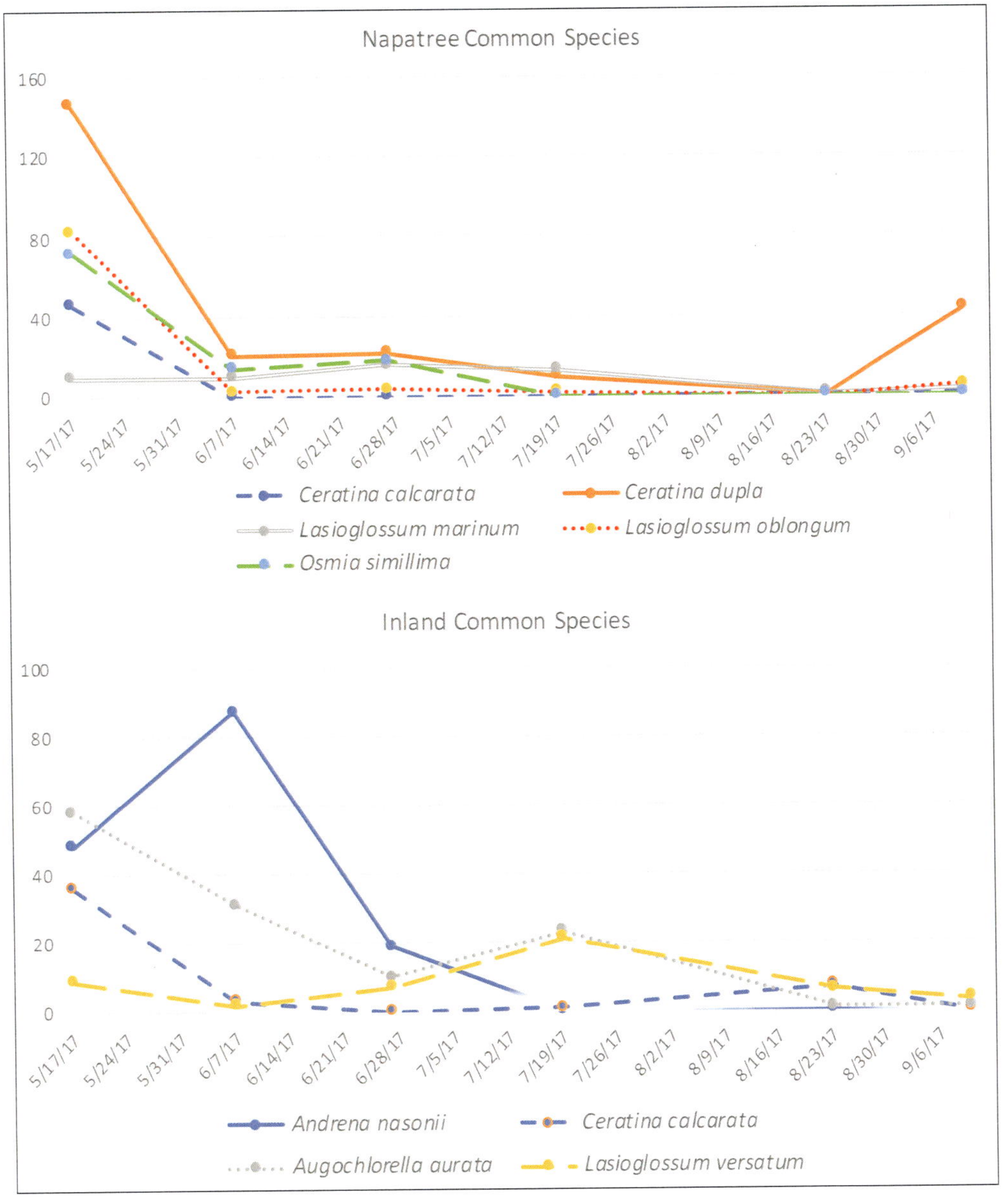


Figure 5. Total number of individuals of common species captured in bee bowls at Napatree Point and at inland sites.

commonly collected species at the inland sites was the soil-nesting *Andrena nasonii*, which peaked in early June. There was also a small July increase in numbers of the social, soil-nesting halictids *Augochlorella aurata* and *Lasioglossum versatum* at the inland sites.

The bee taxa detected by netted samples differed substantially from those from bee-bowl samples at Napatree Point ($\chi^2$ = 338.2, df = 7, $P$ < 0.0001). Bee-bowl samples were biased toward smaller bees (Fig. 6), capturing a significantly lower proportion of large bees than did netting samples ($\chi^2$ = 36.6, df = 1, $P$ < 0.0001). The most commonly netted species included bees that were not found in bee-bowl samples at Napatree, including *Perdita octomaculata* and larger bees such as *Bombus impatiens* and *B. vagans*, as well as some species that were also common in bee-bowl samples, such as the coastal dune species *Lasioglossum marinum* (Fig. 5). We collected no netting samples in May 2017 because there were no flowering species at anthesis in the open secondary dune habitat where we placed the bee-bowl transects. However, site visits in May 2018 revealed several flowering herbs in the woods and thicket habitats surrounding the fort at the western end of Napatree Point, including *Cardamine parviflora* L. (Sand Bittercress), *Arabidopsis thaliana* (L.) Heynhold (Thale Cress), *Barbarea vulgaris* Aiton (Bittercress), *Galium aparine* L. (Cleavers), and Dandelion. In netted samples at Napatree, the Shannon–Weiner index value was 1.27, species richness was 33, and evenness was 0.86. For comparison, the combined bee-bowl samples at Napatree had a Shannon–Weiner index value of 0.916, species richness of 35, and evenness of 0.63.

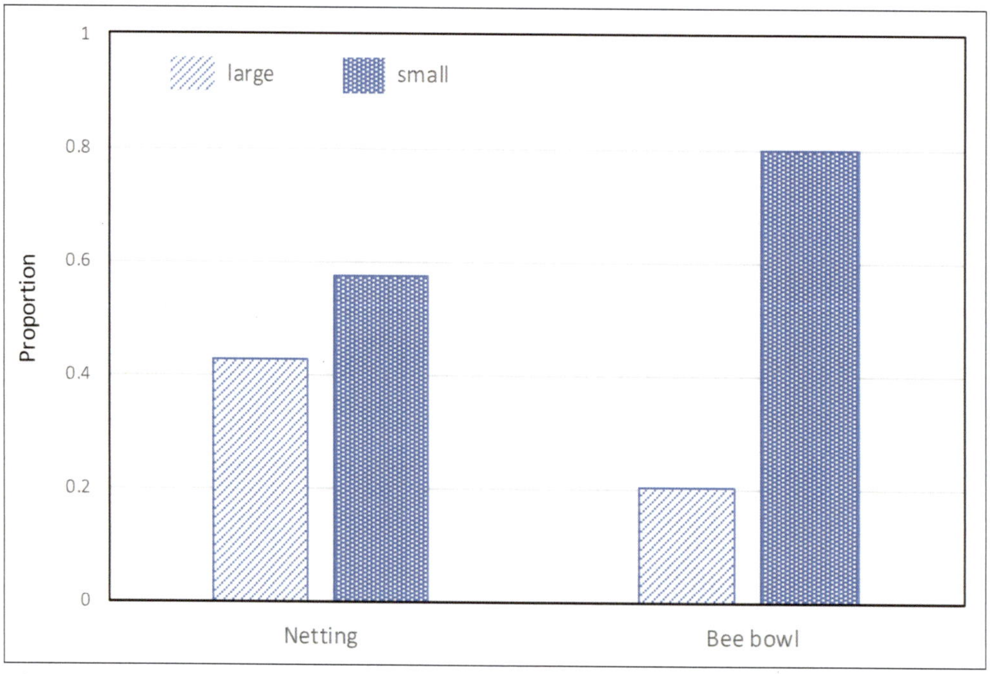


Figure 6. Sizes of bees captured in bee bowls and netting samples at Napatree Point. Proportion of all captures that were large (≥10 mm length) vs. small (<10 mm length) collected by the 2 sampling methods.

## Discussion

Based on our samples, the bee fauna at Napatree differred from that of the 2 inland sites in several ways. Many species were shared between Napatree Point and the 2 inland sites (Tables 1, 2), but species composition of the Napatree bee fauna was clearly distinct from those of the 2 inland sites (Fig. 3). The horizontal axis of the CCA, which reflects similarities of species composition among samples, is essentially an axis of distance from the coast, which is characterized by sandier soils and less forest cover at Napatree than inland, with associated differences in the bee faunas.

Species diversity (as measured by the Shannon–Weiner index), species richness, and evenness, tended to be lower at Napatree than at the inland sites, but there were more novel species at Napatree, including 4 oligolectic native bee species—*Colletes simulans, Andrena asteris, Andrena hirticincta, Perdita octomaculata*—that were not present at the inland sites (Tables 1, 2). *Lasioglossum marinum* and *L. oblongum*, though both considered uncommon bees (Gibbs 2011), were abundant at Napatree Point. Both species are also abundant on Grass Island, CT, another coastal site (Zarillo and Stoner, in press).

Previous bee collections at Great Gull Island and Fishers Island, RI (coastal habitats within 5–23 km [3–14 mi] of Napatree Point) did not yield *L. oblongum* (Ascher et al. 2014). However, some of those surveys were conducted before the name *L. oblongum* was in standard usage, starting ca. 1960 (Gibbs 2010, Mitchell 1960); thus, *L. oblongum* might have been present but not recognized taxonomically in early records. The most recent survey at those islands was in 1976 (Ascher et al. 2014). Pan-trapping became a popular bee-sampling method starting in the 1990s (Cane et al. 2000), and the earlier surveys at Great Gull Island and Fishers Island presumably used netting to sample bees and might have under-sampled smaller-sized bees such as *L. oblongum*. More recent samples from coastal northeastern sites used bee bowls and collected *L. oblongum* (Goldstein and Ascher 2016; Rykken and Farrell 2013; Zarillo and Stoner, in press).

Bee phenology was similar at Napatree and inland sites in general form, with some interesting differences. At Napatree, the numbers of individuals and species captured increased during the last bee-bowl sampling in September, whereas the inland sites showed a decline in numbers on the same date. The number of *Ceratina dupla* captured on Napatree increased in September, possibly because *C. dupla* can be bivoltine at some sites, creating a second brood in later summer (Vickruck et al. 2011). Foraging by late season females and males before overwintering might also be possible.

The most common coastal and inland bee species differed in nest-site associations. Abundant species at Napatree Point included *L. marinum*, which is a coastal dune species, and *L. oblongum*, which nests in rotting logs (Sakagami and Michener 1962) and has been collected from under the bark of fallen logs (Gibbs 2011). This species has been found in forests (Gibbs 2010, 2011; Ulyshen et al. 2010) as well as coastal areas (Zarillo and Stoner, in press), including Gardiners Island, in New York, about 29 km (18 miles) from Napatree Point (Ascher et al. 2014), and on Martha's Vineyard, MA (Goldstein and Ascher 2016). *Osmia simillima*, another common spring bee at Napatree, nests in twigs and cavities, and could presumably find appropriate nesting sites in the woods and possibly in the crumbling walls of the fort. Interestingly, other samples at coastal sites collected relatively few individuals of this species (Ascher et al. 2014; Goldstein and Ascher 2016; Rykken and Farrell 2013; Stage 2009; Zarillo and Stoner, in press). This species has been reported to nest in *Quercus* (oak) apple-galls (Cane et al. 2007) and wood buried in a dune area (Scott 2017). We did not see oaks on Napatree, but driftwood and pine wood are common. In general, *Osmia* are cavity nesters and can use a wide variety of substrates (Bosch 2001). *Bombus* spp. typically nest in larger hollows, including cavities under rock piles (Hatfield et al. 2012). On Napatree, *O. simillima* and the *Bombus* species may have utilized crevices and hollows at the abandoned fort structure located at the west end of Napatree. Alternatively, these strong-flying species could have flown in from nearby mainland nesting sites. *Osmia* species can fly up to 500 m (Biddinger et al. 2013) and *Bombus* species can fly up to several kilometers (Rao and Strange 2012). The sand-nesting species *Perdita octomaculata* was also common at Napatree but not inland, although this species can occur at inland sites with sandy soil (Eickwort 1977).

Abundant inland species included *Andrena nasonii, Augochlorella aurata,* and *Lasioglossum versatum*, which are all ground-nesting bees (Michener 1966, Renauld et al. 2016, Richards et al. 2011, Selfridge et al. 2017). *Ceratina calcarata* was among the common bees found both at coastal Napatree and inland sites. This species is a twig nester that uses brambles, *Rhus* (sumac), and other plants with soft pith for nesting (Ginsberg 1983, Vickruck et al. 2011); these are common plants at the inland sites. *Rhus copallinum* L. (Poison Sumac) has been reported to grow on the west and east end of Napatree (H. Leeson, Rhode Island Natural History Survey, Kingston, RI, pers. comm.), and we observed apparent *Ceratina* nests in twigs of other shrub species that had been clipped or had been broken or browsed by deer.

Bee species collected in netting samples differed substantially from bee-bowl samples at Napatree Point. The differences in phenology may partly have resulted from the relatively late start of netting sampling (about 2 weeks after bee-bowl samples) because we did not detect any flowering activity in the open dune habitats when we took the first bee-bowl sample in May. Site visits in 2018 revealed several herbaceous species flowering in the woods around the fort, where we had not sampled in 2017. The end of the season showed a difference in phenology as well, in that *Bombus* captures increased through the season in netting samples, as is typical for bumble bees (Plowright and Laverty 1984), but they were not captured in the bee bowls. *Bombus* can thermoregulate (Heinrich 1972), and thus can forage at lower fall temperatures than other bees. Many of these late-summer *Bombus* specimens were males or gynes.

Bee size is another factor in sampling effectiveness (Fig. 6). The most common species collected in bee bowls at Napatree included the diminutive *Lasioglossum* and *Ceratina* bees. Netting samples caught high numbers of *L. marinum* but did not catch other bees commonly collected in bee bowls (Table 1). In general, bee bowls catch smaller bees that can be missed by netting (Droege et al. 2010, Selfridge et al. 2017, Westphal et al. 2008). We captured numerous *B. impatiens* and *B. vagans* and smaller numbers of *B. griseocollis* and *Xylocopa virginica* in netting samples, but captured none of these species in bee bowls, likely because these larger bees could climb out of the bee bowls (A. Rothwell, pers. observ.; Westphal et al. 2008).

We captured several *Perdita octomaculata* by netting, while our bee-bowl samples never included this species (Table 1). *Perdita octomaculata* is a small bee which would presumably be effectively sampled by bee bowls. This species emerges and forages on goldenrods late in the summer (Eickwort 1977, Ginsberg 1983), and we netted this bee on 25 September. The last bee-bowl sample was on 10 September, which may have been too early to catch the species. Furthermore, the 2 locations where *P. octomaculata* were caught were at least 44 m from the closest bee-bowl transect (Napatree Site 2), and *P. octomaculata* might have foraged for its preferred host plants, goldenrods, at sites distant from the bee-bowl transect. *Perdita octomaculata* specifically nests in sandy slopes, which is the area where they were netted, and the bees may not have foraged in more distant areas.

Some investigators have reported that results from bee bowls and bee netting were highly correlated (Richards et al. 2011), but there were marked differences in our study. Richards et al. (2011) conducted timed walking samples using sweep nets in a figure-eight motion to collect insects from vegetation, flowers, etc., which differs substantially from our more traditional method of focusing on a single flowering patch for a period of time. The different results between our bee bowl and netting samples suggest that the focused netting method we used helped capture a distinct subset of the bee fauna that bee-bowl sampling missed. We limited our netting collections at flower patches to 15 individuals per sample, which undoubtedly affected the numbers of selected species captured. Netting samples were taken through all open areas of Napatree and thus were not restricted to just 2 transect sites, as were the bee-bowl samples. Therefore, our results suggest that, while repeated samples using objective methods such as bee bowls have great value for comparative samples and monitoring programs, multiple sampling methods provide a more complete view of a local bee fauna for survey purposes.

### Acknowledgments

We thank the organizations involved with the study sites, including the Watch Hill Conservancy and Watch Hill Fire District, the Rhode Island Department of Environmental Management, and the Nature Conservancy. Janice Sassi, Manager of the Napatree Point Conservation Area, and her crew helped transport us through the field sites. The Napatree team also provided information about the field sites and sometimes accompanied us during fieldwork. Hope Leeson, Noah Conway and Robin Baranowski helped identify flowering vegetation. The Rhode Island Department of Environmental Management and Leland Mello helped us set up the study site at the Great Swamp Management Area. The Nature Conservancy and Jeanne Cooper allowed us to set up a study site at Francis C. Carter Memorial Preserve and provided information about the area. We thank Sam Droege for help with sampling design and for much of the specimen handling and identification, and Jason Gibbs for identification of some *Lasioglossum* netted specimens. Steve Alm and Sam Droege provided helpful comments on early drafts of the manuscript. Research was funded, in part, by a Dean's grant from the University of Rhode Island (URI) MESM program, the URI Graduate Student Association and a grant-in aid of research from the URI Coastal Institute. Any use of trade, firm, or product names is for descriptive purposes only and does not imply endorsement by the US Government.

## Literature Cited

Allen-Wardell, G., P. Bernhardt, R. Bittner, A. Burquez, S. Buchmann, J. Cane, P. Cox, P. Feinsinger, M. Ingram, D. Inouye, C. Jones, K. Kennedy, P. Kevan, H. Koopowitz, R. Medellin, S. Medellin-Morales, and G. Naban. 1998. The potential consequences of pollinator declines on the conservation of biodiversity and stability of food crop yields. Conservation Biology 12:8–17.

Ascher, J.S., S. Kornbluth, and R.G. Goelet. 2014. Bees (Hymenoptera: Apoidea: Anthophila) of Gardiners Island, Suffolk County, New York. Northeastern Naturalist 21:47–71.

Biddinger, D.J., N.K. Joshi, E.G. Rajotte, N.O. Halbrendt, C. Pulig, K.J. Naithani, and M. Vaughan. 2013. An immunomarking method to determine the foraging patterns of *Osmia cornifrons* and resulting fruit set in a cherry orchard. Apidologie 44:738–749.

Bosch, J., Y. Maeta, and R. Rust. 2001. A phylogenetic analysis of nesting behavior in the genus *Osmia* (Hymenoptera: Megachilidae). Annals of the Entomological Society of America 94:617–627.

Burnham, K.P., and W.S. Overton. 1979. Robust estimation of population size when capture probabilities vary among animals. Ecology 60:927–936.

Cane, J.H., R.L. Minckley, and L.J. Kervin. 2000. Sampling bees (Hymenoptera: Apiformes) for pollinator community studies: Pitfalls of pan-trapping. Journal of the Kansas Entomological Society 73:225–231.

Droege, S., V.J. Tepedino, G. LeBuhn, W. Link, R.L. Minckley, Q. Chen, and C. Conrad. 2010. Spatial patterns of bee captures in North American bowl-trapping surveys. Insect Conservation and Diversity 3:15–23.

Ehrenfeld, J. 1990. Dynamics and processes of barrier island vegetation. Aquatic Sciences 2:437–480.

Eickwort, G.C. 1977. Aspects of the nesting biology and descriptions of immature stages of *Perdita octomaculata* and *P. halictoides* (Hymenoptera: Andrenidae). Journal of the Kansas Entomological Society 50:577–599.

Environmental Systems Research Institute (ESRI). 2012. "Topographic" [basemap]. 1:15,000. "World Topographic Map". 19 February 2012. Available online at http://www.arcgis.com/home/item.html?id=30e5fe3149c34df1ba922e6f5bbf808f. Accessed 25 May 2017.

ESRI. 2017. ArcGIS Desktop: Release 10.5.1. Redlands, CA.

Fowler, J. 2016. Specialist bees of the Northeast: Host plants and habitat conservation. Northeastern Naturalist 23:305–320.

Gibbs, J. 2010. Revision of the metallic species of *Lasioglossum* (*Dialictus*) in Canada (Hymenoptera, Halictidae, Halictini). Zootaxa 2591:1–382.

Gibbs, J. 2011. Revision of the metallic *Lasioglossum* (*Dialictus*) of eastern North America (Hymenoptera: Halictidae: Halictini). Zootaxa 3073:1–216.

Ginsberg, H.S. 1983. Foraging ecology of bees in an old field. Ecology 64:165–175.

Goldstein, P.Z., and J.S. Ascher. 2016. Taxonomic and behavioral composition of an island fauna: A survey of bees (Hymenoptera: Apoidea: Anthophila) on Martha's vineyard, Massachusetts. Proceedings of the Entomological Society of Washington 118(1):37–92.

Hammer, Ø., D.A.T. Harper, and P.D. Ryan. 2001. PAST: Paleontological statistics software package for education and data analysis. Palaeontologia Electronica 4(1):9.

Hatfield, R., S. Jepsen, E. Mader, S.H. Black, and M. Shepherd. 2012. Conserving bumble bees: Guidelines for creating and managing habitat for America's declining pollinators. The Xerces Society for Invertebrate Conservation, Portland, OR. 32 pp.

Heinrich, B. 1972. Temperature regulation in the bumblebee *Bombus vagans*: A field study. Science 175:185–187.

Howe, M.A., G.T. Knight, and C. Clee. 2010. The importance of coastal sand dunes for terrestrial invertebrates in Wales and the UK, with particular reference to aculeate Hymenoptera (bees, wasps, and ants). Journal of Coastal Conservation 14:91–102.

Mayo, T.W., P.W. Paton, and P.V. August. 2015. Responses of birds to humans at a coastal barrier beach: Napatree Point, Rhode Island. Northeastern Naturalist 22:501–512.

Michener, C.D. 1966. Interaction among workers from different colonies of sweat bees (Hymenoptera, Halictidae). Animal Behaviour 14:126–129.

Mitchell, T.B. 1960. Bees of the Eastern United States: Volume I. NC Agricultural Experimental Station Technical Bulletin 141:1–538.

Mitchell, T.B. 1962. Bees of the Eastern United States: Volume II. NC Agricultural Experimental Station Technical Bulletin 152:1–557.

National Research Council Committee on the Status of Pollinators in North America. 2007. Status of Pollinators in North America. The National Academies Press, Washington, DC. 312 pp.

Peet, R.K. 1974. The measurement of species diversity. Annual Review of Ecology and Systematics 5:285–307.

Plowright, R.C., and T.M. Laverty. 1984. The ecology and sociobiology of bumble bees. Annual Review of Entomology 29:175–199.

Rao, S., and J.P. Strange. 2012. Bumble bee (Hymenoptera: Apidae) foraging distance and colony density associated with a late-season mass flowering crop. Environmental Entomology 41:905–915.

Renauld, M., A. Hutchinson, G. Loeb, K. Poveda, and H. Connelly. 2016. Landscape simplification constrains adult size in a native ground-nesting bee. Plos One 11(3):e0150946.

Richards, M., A. Rutgers-Kelly, J. Gibbs, J. Vickruck, S. Rehan, and C. Sheffield. 2011. Bee diversity in naturalizing patches of Carolinian grasslands in southern Ontario, Canada. The Canadian Entomologist 143:279–299.

Rykken, J.J., and B.D. Farrell. 2013. Boston Harbor Islands all taxa biodiversity inventory: Discovering the "microwilderness" of an urban island park. Natural Resource Technical Report NPS/BOHA/NRTR—2013/746. National Park Service, Fort Collins, CO.

Rykken, J., A. Rodman, S. Droege, and R. Grundel. 2014. Pollinators in peril? A multipark approach to evaluating bee communities in habitats vulnerable to effects from climate change. Park Science 31:84–90.

Sakagami, S.F., and C.D. Michener. 1962. The Nest Architecture of the Sweat Bees (Halictinae). University of Kansas Press, Lawrence, KS.

Scott, V. 2017. Use of communal nest entrances by *Osmia simillima* (Hymenoptera: Megachilidae). The Great Lakes Entomologist 26:10.

Scott, Z., H.S. Ginsberg, and S.R. Alm. 2016. Native bee diversity and pollen-foraging specificity in cultivated Highbush Blueberry (Ericaceae: *Vaccinium corymbosum*) in Rhode Island. Environmental Entomology 45:1432–1438.

Selfridge, J.A., C.T. Frye, J. Gibbs, and R.P. Jean. 2017. The bee fauna of inland sand dune and ridge woodland communities in Worcester County, Maryland. Northeastern Naturalist 24:421–445.

Shannon C.E., and W. Weaver. 1949. The Mathematical Theory of Communication. University of Illinois Press, Urbana IL. 117 pp.

Southwood, T.R.E., and P.A. Henderson. 2000. Ecological Methods. Blackwell Science, Hoboken, NJ. 575 pp.

Stage, G.I. 2009. Survey of the bees (Hymenoptera: Apoidea) of Penikese and Cuttyhunk Islands. Summary of field work, results, and preliminary conclusions. Final Report to Massachusetts Natural Heritage and Endangered Species Program, Westborough, MA.

Tepedino, V.J., and H.S. Ginsberg. 2000. Report of the US Department of Agriculture and US Department of the Interior Joint Workshop on Declining Pollinators, 27–28 May 1999, Logan, UT. US Geological Survey Information and Technology Report USGS/BRD/ITR-2000-0007. Biological Resources Division, Springfield, VA. 9 pp.

Ulyshen, M.D., V. Soon, and J.L. Hanula. 2010. On the vertical distribution of bees in a temperate deciduous forest. Insect Conservation and Diversity 3:222–228.

Vickruck, J.L., S.M. Rehan, C.S. Sheffield, and M.H. Richards. 2011. Nesting biology and DNA barcode analysis of *Ceratina dupla* and *C. mikmaqi*, and comparisons with *C. calcarata* (Hymenoptera: Apidae: Xylocopinae). The Canadian Entomologist 143:254–262.

Westphal, C., R. Bommarco, G. Carré, E. Lamborn, N. Morison, T. Petanidou, S.G. Potts, S.P.M. Roberts, H. Szentgyörgyi, T. Tscheulin, B.E. Vaissière, M. Woyciechowski, J.C. Biesmeijer, W.E. Kunin, J. Settele, and I. Steffan-Dewenter. 2008. Measuring bee diversity in different European habitats and biogeographical regions. Ecological Monographs 78:653–671.

Williams, N.M., R.L. Minckley, and F.A. Silveira. 2001. Variation in native bee faunas and its implications for detecting community changes. Conservation Ecology 5(1):7.

Winfree, R. 2010. The conservation and restoration of wild bees. Annals of the New York Academy of Sciences 1195:169–197.

Zarrillo, T.A., and Stoner, K.A. In press. The bee fauna of an Atlantic coastal plain tidal marsh community in Southern New England, USA. Journal of Mellitology.

*Catocala nebulosi* caught on Napatree. First record for Rhode Island.
Photo Credit: Mark Mello

CHAPTER 16

# A Survey of Moths in Two Rhode Island Barrier Dune Systems, Goosewing Beach Preserve in Little Compton and Napatree Point Conservation Area in Westerly, during 2020

*Mark J. Mello & Jamie Bogart*

Lloyd Center for the Environment

Dartmouth, Massachusetts

**Citation**: M. J. Mello, and J. Bogart. 2021. "A Survey of Moths in Two Rhode Island Barrier Dune Systems, Goosewing Beach Preserve in Little Compton and Napatree Point Conservation Area in Westerly, during 2020." Rhode Island Natural History Survey, Kingston, Rhode Island. Reprinted with permission.

A beautifully illustrated abbreviated version can be found in the Rhode Island Natural History Survey's Rhode Island Naturalist, Volume 16, Number 1, Spring 2021. Available at: https://rinhs.org/wp-content/uploads/2021/09/ri_naturalist_vol16-num1_spring21.pdf.

LLOYD CENTER FOR THE ENVIRONMENT Protecting nature through research, education and outreach.
430 Potomska Road | Dartmouth | Massachusetts | 02748 | tel 508-990-0505 | fax 508-993-7868 | www.lloydcenter.org

# A survey of moths in two Rhode Island barrier dune systems, Goosewing Beach Preserve in Little Compton and Napatree Point Conservation Area in Westerly, during 2020

Mark J. Mello
Jamie Bogart

Report to:

Rhode Island Natural History Survey
&
The Nature Conservancy of Rhode Island

January 2021

Lloyd Center Report #2021-1

## INTRODUCTION

Napatree Point Conservation area in Westerly (Figure 1) and Goosewing Beach Preserve in Little Compton (Figure 2) are small barrier spits in Rhode Island that are representative of many seen in southern New England. Although Rhode Island's barrier beaches are largely protected, dune systems may be developed should it be shown that the development will have no negative effect on resource values such as flood prevention. Also, other states may not necessarily have similar protections (Figure 3). Accelerated sea level rise impacts the primary dunes, and storms impact both primary and secondary dunes. Thus, insects utilizing barrier dune systems may be squeezed by habitat loss on both the seaward and landward ends.

Figure 1. Napatree Point Conservation Area, station locations.

Figure 2. Goosewing Conservation Area, station location.

**Figure 3. Map of barrier dunes** (from Cape San Blas website). "There are boardwalks running throughout the neighborhood that are not pictured on this map!"

"Learn more about our Cape San Blas (Florida) Vacation Rentals in Barrier Dunes."

Staff at the Lloyd Center have conducted surveys within barrier dune systems at Miacomet Beach on Nantucket, MA, Good Harbor and Wingarsheek Beaches in Gloucester, MA (2013-14), Horseneck Reservation in Westport, MA (2014-16; 2019), South Cape Beach in Mashpee (2019), and on Cuttyhunk Island in Gosnold (2019); or identified samples collected in barrier systems from Crane Beach Ipswich, MA (2006) for The Trustees of Reservations by Franz Ingelfinger and Plum Island in Newbury and Rowley, MA (2013-14) by National Wildlife Refuge staff. These surveys have resulted in the documentation of 525 species of macro-moths from 182 light trap and eight bait samples, including five species protected by the Massachusetts Endangered Species Act: *Euchaena madusaria* (2 individuals), *Apamea inebriata* (1), *Sympistis riparia* (17), *Dargida rubripennis* (1), and *Abagrotis benjamini* (18) (Mello & Bogart, in prep.).

An internet search for other barrier dune moth surveys turned up scant records in the Northeast. Goldstein and Nelson (2017) conducted intensive surveys at barrier dune systems in Dartmouth, MA (Allens Pond), and Martha's Vineyard (Lobsterville), for the larval hostplant or plants of *Simpistis riparia* and *Abagrotis benjamini,* two moth species listed as Special Concern in the Massachusetts Endangered Species Act, for which the larval hostplant was unknown. They found that beach plum (*Prunus maritima*) was documented as the sole hostplant for both species at these sites. They also noted several other species that are in decline or are psammophilic species that occur in this habitat, including *Hyalophora cecropia, Eucoptocnemis fimbriaris, Sphinx drupiferarum, Schinia spinosae* and *Drasteria graphica*, as well as other species of *Abagrotis*. Goldstein and Nelson conclude that "dunes may be overlooked as primary natural areas for Lepidoptera". The purpose of this current survey was to expand upon current knowledge of the relationship and importance of barrier dune habitats for moths.

## METHODS

Trap locations for sampling were chosen to provide late spring through early fall coverage of barrier dune habitats within the Napatree (Figure 1) and Goosewing (Figure

2) barrier dune systems. Portable ultraviolet light traps were operated roughly bimonthly for each system from June 3 to October 6 (Table 1), thus, 22 light trap samples were obtained. Light traps charged with ethyl acetate were set prior to dusk and retrieved after dawn the following morning.

Table 1. Sampling schedule for light trap operation conducted in 2020.

| System | | Goosewing Beach Preserve | Napatree Conservation Area | | | | | | | TOTAL # |
|---|---|---|---|---|---|---|---|---|---|---|
| Stations | Day | GW1 | N1 | N2 | N3 | N4 | N5 | N6 | N1-N6 | samples |
| June | 3 | | X | | | | | | 1 | 1 |
| | 17 | | | X | X | | | | 2 | 2 |
| | 23 | X | | | | | | | | 1 |
| | 30 | | | X | | X | | | 2 | 2 |
| July | 14 | X | | | | | | | | 1 |
| | 19 | | | X | | X | | | 2 | 2 |
| | 27 | X | | | | | | | | 1 |
| August | 6 | | | X | | | X | | 2 | 2 |
| | 24 | | | | | | X | | 1 | 1 |
| | 26 | X | | | | | | | | 1 |
| September | 13 | X | | | | | | | | 1 |
| | 16 | | | X | | X | | | 2 | 2 |
| | 24 | X | | | | | | | | 1 |
| October | 6 | X | X | | | X | | X | 2 | 2 |
| Total # samples | | 7 | 2 | 5 | 1 | 4 | 2 | 1 | 15 | 22 |

All macro-moths were identified to species with the exception of the Geometrid genus, *Eupithecia*. "*Eupithecia* sp." may include two or more species, but for the purposes of species "counts", this genus was treated as being comprised of one species unless otherwise specified. At least one specimen per sample of species listed in the Massachusetts Endangered Species Act (MESA) was saved as vouchers. At least one voucher of each other species per system was saved as well. All macrolepidoptera from light trap samples were counted, and the numbers in the tables for these species reflect numbers of individuals present. Microlepidoptera (considered to be the families Eriocraniidae through Pterophoridae, as listed on the Moth Photographers Group website) were also saved, but not counted. A partial list is presented, although members of some families, particularly the Gelechiidae await identification. Unlike for the macro-moths, a complete collection and count of the micro's was not attempted due to their small size, condition and/or unfamiliarity with some of the groups.

The Noctuoidea in the checklist for Moths of North America (Hodges et al., 1983) have been revised by Lafontaine & Schmidt (Zookeys 40:1-239 (2010) and Pohl (2016), which are available on the web (www.pensoftonline.net/zookeys and https://www.researchgate.net/publication/302570819 respectively). I have incorporated the changes in species, genus, and subfamily designations for the Noctuoidea. I have retained the catalog numbers as they appear in the 1983 checklist, save for species described after 1983 that have new numbers; but I have listed and ordered the sequence by the new Zookeys catalog numbers. Many of the MONA numbers will therefore be out of sequence, reflecting the new phylogenetic arrangement. The annotated checklist by Pohl, et al. (2016) has reconciled taxonomy for the remaining families, as well as the Noctuoidea, however, nomenclatural changes since these publications have also been incorporated.

GPS coordinates for each station were recorded with a handheld Garmin 76 unit, and the habitat classifications follow the U.S. National Classification System.

## STUDY AREA

**Napatree Point Conservation Area (**Figure 1) is an 86-acre barrier spit owned and managed by the Watch Hill Fire District and the Watch Hill Conservancy. It extends west from the large parking lot of a private beach club to mouth of Little Narragansett Bay. Most of the barrier spit is dominated by American beachgrass, however the point appear to have been a small glacial island (probably on bedrock) on which an historical fort and associated structures once existed. Non-native plants, especially shrubs are dominant in the uplands and phragmites dominate the freshwater swamp/marsh. A small area of saltmarsh occurs along the border of the lagoon,

Station N1 (N41.31119°; W-71.86295°) is located next to the second access path from the parking area to the beach. American beachgrass, small bayberry, beach rose, and seaside goldenrod predominate within 50 meters of the trap site. Poison ivy and red cedar are also present.

Station N2 (N41.30747°; W-71.88115°) is located immediately adjacent to the west side of the pitch pine grove prior to the large shrub-dominated end of the spit. American beachgrass, pitch pine, seaside goldenrod and small bayberry are predominant. In addition, red cedar, poison ivy, beach rose, and beach pea were present.

Station N3 (N41.30761°; W-71.87747°) is located within the long stretch of the beachgrass-dominated barrier dunes prior to the lagoon inlet. American beachgrass is predominant, but there are small patches of beach rose, bayberry, seaside goldenrod and beach pea. Scattered red cedar are also present.

Station N4 (N41.30612°; W-71.88378°) is located along the footpath leading to the old fort opposite the remains of a groin.

Station. N5 (N41.30610°; W-71.88432°) is located atop the eastern end of the old fort – the highest point on the Napatree Conservation area overlooking maritime shrubland, shrub swamp and phragmites-dominated marsh.

Station N6 (N41.30769°; W-71.88107°) is located immediately adjacent to the east side of the pitch pine grove adjacent to the long stretch of beachgrass and seaside goldenrod. In addition, to the aforementioned species, small bayberry, red cedar, pinweed, pearly everlasting, poison ivy, beach rose, groundsel shrub common mullein, saltmeadow cordgrass and smooth cordgrass were present.

**Goosewing Beach Preserve** (Figure 2) is a 3-acre barrier spit located between the Tunipus and Quicksand Ponds' inlets. It is part of a 75-acre property owned by The Nature Conservancy. The barrier spit is dominated by American beachgrass, seaside goldenrod, small bayberry and beach rose.  Salt marsh borders the inner boundary with quicksand pond.

Station GW1 (N41.49695°; W-71.13099°) is located one-third of the way out between the start of the barrier spit and the inlet to Quicksand Pond. American beachgrass, small bayberry, beach rose, and seaside goldenrod predominate within 50 meters of the trap site. Also present are saltmeadow cordgrass, pearly everlasting, bush honeysuckle, groundsel shrub, poison ivy, beach pea, freshwater cordgrass, switch grass, swamp rose-mallow, phragmites, and other grass species.

Table 2 summarizes the habitat types within roughly 50 meters of each trap station.

Table 2. Dune communities surveyed at Napatree and Goosewing conservation areas during 2020.

GW = Goosewing; N = Napatree; D = dominant habitat; C = Co-dominant habitats; S = secondary habitat; M - marginal habitat

| **Community type*** | Stations | | | | | | |
|---|---|---|---|---|---|---|---|
| | N1 | N2 | N3 | N4 | N5 | N6 | GW1 |
| beachgrass/herbaceous Alliance (GH) | D | C | D | | | C | S |
| maritime shrubland | M | C | M | D | M | M | D |
| pitch pine grove/woodland | | C | | | | S | |
| shrub swamp | | | | | S | | |
| invasive-dominated shrubland | | | | S | D | | |
| salt marsh | | | | | | M | M |

## RESULTS

A total of 145 macro-moth species were identified from seven light trap samples at the one station established on Goosewing and 161 species from 15 samples at the six 6 stations established on Napatree during 2020 (Appendix I) resulting in 216 species of macro-moths recorded by this survey. A summary by station is presented in Table 3.

Table 3, Summary of macro-moths documented at Napatree Point Conservation Area and Goosewing Preserve during 2020.

red = species not previously documented by the author in MA or RI barrier dunes.

| | | | total | total | | | | | | | | total | GW | N | both |
|---|---|---|---|---|---|---|---|---|---|---|---|---|---|---|---|
| | | barrier dune system | GW | N | | | | | | | | N + GW | only | only | |
| Zook. # | MONA # | Station | GW1 | N1 | N2 | N3 | N4 | N5 | N6 | N1-N6 | | | | | |
| | | **THYATIRIDAE** | | | | | | | | | | | | | |
| | 6237 | Pseudothyatira cymatophoroides | | | 1 | | | | | 1 | | 1 | | x | |
| | | **DREPANIDAE** | | | | | | | | | | | | | |
| | 6255 | Oreta rosea | 1 | | | | | | | | | 1 | x | | |
| | | **GEOMETRIDAE** | | | | | | | | | | | | | |
| | 6272 | Eumacaria madopata | 4 | | | | | | | | | 4 | x | | |
| | 6273 | Macaria pustularia | 1 | | | | | | | | | 1 | x | | |
| | 6326 | Macaria aemulataria | 2 | | | | | | | | | 2 | x | | |
| | 6339 | Macaria transitaria | | | 1 | | | | | 1 | | 1 | | x | |
| | 6362 | Digrammia continuata | 3 | | | | | | | | | 3 | x | | |
| | 6443 | Glenoides texanaria | 2 | | 7 | | 2 | | 4 | 13 | | 15 | | | x |
| | 6583 | Iridopsis ephyraria | | | 1 | | | | | 1 | | 1 | | x | |
| | 6588 | Iridopsis larvaria | 1 | | | | | | | | | 1 | x | | |
| | 6598 | Protoboarmia porcelaria | | | 1 | | | 1 | | 2 | | 2 | | x | |
| | 6620 | Melanolophia canadaria | | | | | 1 | | | 1 | | 1 | | x | |
| | 6654 | Hypagyrtis unipunctata | | | | | | 1 | | 1 | | 1 | | x | |
| | 6724 | Euchlaena serrata | 2 | | | | | | | | | 2 | x | | |
| | 6725 | Euchlaena muzaria | 2 | | 2 | | 1 | | | 3 | | 5 | | | x |
| | 6729 | Euchlaena johnsonaria | | | 7 | 8 | 1 | 3 | | 19 | | 19 | | x | |
| | 6753 | Pero honestaria | 1 | | | | | | | | | 1 | x | | |
| | 6754 | Pero ancetaria | | | | | 1 | 6 | | 7 | | 7 | | x | |
| | 6755 | Pero morrisonaria | | 1 | 5 | | | | | 6 | | 6 | | x | |
| | 6805 | Tacparia zalissaria | | 3 | 1 | | | | | 4 | | 4 | | x | |
| | 6822 | Metarranthis duaria | 1 | | 2 | | | | | 2 | | 3 | | | x |
| | 6827 | Metarranthis refractaria | | | 6 | | | | | 6 | | 6 | | x | |
| | 6829.1 | Metarranthis sp.nr. lateritiaria | | | 2 | | | | | 2 | | 2 | | x | |
| | 6941 | Eusarca confusaria | 4 | | | | | 1 | | 1 | | 5 | | | x |
| | 6964 | Tetracis cachexiata | | | 1 | | | | | 1 | | 1 | | x | |
| | 6974 | Patalene olyzonaria | | 1 | 1 | | | | 1 | 3 | | 3 | | x | |
| | 6982 | Prochoerodes lineola | 15 | | 3 | | 2 | 4 | 1 | 10 | | 25 | | | x |
| | 6987 | Antepione thiosaria | 2 | | | | 2 | 1 | | 3 | | 5 | | | x |
| | 7010 | Nematocampa resistaria | 1 | | 1 | | 1 | | | 2 | | 3 | | | x |
| | 7047 | Nemoria rubrifrontaria | 8 | | | | | 1 | | 1 | | 9 | | | x |
| | 7053 | Dichorda iridaria | | | | | | 1 | | 1 | | 1 | | x | |
| | 7058 | Synchlora aerata | 1 | | | | | 3 | | 3 | | 4 | | | x |
| | 7071 | Chlorochlamys chloroleucaria | | 1 | 1 | | | 2 | | 4 | | 4 | | x | |
| | 7094 | Lobocleta ossularia | 14 | | 1 | | | | | 1 | | 15 | | | x |
| | 7114 | Idaea demissaria | | | 1 | | | | | 1 | | 1 | | x | |
| | 7126 | Idaea dimidiata | 1 | | | | 1 | | | 1 | | 2 | | | x |
| | 7132 | Pleuroprucha insulsaria | 5 | | | | 3 | 15 | | 18 | | 23 | | | x |
| | 7159 | Scopula limboundata | 1 | | 1 | | | | | 1 | | 2 | | | x |
| | 7196 | Eulithis diversilineata | 3 | | 1 | | 3 | 8 | | 12 | | 15 | | | x |
| | 7414 | Orthonama obstipata | | 1 | 1 | | 1 | 13 | | 16 | | 16 | | x | |
| | 7416 | Costaconvexa centrostrigaria | 5 | 2 | 3 | | 1 | 3 | | 9 | | 14 | | | x |
| | | Eupithecia sp. | 3 | | 4 | | 3 | 10 | | 17 | | 20 | | | x |
| | | **APATELODIDAE** | | | | | | | | | | | | | |
| | 7663 | Apatelodes torrefacta | 1 | | 1 | | 1 | | | 2 | | 3 | | | x |
| | | **LASIOCAMPIDE** | | | | | | | | | | | | | |
| | 7670 | Tolype velleda | 3 | | | | | | | | | 3 | x | | |
| | 7701 | Malacosoma americana | 1 | | | | | | | | | 1 | x | | |
| | | **SATURNIIDAE** | | | | | | | | | | | | | |
| | 7746 | Automeris io | | | 2 | 1 | 1 | | | 4 | | 4 | | x | |
| | | **SPHINGIDAE** | | | | | | | | | | | | | |
| | 7775 | Manduca sexta | | | | | 1 | | | 1 | | 1 | | x | |
| | 7810 | Sphinx gordius | | | 1 | | | | | 1 | | 1 | | x | |
| | 7824 | Paonias excaecata | | | | | | 1 | | 1 | | 1 | | x | |
| | 7859 | Eumorpha pandorus | | | | | | 1 | | 1 | | 1 | | x | |
| | 7885 | Darapsa myron | | | 3 | | | | | 3 | | 3 | | x | |
| | 7890 | Xylophanes tersa | | | | | | 1 | | 1 | | 1 | | x | |
| | 7894 | Hyles lineata | 1 | | | | 1 | | | 1 | | 2 | | | x |
| | 7786 | Ceratomia amyntor | 1 | | | | | | | | | 1 | x | | |
| | | **NOTODONTIDAE** | | | | | | | | | | | | | |
| 930019 | 7931 | Gluphisia septentrionis | | | | | 1 | | | 1 | | 1 | | x | |
| 930035 | 7904 | Datana drexelii | | | 1 | | | | | 1 | | 1 | | x | |

Table 3. Summary of macro-moths documented at Napatree Point Conservation Area and Goosewing Preserve during 2020.

red = species not previously documented by the author in MA or RI barrier dunes.

| | | | total | total | | | | | | | | total | GW | N | both |
|---|---|---|---|---|---|---|---|---|---|---|---|---|---|---|---|
| | | barrier dune system | GW | N | | | | | | | | N + GW | only | only | |
| Zook. # | MONA # | Station | GW1 | N1 | N2 | N3 | N4 | N5 | N6 | N1-N6 | | | | | |
| 930075 | 7983 | Heterocampa obliqua | 1 | | | | | | | | | 1 | x | | |
| 930087 | 7995 | Heterocampa biundata | | | 4 | | | 1 | | 5 | | 5 | | x | |
| 930100 | 8007 | Schizura unicornis | | 1 | 1 | | | | | 2 | | 2 | | x | |
| | | **EREBIDAE** | | | | | | | | | | | | | |
| | | **Lymantriinae** | | | | | | | | | | | | | |
| 930141 | 8318 | Lymantria dispar | 3 | | | | | | | | | 3 | x | | |
| 930168 | 8316 | Orgyia leucostigma | 56 | 2 | 4 | | | | | 6 | | 62 | | | x |
| | | **Arctiinae** | | | | | | | | | | | | | |
| 930189 | 8072 | Cisthene packardii | | | 2 | | | 5 | | 7 | | 7 | | x | |
| 930219 | 8045.1 | Crambidia pallida | 1 | | 1 | | | | | 1 | | 2 | | | x |
| 930244 | 8197 | Grammia virgo | 10 | | | | | | | | | 10 | x | | |
| 930278 | 8169 | Apantesis phalerata | 3 | 1 | 1 | | 1 | 1 | | 4 | | 7 | | | x |
| 930297 | 8118 | Virbia opella | 1 | | | | | | | | | 1 | x | | |
| 930299 | 8121 | Virbia aurantiaca | 56 | | 2 | 6 | | 5 | | 13 | | 69 | | | x |
| 930309 | 8134 | Spilosoma congrua | | | 5 | | | | | 5 | | 5 | | x | |
| 930311 | 8133 | Spilosoma latipennis | | | 1 | | | | | 1 | | 1 | | x | |
| 930316 | 8137 | Spilosoma virginica | | | | 1 | | 4 | | 5 | | 5 | | x | |
| 930319 | 8140 | Hyphantria cunea | 1 | | | | | 5 | | 5 | | 6 | | | x |
| 930332 | 8156 | Phragmatobia fuliginosa | 7 | | | | | | | | | 7 | x | | |
| 930335 | 8129 | Pyrrharctia isabella | 1 | | 17 | 1 | 3 | 2 | | 23 | | 24 | | | x |
| 930360 | 8203 | Halysidota tessellaris | 2 | | 16 | 1 | 6 | | | 23 | | 25 | | | x |
| 930370 | 8211 | Lophocampa caryae | | 1 | 2 | | | | | 3 | | 3 | | x | |
| 930404 | 8230 | Cycnia tenera | 1 | | | | 1 | 3 | | 4 | | 5 | | | x |
| 930440 | 8267 | Cisseps fulvicollis | 4 | | 1 | | 1 | | | 2 | | 6 | | | x |
| | | **Herminiinae** | | | | | | | | | | | | | |
| 930469 | 8322 | Idia americalis | | | 1 | | | | | 1 | | 1 | | x | |
| 930487 | 8338 | Phalaenophana pyramusalis | | | 1 | | | 2 | | 3 | | 3 | | x | |
| 930489 | 8340 | Zanclognatha lituralis | | | 3 | | | | | 3 | | 3 | | x | |
| 930490 | 8341 | Zanclognatha theralis | 1 | | | | | | | | | 1 | x | | |
| 930500 | 8353 | Zanclognatha jacchusalis | | | 1 | | | | | 1 | | 1 | | x | |
| 930502 | 8355.1 | Chytolita morbidalis "petrealis" | 5 | | 1 | | 2 | | | 3 | | 8 | | | x |
| 930511 | 8360 | Macrochilo orciferalis | 11 | | 3 | 22 | | | | 25 | | 36 | | | x |
| 930516 | 8366 | Tetanolita mynesalis | 6 | | 2 | | 1 | | | 3 | | 9 | | | x |
| 930518 | 8368 | Tetanolita floridana | 7 | | | | | | | | | 7 | x | | |
| 930520 | 8370 | Bleptina caradrinalis | 50 | | 78 | 9 | 21 | 1 | | 109 | | 159 | | | x |
| 930530 | 8379 | Renia factiosalis | 1 | | | | | | | | | 1 | x | | |
| 930536 | 8384.1 | Renia flavipunctalis | 25 | | 1 | | 1 | 8 | | 10 | | 35 | | | x |
| 930538 | 8386 | Renia adspergillus | | | 11 | | | 5 | | 16 | | 16 | | x | |
| 930539 | 8387 | Renia sobrialis | 7 | | | | | | | | | 7 | x | | |
| | | Renia sp. | 3 | | | | 1 | | | 1 | | 4 | | | x |
| 930547 | 8393 | Lascoria ambigualis | 7 | | 2 | | 2 | | | 4 | | 11 | | | x |
| 930551 | 8397 | Palthis angulalis | 19 | | 1 | | 1 | 3 | | 5 | | 24 | | | x |
| 930552 | 8398 | Palthis asopialis | 1 | | | | | 2 | | 2 | | 3 | | | x |
| | | **Hypeninae** | | | | | | | | | | | | | |
| 930561 | 8441 | Hypena manalis | 1 | | | | | | | | | 1 | x | | |
| 930565 | 8444 | Hypena palparia | | | | | 1 | | | 1 | | 1 | | x | |
| 930588 | 8465 | Hypena scabra | 14 | | 7 | | 1 | 43 | | 51 | | 65 | | | x |
| | | **Rivulinae** | | | | | | | | | | | | | |
| 930592 | 8404 | Rivula propinqualis | 3 | | | | | | | | | 3 | x | | |
| | | **Boletobiinae** | | | | | | | | | | | | | |
| 930679 | 8499 | Metalectra discalis | | | | | | 1 | | 1 | | 1 | | x | |
| | | **Catocalinae** | | | | | | | | | | | | | |
| 930767 | 8777 | Catocala badia | 3 | | | | | | | | | 3 | x | | |
| **930787** | **8796** | **Catocala nebulosa** | | | | | | 1 | | 1 | | 1 | | x | |
| 930833 | 8847 | Catocala gracilis | 1 | | | | | | | | | 1 | x | | |
| 930841 | 8857 | Catocala ultronia | 2 | | | | | 1 | | 1 | | 3 | | | x |
| 930923 | 8738 | Caenurgina crassiuscula | 3 | | | | 1 | 1 | | 2 | | 5 | | | x |
| 930924 | 8739 | Caenurgina erechtea | 1 | | | | | | | | | 1 | x | | |
| 930927 | 8767 | Doryodes spadaria | 1 | | | 2 | | | | 2 | | 3 | | | x |
| 930944 | 8745 | Mocis texana | | | | | | 1 | | 1 | | 1 | | x | |
| 930962 | 8721 | Allotria elonympha | 1 | | | | | | | | | 1 | x | | |
| 931023 | 8689 | Zale lunata | 2 | | | | | 1 | | 1 | | 3 | | | x |
| 931042 | 8707 | Zale metatoides | | | 1 | | | | | 1 | | 1 | | x | |
| 931053 | 8717 | Zale horrida | 1 | | | | | | | | | 1 | x | | |

Table 3. Summary of macro-moths documented at Napatree Point Conservation Area and Goosewing Preserve during 2020.

red = species not previously documented by the author in MA or RI barrier dunes.

| | | | total | total | | | | | | | | total | GW | N | both |
|---|---|---|---|---|---|---|---|---|---|---|---|---|---|---|---|
| | | barrier dune system | GW | N | | | | | | | | N + GW | only | only | |
| Zook. # | MONA # | Station | GW1 | N1 | N2 | N3 | N4 | N5 | N6 | N1-N6 | | | | | |
| 931060 | 9818 | Amolita fessa | 1 | | | | | | | | | 1 | x | | |
| | | **Eulepidotinae** | | | | | | | | | | | | | |
| 931077 | 8574 | Anticarsia gemmatalis | | | 1 | | | | | 1 | | 1 | | x | |
| | | **EUTELLIIDAE** | | | | | | | | | | | | | |
| 931106 | 8957 | Paectes oculatrix | 6 | | 2 | | | 1 | | 3 | | 9 | | | x |
| | | **NOLIDAE** | | | | | | | | | | | | | |
| 931123 | 8983.2 | Meganola spodia | | | 1 | | | | | 1 | | 1 | | x | |
| 931131 | 8991 | Nola cereella | | | 1 | | | | | 1 | | 1 | | x | |
| | | **NOCTUIDAE** | | | | | | | | | | | | | |
| | | **Plusiinae** | | | | | | | | | | | | | |
| 931170 | 8890 | Chrysodeixis includens | 1 | | | | 3 | | | 3 | | 4 | | | x |
| 931191 | 8908 | Autographa precationis | | | | | 1 | | 1 | 2 | | 2 | | x | |
| 931234 | 8924 | Anagrapha falcifera | 1 | | | | | 1 | | 1 | | 2 | | | x |
| | | **Eustrotiinae** | | | | | | | | | | | | | |
| 931284 | 9044 | Marimatha nigrofimbria | | | | | | 9 | | 9 | | 9 | | x | |
| 931295 | 9049 | Maliattha synochitis | | | 3 | | | | | 3 | | 3 | | x | |
| | | **Acronictinae** | | | | | | | | | | | | | |
| 931421 | 9200 | Acronicta americana | 1 | | | | | | | | | 1 | x | | |
| 931478 | 9264 | Acronicta longa | 1 | | 1 | | | | | 1 | | 2 | | | x |
| 931493 | 9280 | Simyra insularis | 2 | | | | | | | | | 2 | x | | |
| | | **Cuculliinae** | | | | | | | | | | | | | |
| 931504 | 10200 | Cucullia asteroides | | | 1 | | 1 | | | 2 | | 2 | | x | |
| | | **Amphipyrinae** | | | | | | | | | | | | | |
| 931544 | 9638 | Amphipyra pyramidoides | 1 | | | | | 1 | | 1 | | 2 | | | x |
| | | **Oncocneminae** | | | | | | | | | | | | | |
| 931875 | 10135 | Sympistis riparia | 1 | | | | | | | | | 1 | x | | |
| | | **Agaristinae** | | | | | | | | | | | | | |
| 931964 | 9299 | Eudryas unio | | | 1 | | | | | 1 | | 1 | | x | |
| 931966 | 9301 | Eudryas grata | 2 | | 6 | | 3 | | | 9 | | 11 | | | x |
| | | **Condicinae** | | | | | | | | | | | | | |
| 931989 | 9690 | Condica videns | 10 | | 3 | 7 | | 1 | | 11 | | 21 | | | x |
| 931995 | 9696 | Condica vecors | 7 | | | | | | | | | 7 | x | | |
| 931998 | 9699 | Condica sutor | | | | | 1 | | | 1 | | 1 | | x | |
| 932015 | 9714 | Condica confederata | 1 | | | | | | | | | 1 | x | | |
| 932018 | 9720 | Ogdoconta cinereola | 1 | | | | | 5 | | 5 | | 6 | | | x |
| 932026 | 9065 | Leuconycta diphteroides | 1 | | 1 | | | | | 1 | | 2 | | | x |
| | | **Heliothinae** | | | | | | | | | | | | | |
| 932039 | 11055 | Derrima stellata | | | 1 | | | | | 1 | | 1 | | x | |
| 932045 | 11068 | Helicoverpa zea | 2 | | 2 | | 4 | 1 | 3 | 10 | | 12 | | | x |
| 932120 | 11117 | Schinia lynx | | | 3 | | | | | 3 | | 3 | | x | |
| 932134 | 11128 | Schinia arcigera | 2 | | | | | 2 | | 2 | | 4 | | | x |
| 932156 | 11177 | Schinia nundina | | | | | | 4 | | 4 | | 4 | | x | |
| | | **Noctuinae** | | | | | | | | | | | | | |
| 932205 | 9053 | Pseudeustrotia carneola | 1 | | | | 1 | 5 | | 6 | | 7 | | | x |
| 932209 | 9619 | Phosphila miselioides | 2 | | | | | | | | | 2 | x | | |
| 932216 | 9666 | Spodoptera frugiperda | 1 | | | | | | | | | 1 | x | | |
| 932219 | 9669 | Spodoptera ornithogalli | 12 | | 2 | | 3 | | | 5 | | 17 | | | x |
| 932228 | 9678 | Elaphria versicolor | 1 | | 1 | | | 8 | | 9 | | 10 | | | x |
| 932230 | 9679 | Elaphria chalcedonia | 5 | | 2 | | | | | 2 | | 7 | | | x |
| 932234 | 9681.1 | Elaphria alapallida | | | 3 | | | | | 3 | | 3 | | x | |
| 932238 | 9684 | Elaphria grata | 1 | | 14 | | 25 | 37 | 1 | 77 | | 78 | | | x |
| 932249 | 9688 | Galgula partita | 1 | | | | | | | | | 1 | x | | |
| 932266 | 9647 | Proxenus miranda | 10 | 3 | 18 | 53 | 1 | 1 | | 76 | | 86 | | | x |
| 932269 | 9650 | Athetis tarda | | | | | | 4 | | 4 | | 4 | | x | |
| 932272 | 9560 | Dypterygia rozmani | | | 4 | 1 | | 2 | | 7 | | 7 | | x | |
| 932282 | 9637.1 | Magusa divaricata | | | 1 | 1 | 2 | | | 4 | | 4 | | x | |
| 932290 | 9545 | Euplexia benesimilis | | | 1 | | 1 | 2 | | 4 | | 4 | | x | |
| 932291 | 9546 | Phlogophora iris | | | | 1 | | | | 1 | | 1 | | x | |
| 932308 | - | Apamea unanimis | | | 1 | | | | | 1 | | 1 | | x | |
| 932314 | 9364 | Apamea sordens | | | | 1 | | | | 1 | | 1 | | x | |
| 932339 | 9378 | Apamea burgessi | | | 1 | | | | | 1 | | 1 | | x | |
| 932356 | 9395 | Apamea lintneri | 35 | 5 | 53 | | 65 | | | 123 | | 158 | | | x |
| 932358 | 9454 | Loscopia velata | 1 | | | | | | | | | 1 | x | | |
| 932369 | 9393 | Resapamea stipata | 2 | | | | | | | | | 2 | x | | |

Table 3. Summary of macro-moths documented at Napatree Point Conservation Area and Goosewing Preserve during 2020.

red = species not previously documented by the author in MA or RI barrier dunes.

| | | | total | total | | | | | | | | total | GW | N | both |
|---|---|---|---|---|---|---|---|---|---|---|---|---|---|---|
| | | barrier dune system | GW | N | | | | | | | | N + GW | only | only | |
| Zook. # | MONA # | Station | GW1 | N1 | N2 | N3 | N4 | N5 | N6 | N1-N6 | | | | | |
| 932370 | 9406 | Mesapamea fractilinea | 1 | | | | | | | | | 1 | x | | |
| 932375 | 9415.1 | Oligia strigilis | 1 | | | | | | | | | 1 | x | | |
| 932376 | 9415.2 | Oligia latruncula | | | 2 | | | | | 2 | | 2 | | x | |
| 932419 | 9435 | Photedes inops | 6 | | | | | | | | | 6 | x | | |
| 932439 | 9450 | Globia subflava | 7 | | | | | | | | | 7 | x | | |
| 932443 | 9453 | Helotropha reniformis | 1 | | | | | | | | | 1 | x | | |
| 932444 | 9520 | Achatodes zeae | | | | | | 2 | | 2 | | 2 | | x | |
| 932446 | 9456 | Amphipoea interoceanica | 4 | | | | | | | | | 4 | x | | |
| 932496 | 9465 | Papaipema duovata | 2 | | | | 1 | | 1 | 2 | | 4 | | | x |
| 932609 | 9952 | Eucirroedia pampina | 1 | | | | | | | | | 1 | x | | |
| 932616 | 9957 | Sunira bicolorago | 3 | 1 | 1 | | | | | 2 | | 5 | | | x |
| 932810 | 10524 | Nephelodes minians | 5 | | 1 | | | | | 1 | | 6 | | | x |
| 932882 | 10300 | Lacanobia grandis | 1 | | 6 | 2 | | | | 8 | | 9 | | | x |
| 932931 | 10434 | Dargida rubripennis | 1 | | | | | 1 | | 1 | | 2 | | | x |
| 932933 | 10436 | Mythimna oxygala | 3 | 1 | 3 | 5 | | | | 9 | | 12 | | | x |
| 932935 | 10438 | Mythimna unipuncta | 73 | | 21 | 1 | 79 | 19 | 2 | 122 | | 195 | | | x |
| 932937 | 10439 | Leucania extincta | 21 | 18 | 29 | 41 | | 5 | | 93 | | 114 | | | x |
| 932938.1 | 10440.1 | Leucania amygdalina | 1 | 1 | 3 | | | 3 | | 7 | | 8 | | | x |
| 932943 | 10444 | Leucania phragmitidicola | 37 | | 3 | 5 | | | | 8 | | 45 | | | x |
| 932944 | 10445 | Leucania linda | | | 2 | 2 | | | | 4 | | 4 | | x | |
| 932946 | 10446.1 | Leucania lapidaria | 1 | | | | | | | | | 1 | x | | |
| 932960 | 10456 | Leucania adjuta | 5 | 2 | 1 | | 1 | | | 4 | | 9 | | | x |
| 932965 | 10461 | Leucania ursula | 1 | | | | | | | | | 1 | x | | |
| 933044 | 10397 | Lacinipolia renigera | 63 | 1 | 11 | 3 | 8 | | | 23 | | 86 | | | x |
| 933052 | 10405 | Lacinipolia lorea | 1 | | | | | | | | | 1 | x | | |
| 933067 | 10413 | Lacinipolia explicata | 1 | | | | | | | | | 1 | x | | |
| 933136 | 10585 | Orthodes majuscula | 2 | | 6 | | | 11 | | 17 | | 19 | | | x |
| 933138 | 10587 | Orthodes cynica | 1 | | 98 | | 7 | | | 105 | | 106 | | | x |
| 933146 | 10288 | Orthodes detracta | | | 1 | | | | | 1 | | 1 | | x | |
| 933193 | 10627 | Tricholita signata | 36 | | 3 | | 3 | | | 6 | | 42 | | | x |
| 933211 | 10915 | Peridroma saucia | 1 | | 1 | | 1 | 1 | | 3 | | 4 | | | x |
| 933212 | 10911 | Anicla infecta | 4 | | 2 | | | | | 2 | | 6 | | | x |
| 933329 | 10715 | Euxoa scandens | 9 | | 7 | | 1 | 1 | | 9 | | 18 | | | x |
| 933346 | 10727 | Euxoa pleuritica | 1 | | | | | | | | | 1 | x | | |
| 933461 | 10838 | Euxoa detersa | 364 | 20 | 14 | | 25 | | 16 | 75 | | 439 | | | x |
| 933495 | 10680 | Feltia geniculata | 21 | 9 | | | | | | 9 | | 30 | | | x |
| 933498 | 10670 | Feltia jaculifera | 2 | | 2 | | 1 | | | 3 | | 5 | | | x |
| 933501 | 10674 | Feltia subgothica | 7 | | | | | | | | | 7 | x | | |
| 933503 | 10676 | Feltia herilis | 13 | | | | 5 | 2 | | 7 | | 20 | | | x |
| 933504 | 10664 | Feltia subterranea | | | | | | 1 | 1 | 2 | | 2 | | x | |
| 933506 | 10641 | Agrotis vetusta | 111 | 21 | 32 | | 8 | | 3 | 64 | | 175 | | | x |
| 933515 | 10650 | Agrotis gladiaria | 1 | | | | | | | | | 1 | x | | |
| 933528 | 10663 | Agrotis ipsilon | 3 | 2 | 1 | | 3 | 10 | | 16 | | 19 | | | x |
| 933529 | 10891 | Ochropleura implecta | 2 | | | | | 3 | | 3 | | 5 | | | x |
| 933551 | 11003.1 | Noctua pronuba | 38 | 1 | 11 | 5 | 2 | 4 | | 23 | | 61 | | | x |
| 933554 | 10926 | Spaelotis clandestina | 2 | | | | | | | | | 2 | x | | |
| 933572 | 10944 | Xestia smithii | 3 | | | | | | | | | 3 | x | | |
| 933588 | 10942 | Xestia c-nigrum | 1 | | | | | | | | | 1 | x | | |
| 933589 | 10942.1 | Xestia dolosa | 5 | | 2 | 1 | 2 | 3 | | 8 | | 13 | | | x |
| 933626 | 10955 | Agnorisma badinodis | 3 | | | | 10 | | 1 | 11 | | 14 | | | x |
| 933649 | 11006 | Protolampra brunneicollis | 3 | | 6 | | | | | 6 | | 9 | | | x |
| 933663.1 | 11024.1 | Abagrotis benjamini | 13 | 9 | 13 | | 6 | 1 | | 29 | | 42 | | | x |
| 933680 | 11029 | Abagrotis alternata | | 1 | | | | | | 1 | | 1 | | x | |
| 933686 | 11043.2 | Abagrotis magnicupida | 10 | | 4 | | 1 | | | 5 | | 15 | | | x |
| | | | | | | | | | | | | | | | |
| | | Total # species | **145** | **25** | **110** | **24** | **65** | **71** | **12** | **161** | | **216** | **55** | **71** | **90** |
| | | Total # individuals | **1422** | **109** | **647** | **180** | **347** | **323** | **35** | **1641** | | **3063** | | | |
| | | Number samples | **8** | **2** | **5** | **1** | **4** | **2** | **1** | **15** | | **23** | | | |

Ninety species were found at both sites, 55 only at Goosewing and 71 only at Napatree. Thus 126, or 58% of the species, were unique to one site. Figure 4 depicts a randomized species accumulation curve for each site. Neither site is leveling off, indicating that increased sampling would continue to increase the number of new species for each site, thus the sampling effort at each site was not sufficient to provide a complete list of species present at each site. This is consistent with other season-long inventories I have conducted. Multi-year single site surveys suggest that even with a more intensive sampling regime, a single year produces only between 35 – 50 percent of the species utilizing the site. The data suggest that per sample, Goosewing species richness may be greater than that at Napatree despite the increased number of samples and variety of habitats at Napatree. However, additional sampling might bring these curves closer together.

Figure 4. Randomized species accumulation curves for Goosewing (red) and Napatree (blue) in 2020.

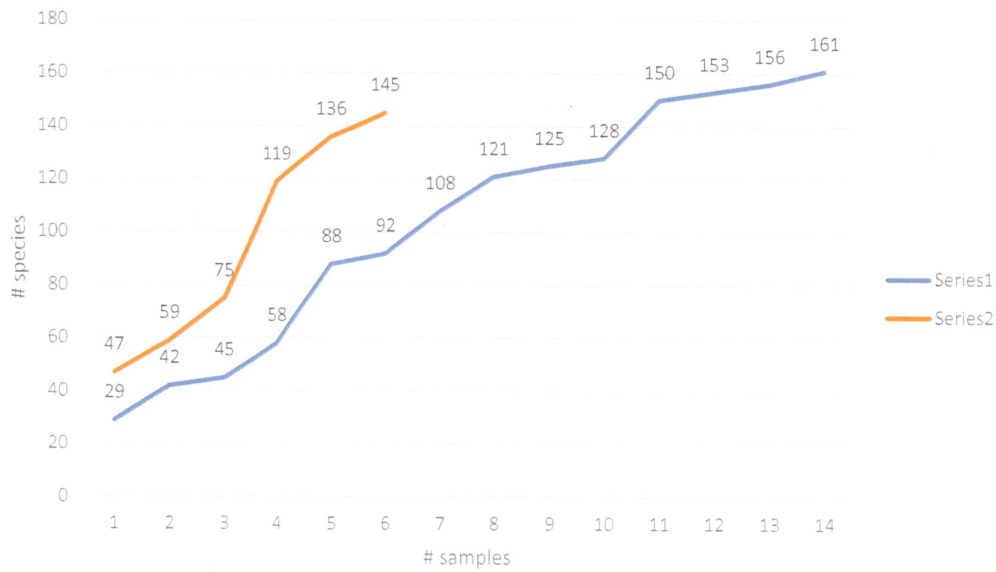


Three dune affiliate/obligate species listed in the Massachusetts Endangered Act (MESA) plus an additional 15 uncommon dune affiliates were documented during a much more intensive survey at Dead Neck in Mashpee, and Horseneck Beach Reservation in Wesport during 2019 (Mello and Bogart, in prep.). In the current survey, 13 dune affiliates were documented, 11 at Goosewing and 9 at Napatree (Table 4). A comparison of the Massachusetts 2019 data and the current Rhode Island survey grouped by the larval hostplant affinities is presented in Table 5. The 2019

Table 4. Dune-affiliated macromoths documented at Napatree Point Conservation Area and Goosewing Preserve during 2020.

| | | | total | total | | | | | | | total |
|---|---|---|---|---|---|---|---|---|---|---|---|
| | | barrier dune system | GW | N | | | | | | | N + GW |
| Zook. # | MONA # | Station | GW1 | N1 | N2 | N3 | N4 | N5 | N6 | N1-N6 | |
| | | **EREBIDAE** | | | | | | | | | |
| 930767 | 8777 | Catocala badia | 3 | | | | | | | | 3 |
| 930927 | 8767 | Doryodes spadaria | 1 | | | 2 | | | | 2 | 3 |
| | | **NOCTUIDAE** | | | | | | | | | |
| 931875 | 10135 | Sympistis riparia | 1 | | | | | | | | 1 |
| 932039 | 11055 | Derrima stellata | | | 1 | | | | | 1 | 1 |
| 932339 | 9378 | Apamea burgessi | | | 1 | | | | | 1 | 1 |
| 932356 | 9395 | Apamea lintneri | 35 | 5 | 53 | | 65 | | | 123 | 158 |
| 932419 | 9435 | Photedes inops | 6 | | | | | | | | 6 |
| 932496 | 9465 | Papaipema duovata | 2 | | | | 1 | | 1 | 2 | 4 |
| 932931 | 10434 | Dargida rubripennis | 1 | | | | | 1 | | 1 | 2 |
| 932937 | 10439 | Leucania extincta | 21 | 18 | 29 | 41 | | 5 | | 93 | 114 |
| 933346 | 10727 | Euxoa pleuritica | 1 | | | | | | | | 1 |
| 933461 | 10838 | Euxoa detersa | 364 | 20 | 14 | | 25 | | 16 | 75 | 439 |
| 933663.1 | 11024.1 | Abagrotis benjamini | 13 | 9 | 13 | | 6 | 1 | | 29 | 42 |
| | | | | | | | | | | | |
| | | Total # species | **11** | **4** | **6** | **2** | **4** | **3** | **2** | **9** | **13** |
| | | Total # individuals | **448** | **52** | **111** | **43** | **97** | **7** | **17** | **327** | **775** |
| | | Number samples | **8** | **2** | **5** | **1** | **4** | **2** | **1** | **15** | **23** |

Table 5. Comparison of dune-affiliated macro-moths documented from Massachusetts (2019) and Rhode Island (2020) surveys.

| | Rhode Island | | | Massachusetts | | |
|---|---|---|---|---|---|---|
| barrier dune system | Goosewing | Napatree | **RI total** | Dead Neck | Horseneck | **MA** |
| **larvae on grass species** | | | | | | |
| *Doryodes spadaria* (Dull Doryodes) | 1 | 2 | **3** | 8 | 6 | **14** |
| *Apamea inorndinata* | | | | 166 | 1 | **167** |
| *Apamea burgessi* | | 1 | **1** | 2 | | **2** |
| *Apamea lintneri* (Sand Wainscot) | 35 | 123 | **158** | 4 | 5 | **9** |
| *Photedes inops* (Cordgrass Borer) | 6 | | **6** | | | |
| *Dargida rubripennis* (The Pink Streak) | 1 | 1 | **2** | | 1 | **1** |
| *Leucania extincta* | 21 | 93 | **114** | 24 | 14 | **38** |
| *Eucoctopnemis fimbriaris* (Fringed Dart) | | | | 2 | | **2** |
| *Euxoa pleuritica* (Fawn Brown Dart) | 1 | | **1** | 34 | 2 | **36** |
| *Euxoa violaris* (Violet Dart) | | | | 3 | | **3** |
| *Euxoa detersa* (Rubbed Dart) | 364 | 75 | **439** | 15 | 307 | **322** |
| **larvae on forbs species** | | | | | | |
| *Drasteria graphica* (Graphic Moth) | | | | 3 | 25 | **28** |
| *Derrima stellata* (Pink Star Moth) | | 1 | **1** | 2 | 2 | **4** |
| *Schinia spinosae* (Spinose Flower Moth) | | | | 64 | 11 | **75** |
| *Papaipema duovata* (Seaside Goldenrod Borer) | 2 | 2 | **4** | 3 | 26 | **29** |
| **larvae on shrub species** | | | | | | |
| *Catocala badia* (Bay Underwing) | 3 | | **3** | 6 | 1 | **7** |
| *Argyrostrotis anilis* (Short-lined Chocolate Moth) | | | | 124 | 44 | **168** |
| *Sympistis riparia* (*Dune sympistis)* | 1 | | **1** | 126 | 120 | **246** |
| *Abagrotis benjamini* (Coastal Heathland Cutworm) | 13 | 29 | **42** | 25 | 17 | **42** |
| | | | | | | |
| Total # species | 4 | 9 | **13** | 17 | 15 | **18** |
| Total # individuals | 19 | 327 | **775** | 437 | 575 | **1012** |
| Number samples | 8 | 15 | **23** | 28 | 40 | **68** |

Massachusetts dune survey produced 18 barrier dune affiliated species, five more than the Rhode Island survey. There were nearly three times the number of samples collected in Massachusetts than in the Rhode Island survey. Nevertheless, there were distinct differences for some of the dominant species found in each survey that would be explained by more than the difference in sampling effort. The most dramatic differences in dominant species can be explained by larval hostplant presence or lack thereof. One species, *Abagrotis benjamini* (Figure 5), that is listed in *Rare Native Animals of Rhode Island* (Enser, 2006) was found at both Goosewing (13 individuals) and Napatree (29 individuals) despite the near lack of its documented larval hostplant, beach plum (Goldstein and Nelson (2017). It is possible that other *Prunus* species such as black cherry (*P. serotina*) may be an alternate hostplant on barrier dunes. The remaining 12 species (Figure 6) that I've identified as dune affiliates share these barrier dunes and may be as dune restricted as *A. benjamini*, at least in the Northeast.

Figure 5. *Abagrotis benjamini*

Figure 6. Dune affiliates documented at Napatree and/or Goosewing during 2020.

| | | |
|---|---|---|
| *Catocala badia* | *Apamea burgessi* | *Dargida rubripennis* |
| *Doryodes spadaria* | *Apamea lintneni* | *Leucania extincta* |
| *Sympistis riparia* | *Photedes inops* | *Euxoa pleuritica* |
| *Derrima stellata* | *Papaipema duovata* | *Euxoa detersa* |

*Sympistis riparia* and *Argyrostrotis anilis*, two other beach plum feeders, were common at both sites surveyed in Massachusetts during 2019 where beach plum was a major part of the shrub community but were represented by one individual (at Goosewing, *S. riparia)* in the current survey. Thus, these two species seem to be more tied to beach plum than *Abagrotis benjamini*.

Because moths can fly, any light trap sample may intercept moths that are wind-driven (especially along the coast), or simply flying by on route from one habitat patch to another. Two examples of this were seen from samples collected at Napatree. A single *Catocala nebulosa* (Clouded Underwing) occurred at Napatree, station N5 on August 24 (Figure 6). This is the first record for this species from Rhode Island. It is an example of a species that seems due to climate change to be expanding its distribution eastward and northward, as it has recently been reported from Connecticut (Larry Gall, pers. com.) and Massachusetts (MASSMOTHS, in prep.). The larvae feed on hickory, which is absent from Napatree and at least 1.5 miles from the nearest upland likely to contain hickory. Thus, this individual clearly did not originate nor could it breed on Napatree. An example of a southern species, *Xylophanes tersa* (Tersa Sphinx) also was found on Napatree station N5 on August 6 (Figure 7). This species could not survive during even mild southern New England's winters in any of its life stages, but is a sporadic late summer northward wanderer as well as a strong flier.

Figure 6. *Catocala nebulosa*

Figure 7. *Xylophanes tersa*

Table 6 lists the 12 species that were found at either Napatree or Goosewing that were not previously recorded in previous dune surveys in Rhode Island or Massachusetts. Most occurred as single individuals representing southern species that likely were carried by wind currents or were emigrants from south or west of Rhode Island. The 77 individuals of *Elaphria grata* however, likely suggest that a breeding colony is present at Napatree Point Conservation Area. The larvae have been reared by McCabe on green algae and fungi, but larval choices in the field are undetermined (Wagner et al., 2011).

Appendix II lists the 94 microlepidoptera that were documented during this survey. Because approximately two dozen or more species await identification, and the smallest micro's were not collected due their poor condition, this is only a partial list of those that were captured. At least one species, *Chionodes arenella* (Gelechiidae), is a dunes affiliate. There may be other dunes affiliates but published records are so spotty that habitat determinations were not possible.

| Table 6. Species found during RI 2020 survey that were new records for dunes in MA or RI. | | | | | | | | | | |
|---|---|---|---|---|---|---|---|---|---|---|
| | | N = Napatree | GW = Goosewing (one station) | | | | | | | |
| | | | | | | | | total | total | total |
| | | Station | N1 | N2 | N4 | N5 | N6 | **N** | **GW** | **N + GW** |
| | | GEOMETRIDAE | | | | | | | | |
| | 6805 | Tacparia zalissaria | 3 | 1 | | | | **4** | | **4** |
| | | SPHINGIDAE | | | | | | | | |
| | 7775 | Manduca sexta | | | 1 | | | **1** | | **1** |
| | 7890 | Xylophanes tersa | | | | 1 | | **1** | | **1** |
| | 7786 | Ceratomia amyntor | | | | | | | **1** | **1** |
| | | EREBIDAE | | | | | | | | |
| 930787 | 8796 | Catocala nebulosa | | | | 1 | | **1** | | **1** |
| | | NOCTUIDAE | | | | | | | | |
| 931421 | 9200 | Acronicta americana | | | | | | | **1** | **1** |
| 931998 | 9699 | Condica sutor | | | 1 | | | **1** | | **1** |
| 932015 | 9714 | Condica confederata | | | | | | | **1** | **1** |
| 932238 | 9684 | Elaphria grata | | 14 | 25 | 37 | 1 | **77** | **1** | **78** |
| 932439 | 9450 | Globia subflava | | | | | | | **7** | **7** |
| 932446 | 9456 | Amphipoea interoceanica | | | | | | | **4** | **4** |
| 933554 | 10926 | Spaelotis clandestina | | | | | | | **2** | **2** |
| | | Total # species | 1 | 2 | 3 | 3 | 1 | **6** | **7** | **12** |
| | | Total # individuals | 3 | 15 | 27 | 39 | 1 | **85** | **17** | **102** |

## CONCLUSIONS

Over 200 species of macro-moths and micro-moths were recorded from Napatree Point Conservation Area and/or Goosewing Beach Preserve in 2020, a significant fraction of which are tied to barrier dune/sand barrens habitats. This study confirms that even relatively small barrier dune systems can support a community of barrens affiliates that would either be extirpated or become critically threatened if this habitat should disappear. Twin threats of development and rising sea level put the squeeze on this vulnerable habitat throughout the northeast. Although the current configuration of the Napatree and Goosewing barrier dunes are protected from development, rising sea level and the specter of increasing violent storms is unabated. Dune restoration and protection from trampling of vegetation is underway at both these sites. Goosewing and Napatree Point suffer from takeover by non-native plant species. Beach plum, abundant at the two Massachusetts dune systems surveyed in 2019 is nearly absent at Napatree and scant at Goosewing. When preparing future restoration plans, it would be beneficial to include beach plum plantings as a major component. Not only would several dune affiliate moths benefit, but it would improve storm damage prevention. Like beachgrass, beach plum can be buried in sand and continue to raise the elevation of the dune on which it sits.

## ACNOWLEDGEMENTS

This project was supported by a grant from the Godzala Research Fund of the RI Natural History Survey and support from the Lloyd Center for the Environment. I thank David Gregg, Executive Director of RINHS for his support of the project and providing contact information for access; Janice M. Sassi, Manager of the Watch Hill Conservancy and Jeanne Cooper, TNC Goosewing Program Manager for arranging permission for access and logistics for this survey. I thank Emily and Conner, who provided jeep access to the Napatree stations, saving me a mile-plus hike or canoe paddle to the light trap stations. Finally, I thank Jamie Bogart, Lloyd Center Research Associate, for conducting the field work at Goosewing.

## CITATIONS

Enser, R.W., 2006. *Rare Native Animals of Rhode Island*. Rhode Island Natural Heritage Program. Rhode Island Department of Environmental Management. Providence, RI

Goldstein, P. Z. and Nelson, M. W. 2017. "Two psammophilic noctuids newly associated with beach plum *Prunus maritima*: The Dune Noctuid (*Sympistis riparia*) and Coastal Heathland Cutworm (Abagrotis nefascia) in Northeastern North America (Lepidoptera: Noctuidae)." ZooKeys. 661:62–89. https://doi.org/10.3897/zookeys.661.10867

Lafontaine, D. and C. Schmidt, 2010. Annotated check list of the Noctuoidea (Insecta, Lepidoptera) of North America north of Mexico. Zookeys 40:1-239. https://doi.org/10.3897/zookeys.40.414 (19 March 2010)

(MONA) Hodges, R.W., et al., 1983. *Check List of the Lepidoptera of America North of Mexico*. E.W. Classey Ltd. and The Wedge Entomological Research Foundation, London. 284 p.

Pohl,G.R., B. Patterson and J.P. Pelham, 2016. Annotated taxonomic checklist of the Lepidoptera of North America, North of Mexico. Working paper published online by the authors at ResearchGate.net. 766 pp.

Wagner, D.L., D.F. Schweitzer, J.B, Sullivan & R.C, Reardon, 2011. *Owlet Caterpillars of Eastern North America*. Princeton University Press. Princeton and Oxford. 576 p.

Appendix I. Raw light trap data from two Rhode Island barrier dune systems (Napatree Point Conservation Area and Goosewing Beach Preserve) recorded in 2020.

# = counts of individuals

red = species that have not been previously found by the author in MA or RI barrier dune systems

| | | Town | Westerly (Watch Hill) | | | | | | | | | | | | | | | | Little Compton | | | | | | | | |
|---|---|---|---|---|---|---|---|---|---|---|---|---|---|---|---|---|---|---|---|---|---|---|---|---|---|---|---|
| | | System | Napatree (N) | | | | | | | | | | | | | | | | Goosewing (GW) | | | | | | | | |
| | | Month | June | June | June | June | June | July | July | Aug | Aug | Aug | Sept | Sept | Sept | Oct | Oct | total | June | July | July | Aug | Sept | Sept | Oct | total | total |
| | | day | 3 | 17 | 17 | 30 | 30 | 19 | 19 | 6 | 6 | 24 | 16 | 16 | 24 | 6 | 6 | | 23 | 14 | 27 | 26 | 13 | 24 | 6 | | N + GW |
| Zook. # | MONA # | Station | N1 | N2 | N3 | N2 | N4 | N2 | N4 | N2 | N5 | N5 | N2 | N4 | N1 | N4 | N6 | | GW1 | GW1 | GW1 | GW1 | GW1 | GW1 | GW1 | GW1 | |
| | | **THYATIRIDAE** | | | | | | | | | | | | | | | | | | | | | | | | | |
| | 6237 | Pseudothyatira cymatophoroides | | 1 | | | | | | | | | | | | | | 1 | | | | | | | | | 1 |
| | | **DREPANIDAE** | | | | | | | | | | | | | | | | | | | | | | | | | |
| | 6255 | Oreta rosea | | | | | | | | | | | | | | | | | | | 1 | | | | | 1 | 1 |
| | | **GEOMETRIDAE** | | | | | | | | | | | | | | | | | | | | | | | | | |
| | 6272 | Eumacaria madopata | | | | | | | | | | | | | | | | | | | 4 | | | | | 4 | 4 |
| | 6273 | Macaria pustularia | | | | | | | | | | | | | | | | | | 1 | | | | | | 1 | 1 |
| | 6326 | Macaria aemulataria | | | | | | | | | | | | | | | | | | | 2 | | | | | 2 | 2 |
| | 6339 | Macaria transitaria | | 1 | | | | | | | | | | | | | | 1 | | | | | | | | | 1 |
| | 6362 | Digrammia continuata | | | | | | | | | | | | | | | | | | 1 | 1 | 1 | | | | 3 | 3 |
| | 6443 | Glenoides texanaria | | | | | | 2 | 1 | | | | 5 | | | 1 | 4 | 13 | | | 2 | | | | | 2 | 15 |
| | 6583 | Iridopsis ephyraria | | 1 | | | | | | | | | | | | | | 1 | | | | | | | | | 1 |
| | 6588 | Iridopsis larvaria | | | | | | | | | | | | | | | | | | | 1 | | | | | 1 | 1 |
| | 6598 | Protoboarmia porcelaria | | | | 1 | | | | | | 1 | | | | | | 2 | | | | | | | | | 2 |
| | 6620 | Melanolophia canadaria | | | | | | | 1 | | | | | | | | | 1 | | | | | | | | | 1 |
| | 6654 | Hypagyrtis unipunctata | | | | | | | | | | 1 | | | | | | 1 | | | | | | | | | 1 |
| | 6724 | Euchlaena serrata | | | | | | | | | | | | | | | | | 2 | | | | | | | 2 | 2 |
| | 6725 | Euchlaena muzaria | | | | 2 | 1 | | | | | | | | | | | 3 | 2 | | | | | | | 2 | 5 |
| | 6729 | Euchlaena johnsonaria | | 3 | 8 | 4 | 1 | | | | | 3 | | | | | | 19 | | | | | | | | | 19 |
| | 6753 | Pero honestaria | | | | | | | | | | | | | | | | | 1 | | | | | | | 1 | 1 |
| | 6754 | Pero ancetaria | | | | | | | 1 | | 6 | | | | | | | 7 | | | | | | | | | 7 |
| | 6755 | Pero morrisonaria | 1 | 3 | | | | | | 2 | | | | | | | | 6 | | | | | | | | | 6 |
| | 6805 | Tacparia zalissaria | 3 | 1 | | | | | | | | | | | | | | 4 | | | | | | | | | 4 |
| | 6822 | Metarranthis duaria | | 2 | | | | | | | | | | | | | | 2 | 1 | | | | | | | 1 | 3 |
| | 6827 | Metarranthis refractaria | | 4 | | 2 | | | | | | | | | | | | 6 | | | | | | | | | 6 |
| | 6829.1 | Metarranthis sp.nr. lateritiaria | | 2 | | | | | | | | | | | | | | 2 | | | | | | | | | 2 |
| | 6941 | Eusarca confusaria | | | | | | | | | | 1 | | | | | | 1 | 2 | | 1 | 1 | | | | 4 | 5 |
| | 6964 | Tetracis cachexiata | | 1 | | | | | | | | | | | | | | 1 | | | | | | | | | 1 |
| | 6974 | Patalene olyzonaria | | | | | | 1 | | | | | | | 1 | | 1 | 3 | | | | | | | | | 3 |
| | 6982 | Prochoerodes lineola | | | | | | | 1 | | 3 | 1 | 3 | 1 | | | 1 | 10 | | 1 | 11 | | 2 | 1 | | 15 | 25 |
| | 6987 | Antepione thiosaria | | | | | | | 2 | | 1 | | | | | | | 3 | | | 2 | | | | | 2 | 5 |
| | 7010 | Nematocampa resistaria | | | | | | 1 | 1 | | | | | | | | | 2 | | | 1 | | | | | 1 | 3 |
| | 7047 | Nemoria rubrifrontaria | | | | | | | | | 1 | | | | | | | 1 | | | 8 | | | | | 8 | 9 |
| | 7053 | Dichorda iridaria | | | | | | | | | 1 | | | | | | | 1 | | | | | | | | | 1 |
| | 7058 | Synchlora aerata | | | | | | | | | | 3 | | | | | | 3 | | | | | 1 | | | 1 | 4 |
| | 7071 | Chlorochlamys chloroleucaria | 1 | 1 | | | | | | | | 2 | | | | | | 4 | | | | | | | | | 4 |
| | 7094 | Lobocleta ossularia | | | | 1 | | | | | | | | | | | | 1 | 14 | | | | | | | 14 | 15 |
| | 7114 | Idaea demissaria | | | | | | | | 1 | | | | | | | | 1 | | | | | | | | | 1 |
| | 7126 | Idaea dimidiata | | | | | 1 | | | | | | | | | | | 1 | | 1 | | | | | | 1 | 2 |
| | 7132 | Pleuroprucha insulsaria | | | | | | | 3 | | 6 | 9 | | | | | | 18 | | | 4 | | | 1 | | 5 | 23 |
| | 7159 | Scopula limboundata | | | | 1 | | | | | | | | | | | | 1 | 1 | | | | | | | 1 | 2 |
| | 7196 | Eulithis diversilineata | | | | | 1 | 1 | 2 | | 8 | | | | | | | 12 | | | 3 | | | | | 3 | 15 |
| | 7414 | Orthonama obstipata | 1 | 1 | | | 1 | | | | 13 | | | | | | | 16 | | | | | | | | | 16 |
| | 7416 | Costaconvexa centrostrigaria | 2 | 1 | | 1 | 1 | 1 | | | 1 | 2 | | | | | | 9 | 1 | | 3 | 1 | | | | 5 | 14 |
| | | Eupithecia sp. | | | | 2 | 1 | 1 | 1 | 1 | 8 | 2 | | | | 1 | | 17 | | 1 | | 2 | | | | 3 | 20 |
| | | **APATELODIDAE** | | | | | | | | | | | | | | | | | | | | | | | | | |
| | 7663 | Apatelodes torrefacta | | 1 | | | 1 | | | | | | | | | | | 2 | 1 | | | | | | | 1 | 3 |
| | | **LASIOCAMPIDE** | | | | | | | | | | | | | | | | | | | | | | | | | |
| | 7670 | Tolype velleda | | | | | | | | | | | | | | | | | | | | | | 3 | | 3 | 3 |
| | 7701 | Malacosoma americana | | | | | | | | | | | | | | | | | | 1 | | | | | | 1 | 1 |
| | | **SATURNIIDAE** | | | | | | | | | | | | | | | | | | | | | | | | | |

Appendix I. Raw light trap data from two Rhode Island barrier dune systems (Napatree Point Conservation Area and Goosewing Beach Preserve) recorded in 2020.

# = counts of individuals — red = species that have not been previously found by the author in MA or RI barrier dune systems

| | | Town | Westerly (Watch Hill) | | | | | | | | | | | | | | | | Little Compton | | | | | | | | |
|---|---|---|---|---|---|---|---|---|---|---|---|---|---|---|---|---|---|---|---|---|---|---|---|---|---|---|---|
| | | System | Napatree (N) | | | | | | | | | | | | | | | | Goosewing (GW) | | | | | | | | |
| | | Month | June | June | June | June | June | July | July | Aug | Aug | Aug | Sept | Sept | Sept | Oct | Oct | total | June | July | July | Aug | Sept | Sept | Oct | total | total |
| | | day | 3 | 17 | 17 | 30 | 30 | 19 | 19 | 6 | 6 | 24 | 16 | 16 | 24 | 6 | 6 | | 23 | 14 | 27 | 26 | 13 | 24 | 6 | | N + GW |
| Zook. # | MONA # | Station | N1 | N2 | N3 | N2 | N4 | N2 | N4 | N2 | N5 | N5 | N2 | N4 | N1 | N4 | N6 | | GW1 | GW1 | GW1 | GW1 | GW1 | GW1 | GW1 | GW1 | |
| | 7746 | Automeris io | | 2 | 1 | | 1 | | | | | | | | | | | 4 | | | | | | | | | 4 |
| | | **SPHINGIDAE** | | | | | | | | | | | | | | | | | | | | | | | | | |
| | **7775** | **Manduca sexta** | | | | | | | | | | | | 1 | | | | 1 | | | | | | | | | 1 |
| | 7810 | Sphinx gordius | | | | 1 | | | | | | | | | | | | 1 | | | | | | | | | 1 |
| | 7824 | Paonias excaecata | | | | | | | | | 1 | | | | | | | 1 | | | | | | | | | 1 |
| | 7859 | Eumorpha pandorus | | | | | | | | | 1 | | | | | | | 1 | | | | | | | | | 1 |
| | 7885 | Darapsa myron | | 1 | | | | 2 | | | | | | | | | | 3 | | | | | | | | | 3 |
| | **7890** | **Xylophanes tersa** | | | | | | | | | 1 | | | | | | | 1 | | | | | | | | | 1 |
| | 7894 | Hyles lineata | | | | | | | | | | | | 1 | | | | 1 | | | | | | 1 | | 1 | 2 |
| | **7786** | **Ceratomia amyntor** | | | | | | | | | | | | | | | | | | 1 | | | | | | 1 | 1 |
| | | **NOTODONTIDAE** | | | | | | | | | | | | | | | | | | | | | | | | | |
| 930019 | 7931 | Gluphisia septentrionis | | | | | | | 1 | | | | | | | | | 1 | | | | | | | | | 1 |
| 930035 | 7904 | Datana drexelii | | 1 | | | | | | | | | | | | | | 1 | | | | | | | | | 1 |
| 930075 | 7983 | Heterocampa obliqua | | | | | | | | | | | | | | | | | | 1 | | | | | | 1 | 1 |
| 930087 | 7995 | Heterocampa biundata | | 4 | | | | | | | 1 | | | | | | | 5 | | | | | | | | | 5 |
| 930100 | 8007 | Schizura unicornis | 1 | | | | | | | 1 | | | | | | | | 2 | | | | | | | | | 2 |
| | | **EREBIDAE** | | | | | | | | | | | | | | | | | | | | | | | | | |
| | | **Lymantriinae** | | | | | | | | | | | | | | | | | | | | | | | | | |
| 930141 | 8318 | Lymantria dispar | | | | | | | | | | | | | | | | | | | 3 | | | | | 3 | 3 |
| 930168 | 8316 | Orgyia leucostigma | | | | 1 | | | | | | | 3 | | 2 | | | 6 | | | 49 | | 1 | 5 | 1 | 56 | 62 |
| | | **Arctiinae** | | | | | | | | | | | | | | | | | | | | | | | | | |
| 930189 | 8072 | Cisthene packardii | | 2 | | | | | | | | 5 | | | | | | 7 | | | | | | | | | 7 |
| 930219 | 8045.1 | Crambidia pallida | | | | | | | | | | | 1 | | | | | 1 | | | | 1 | | | | 1 | 2 |
| 930244 | 8197 | Grammia virgo | | | | | | | | | | | | | | | | | | | 10 | | | | | 10 | 10 |
| 930278 | 8169 | Apantesis phalerata | | | | | 1 | | | 1 | | 1 | | | 1 | | | 4 | 1 | | 1 | | 1 | | | 3 | 7 |
| 930297 | 8118 | Virbia opella | | | | | | | | | | | | | | | | | | 1 | | | | | | 1 | 1 |
| 930299 | 8121 | Virbia aurantiaca | | 2 | 6 | | | | | | | 5 | | | | | | 13 | 23 | | | 32 | 1 | | | 56 | 69 |
| 930309 | 8134 | Spilosoma congrua | | 5 | | | | | | | | | | | | | | 5 | | | | | | | | | 5 |
| 930311 | 8133 | Spilosoma latipennis | | 1 | | | | | | | | | | | | | | 1 | | | | | | | | | 1 |
| 930316 | 8137 | Spilosoma virginica | | | 1 | | | | | | 4 | | | | | | | 5 | | | | | | | | | 5 |
| 930319 | 8140 | Hyphantria cunea | | | | | | | | | 5 | | | | | | | 5 | | | 1 | | | | | 1 | 6 |
| 930332 | 8156 | Phragmatobia fuliginosa | | | | | | | | | | | | | | | | | | 2 | 2 | 2 | 1 | | | 7 | 7 |
| 930335 | 8129 | Pyrrharctia isabella | | 16 | 1 | 1 | 3 | | | | 1 | 1 | | | | | | 23 | 1 | | | | | | | 1 | 24 |
| 930360 | 8203 | Halysidota tessellaris | | | 1 | 13 | 6 | 3 | | | | | | | | | | 23 | 1 | 1 | | | | | | 2 | 25 |
| 930370 | 8211 | Lophocampa caryae | 1 | 2 | | | | | | | | | | | | | | 3 | | | | | | | | | 3 |
| 930404 | 8230 | Cycnia tenera | | | | | 1 | | | | 3 | | | | | | | 4 | | | 1 | | | | | 1 | 5 |
| 930440 | 8267 | Cisseps fulvicollis | | | | | | 1 | 1 | | | | | | | | | 2 | | 2 | 2 | | | | | 4 | 6 |
| | | **Herminiinae** | | | | | | | | | | | | | | | | | | | | | | | | | |
| 930469 | 8322 | Idia americalis | | | | 1 | | | | | | | | | | | | 1 | | | | | | | | | 1 |
| 930487 | 8338 | Phalaenophana pyramusalis | | 1 | | | | | | | 2 | | | | | | | 3 | | | | | | | | | 3 |
| 930489 | 8340 | Zanclognatha lituralis | | 2 | | 1 | | | | | | | | | | | | 3 | | | | | | | | | 3 |
| 930490 | 8341 | Zanclognatha theralis | | | | | | | | | | | | | | | | | | | 1 | | | | | 1 | 1 |
| 930500 | 8353 | Zanclognatha jacchusalis | | | | | | 1 | | | | | | | | | | 1 | | | | | | | | | 1 |
| 930502 | 8355.1 | Chytolita morbidalis "petrealis" | | | | 1 | 2 | | | | | | | | | | | 3 | 5 | | | | | | | 5 | 8 |
| 930511 | 8360 | Macrochilo orciferalis | | 2 | 22 | 1 | | | | | | | | | | | | 25 | 7 | | 1 | 2 | 1 | | | 11 | 36 |
| 930516 | 8366 | Tetanolita mynesalis | | | | | | | | | | | 2 | 1 | | | | 3 | | | | | 1 | 4 | 1 | 6 | 9 |
| 930518 | 8368 | Tetanolita floridana | | | | | | | | | | | | | | | | | | | 2 | 2 | 1 | 2 | | 7 | 7 |
| 930520 | 8370 | Bleptina caradrinalis | | 44 | 9 | 30 | 18 | 1 | 2 | | | 1 | 3 | 1 | | | | 109 | 22 | 2 | 4 | 13 | 6 | 3 | | 50 | 159 |
| 930530 | 8379 | Renia factiosalis | | | | | | | | | | | | | | | | | | | 1 | | | | | 1 | 1 |
| 930536 | 8384.1 | Renia flavipunctalis | | | | | 1 | 1 | | | 8 | | | | | | | 10 | | 2 | 23 | | | | | 25 | 35 |
| 930538 | 8386 | Renia adspergillus | | 8 | | 2 | | | | | | 5 | 1 | | | | | 16 | | | | | | | | | 16 |

Appendix I. Raw light trap data from two Rhode Island barrier dune systems (Napatree Point Conservation Area and Goosewing Beach Preserve) recorded in 2020.

| | | # = counts of individuals | red = species that have not been previously found by the author in MA or RI barrier dune systems | | | | | | | | | | | | | | | | | | | | | | | |
|---|---|---|---|---|---|---|---|---|---|---|---|---|---|---|---|---|---|---|---|---|---|---|---|---|---|---|
| | | Town | Westerly (Watch Hill) | | | | | | | | | | | | | | | Little Compton | | | | | | | | |
| | | System | Napatree (N) | | | | | | | | | | | | | | | Goosewing (GW) | | | | | | | | |
| | | Month | June | June | June | June | June | July | July | Aug | Aug | Aug | Sept | Sept | Sept | Oct | Oct | total | June | July | July | Aug | Sept | Sept | Oct | total | total N + GW |
| | | day | 3 | 17 | 17 | 30 | 30 | 19 | 19 | 6 | 6 | 24 | 16 | 16 | 24 | 6 | 6 | | 23 | 14 | 27 | 26 | 13 | 24 | 6 | | |
| Zook. # | MONA # | Station | N1 | N2 | N3 | N2 | N4 | N2 | N4 | N2 | N5 | N5 | N2 | N4 | N1 | N4 | N6 | | GW1 | GW1 | GW1 | GW1 | GW1 | GW1 | GW1 | GW1 | |
| 930539 | 8387 | Renia sobrialis | | | | | | | | | | | | | | | | | | 5 | 2 | | | | | 7 | 7 |
| | | Renia sp. | | | | | 1 | | | | | | | | | | | 1 | | | 3 | | | | | 3 | 4 |
| 930547 | 8393 | Lascoria ambigualis | | 2 | | | | | 2 | | | | | | | | | 4 | | | 7 | | | | | 7 | 11 |
| 930551 | 8397 | Palthis angulalis | | | | | | | 1 | 1 | 2 | 1 | | | | | | 5 | | | 18 | | | 1 | | 19 | 24 |
| 930552 | 8398 | Palthis asopialis | | | | | | | | | | 2 | | | | | | 2 | | | | | | 1 | | 1 | 3 |
| | | **Hypeninae** | | | | | | | | | | | | | | | | | | | | | | | | | |
| 930561 | 8441 | Hypena manalis | | | | | | | | | | | | | | | | | | | 1 | | | | | 1 | 1 |
| 930565 | 8444 | Hypena palparia | | | | | | | 1 | | | | | | | | | 1 | | | | | | | | | 1 |
| 930588 | 8465 | Hypena scabra | | | | | | 2 | 1 | 4 | 23 | 20 | 1 | | | | | 51 | | 1 | 9 | 4 | | | | 14 | 65 |
| | | **Rivulinae** | | | | | | | | | | | | | | | | | | | | | | | | | |
| 930592 | 8404 | Rivula propinqualis | | | | | | | | | | | | | | | | | | | 3 | | | | | 3 | 3 |
| | | **Boletobiinae** | | | | | | | | | | | | | | | | | | | | | | | | | |
| 930679 | 8499 | Metalectra discalis | | | | | | | | | | 1 | | | | | | 1 | | | | | | | | | 1 |
| | | **Catocalinae** | | | | | | | | | | | | | | | | | | | | | | | | | |
| 930767 | 8777 | Catocala badia | | | | | | | | | | | | | | | | | | | 3 | | | | | 3 | 3 |
| **930787** | **8796** | **Catocala nebulosa** | | | | | | | | | | 1 | | | | | | 1 | | | | | | | | | 1 |
| 930833 | 8847 | Catocala gracilis | | | | | | | | | | | | | | | | | | | 1 | | | | | 1 | 1 |
| 930841 | 8857 | Catocala ultronia | | | | | | | | | 1 | | | | | | | 1 | | | 2 | | | | | 2 | 3 |
| 930923 | 8738 | Caenurgina crassiuscula | | | | | 1 | | | | | 1 | | | | | | 2 | 1 | 1 | | 1 | | | | 3 | 5 |
| 930924 | 8739 | Caenurgina erechtea | | | | | | | | | | | | | | | | | | | | 1 | | | | 1 | 1 |
| 930927 | 8767 | Doryodes spadaria | | | 2 | | | | | | | | | | | | | 2 | | | | | 1 | | | 1 | 3 |
| 930944 | 8745 | Mocis texana | | | | | | | | | 1 | | | | | | | 1 | | | | | | | | | 1 |
| 930962 | 8721 | Allotria elonympha | | | | | | | | | | | | | | | | | | | 1 | | | | | 1 | 1 |
| 931023 | 8689 | Zale lunata | | | | | | | | | 1 | | | | | | | 1 | | | 2 | | | | | 2 | 3 |
| 931042 | 8707 | Zale metatoides | | 1 | | | | | | | | | | | | | | 1 | | | | | | | | | 1 |
| 931053 | 8717 | Zale horrida | | | | | | | | | | | | | | | | | | | 1 | | | | | 1 | 1 |
| 931060 | 9818 | Amolita fessa | | | | | | | | | | | | | | | | | | 1 | | | | | | 1 | 1 |
| | | **Eulepidotinae** | | | | | | | | | | | | | | | | | | | | | | | | | |
| 931077 | 8574 | Anticarsia gemmatalis | | | | | | | | | | | 1 | | | | | 1 | | | | | | | | | 1 |
| | | **EUTELLIIDAE** | | | | | | | | | | | | | | | | | | | | | | | | | |
| 931106 | 8957 | Paectes oculatrix | | | | | | | | 2 | 1 | | | | | | | 3 | 1 | | 5 | | | | | 6 | 9 |
| | | **NOLIDAE** | | | | | | | | | | | | | | | | | | | | | | | | | |
| 931123 | 8983.2 | Meganola spodia | | | | 1 | | | | | | | | | | | | 1 | | | | | | | | | 1 |
| 931131 | 8991 | Nola cereella | | 1 | | | | | | | | | | | | | | 1 | | | | | | | | | 1 |
| | | **NOCTUIDAE** | | | | | | | | | | | | | | | | | | | | | | | | | |
| | | **Plusiinae** | | | | | | | | | | | | | | | | | | | | | | | | | |
| 931170 | 8890 | Chrysodeixis includens | | | | | | | | | | | | 3 | | | | 3 | | | | | | | 1 | 1 | 4 |
| 931191 | 8908 | Autographa precationis | | | | | | | | | | | | | | 1 | 1 | 2 | | | | | | | | | 2 |
| 931234 | 8924 | Anagrapha falcifera | | | | | | | | | 1 | | | | | | | 1 | | | | 1 | | | | 1 | 2 |
| | | **Eustrotiinae** | | | | | | | | | | | | | | | | | | | | | | | | | |
| 931284 | 9044 | Marimatha nigrofimbria | | | | | | | | | 9 | | | | | | | 9 | | | | | | | | | 9 |
| 931295 | 9049 | Maliattha synochitis | | 3 | | | | | | | | | | | | | | 3 | | | | | | | | | 3 |
| | | **Acronictinae** | | | | | | | | | | | | | | | | | | | | | | | | | |
| **931421** | **9200** | **Acronicta americana** | | | | | | | | | | | | | | | | | | | 1 | | | | | 1 | 1 |
| 931478 | 9264 | Acronicta longa | | | | 1 | | | | | | | | | | | | 1 | | | 1 | | | | | 1 | 2 |
| 931493 | 9280 | Acronicta insularis | | | | | | | | | | | | | | | | | | | 1 | 1 | | | | 2 | 2 |
| | | **Cuculliinae** | | | | | | | | | | | | | | | | | | | | | | | | | |
| 931504 | 10200 | Cucullia asteroides | | | | | | | | | | | 1 | 1 | | | | 2 | | | | | | | | | 2 |
| | | **Amphipyrinae** | | | | | | | | | | | | | | | | | | | | | | | | | |
| 931544 | 9638 | Amphipyra pyramidoides | | | | | | | | | 1 | | | | | | | 1 | | | | | | 1 | | 1 | 2 |
| | | **Oncocneminae** | | | | | | | | | | | | | | | | | | | | | | | | | |

Appendix I. Raw light trap data from two Rhode Island barrier dune systems (Napatree Point Conservation Area and Goosewing Beach Preserve) recorded in 2020.

# = counts of individuals; red = species that have not been previously found by the author in MA or RI barrier dune systems

| | | Town | Westerly (Watch Hill) | | | | | | | | | | | | | | | | Little Compton | | | | | | | | |
|---|---|---|---|---|---|---|---|---|---|---|---|---|---|---|---|---|---|---|---|---|---|---|---|---|---|---|
| | | System | Napatree (N) | | | | | | | | | | | | | | | | Goosewing (GW) | | | | | | | | total N + GW |
| | | Month | June | June | June | June | June | July | July | Aug | Aug | Aug | Sept | Sept | Sept | Oct | Oct | total | June | July | July | Aug | Sept | Sept | Oct | total | |
| | | day | 3 | 17 | 17 | 30 | 30 | 19 | 19 | 6 | 6 | 24 | 16 | 16 | 24 | 6 | 6 | | 23 | 14 | 27 | 26 | 13 | 24 | 6 | | |
| Zook. # | MONA # | Station | N1 | N2 | N3 | N2 | N4 | N2 | N4 | N2 | N5 | N5 | N2 | N4 | N1 | N4 | N6 | | GW1 | GW1 | GW1 | GW1 | GW1 | GW1 | GW1 | GW1 | |
| 931875 | 10135 | Sympistis riparia | | | | | | | | | | | | | | | | | 1 | | | | | | | 1 | 1 |
| | | **Agaristinae** | | | | | | | | | | | | | | | | | | | | | | | | | |
| 931964 | 9299 | Eudryas unio | | 1 | | | | | | | | | | | | | | 1 | | | | | | | | | 1 |
| 931966 | 9301 | Eudryas grata | | | | | | 5 | 3 | 1 | | | | | | | | 9 | 2 | | | | | | | 2 | 11 |
| | | **Condicinae** | | | | | | | | | | | | | | | | | | | | | | | | | |
| 931989 | 9690 | Condica videns | | 2 | 7 | | | | | | | 1 | 1 | | | | | 11 | 2 | | 7 | | | | 1 | 10 | 21 |
| 931995 | 9696 | Condica vecors | | | | | | | | | | | | | | | | | | | 6 | 1 | | | | 7 | 7 |
| **931998** | **9699** | **Condica sutor** | | | | | | | | | | | | | | 1 | | 1 | | | | | | | | | 1 |
| **932015** | **9714** | **Condica confederata** | | | | | | | | | | | | | | | | | | | | | 1 | | | 1 | 1 |
| 932018 | 9720 | Ogdoconta cinereola | | | | | | | | | | 5 | | | | | | 5 | | | | | 1 | | | 1 | 6 |
| 932026 | 9065 | Leuconycta diphteroides | | 1 | | | | | | | | | | | | | | 1 | 1 | | | | | | | 1 | 2 |
| | | **Heliothinae** | | | | | | | | | | | | | | | | | | | | | | | | | |
| 932039 | 11055 | Derrima stellata | | | | | | | | 1 | | | | | | | | 1 | | | | | | | | | 1 |
| 932045 | 11068 | Helicoverpa zea | | | | | | | | | | 1 | 2 | 2 | | 2 | 3 | 10 | | | | | | 2 | | 2 | 12 |
| 932120 | 11117 | Schinia lynx | | | | | | | | 3 | | | | | | | | 3 | | | | | | | | | 3 |
| 932134 | 11128 | Schinia arcigera | | | | | | | | | 1 | 1 | | | | | | 2 | | | | 2 | | | | 2 | 4 |
| 932156 | 11177 | Schinia nundina | | | | | | | | | | 4 | | | | | | 4 | | | | | | | | | 4 |
| | | **Noctuinae** | | | | | | | | | | | | | | | | | | | | | | | | | |
| 932205 | 9053 | Pseudeustrotia carneola | | | | | 1 | | | | 5 | | | | | | | 6 | | | 1 | | | | | 1 | 7 |
| 932209 | 9619 | Phosphila miselioides | | | | | | | | | | | | | | | | | | 1 | 1 | | | | | 2 | 2 |
| 932216 | 9666 | Spodoptera frugiperda | | | | | | | | | | | | | | | | | 1 | | | | | | | 1 | 1 |
| 932219 | 9669 | Spodoptera ornithogalli | | | | | | | | | | | 2 | 3 | | | | 5 | | | | 1 | | 10 | 1 | 12 | 17 |
| 932228 | 9678 | Elaphria versicolor | | 1 | | | | | | | 8 | | | | | | | 9 | | | 1 | | | | | 1 | 10 |
| 932230 | 9679 | Elaphria chalcedonia | | 1 | | | | | | | | | 1 | | | | | 2 | | | | | 1 | 4 | | 5 | 7 |
| 932234 | 9681.1 | Elaphria alapallida | | 3 | | | | | | | | | | | | | | 3 | | | | | | | | | 3 |
| **932238** | **9684** | **Elaphria grata** | | | | | | | | 11 | 37 | | 3 | 8 | | 17 | 1 | 77 | | | | | | 1 | | 1 | 78 |
| 932249 | 9688 | Galgula partita | | | | | | | | | | | | | | | | | | | | | | | 1 | 1 | 1 |
| 932266 | 9647 | Proxenus miranda | 3 | 10 | 53 | | 1 | | | 8 | | 1 | | | | | | 76 | 2 | | 7 | 1 | | | | 10 | 86 |
| 932269 | 9650 | Athetis tarda | | | | | | | | | 4 | | | | | | | 4 | | | | | | | | | 4 |
| 932272 | 9560 | Dypterygia rozmani | | 1 | 1 | 1 | | | | 2 | 2 | | | | | | | 7 | | | | | | | | | 7 |
| 932282 | 9637.1 | Magusa divaricata | | | 1 | | | | | | | | 1 | 1 | | 1 | | 4 | | | | | | | | | 4 |
| 932290 | 9545 | Euplexia benesimilis | | | | | | 1 | 1 | | 2 | | | | | | | 4 | | | | | | | | | 4 |
| 932291 | 9546 | Phlogophora iris | | | 1 | | | | | | | | | | | | | 1 | | | | | | | | | 1 |
| 932308 | - | Apamea unanimis | | 1 | | | | | | | | | | | | | | 1 | | | | | | | | | 1 |
| 932314 | 9364 | Apamea sordens | | | 1 | | | | | | | | | | | | | 1 | | | | | | | | | 1 |
| 932339 | 9378 | Apamea burgessi | | | | | | | | | | | 1 | | | | | 1 | | | | | | | | | 1 |
| 932356 | 9395 | Apamea lintneri | | | | | | | | | | | 53 | 65 | 5 | | | 123 | | | | 1 | 24 | 10 | | 35 | 158 |
| 932358 | 9454 | Loscopia velata | | | | | | | | | | | | | | | | | | 1 | | | | | | 1 | 1 |
| 932369 | 9393 | Resapamea stipata | | | | | | | | | | | | | | | | | | | 2 | | | | | 2 | 2 |
| 932370 | 9406 | Mesapamea fractilinea | | | | | | | | | | | | | | | | | | | | 1 | | | | 1 | 1 |
| 932375 | 9415.1 | Oligia strigilis | | | | | | | | | | | | | | | | | 1 | | | | | | | 1 | 1 |
| 932376 | 9415.2 | Oligia latruncula | | 2 | | | | | | | | | | | | | | 2 | | | | | | | | | 2 |
| 932419 | 9435 | Photedes inops | | | | | | | | | | | | | | | | | | | | 1 | 5 | | | 6 | 6 |
| **932439** | **9450** | **Globia subflava** | | | | | | | | | | | | | | | | | | | 7 | | | | | 7 | 7 |
| 932443 | 9453 | Helotropha reniformis | | | | | | | | | | | | | | | | | | | 1 | | | | | 1 | 1 |
| 932444 | 9520 | Achatodes zeae | | | | | | | | | 2 | | | | | | | 2 | | | | | | | | | 2 |
| **932446** | **9456** | **Amphipoea interoceanica** | | | | | | | | | | | | | | | | | | 2 | 2 | | | | | 4 | 4 |
| 932496 | 9465 | Papaipema duovata | | | | | | | | | | | | | | 1 | 1 | 2 | | | | | | | 2 | 2 | 4 |
| 932609 | 9952 | Eucirroedia pampina | | | | | | | | | | | | | | | | | | | | | | 1 | | 1 | 1 |
| 932616 | 9957 | Sunira bicolorago | | | | | | | | | | | 1 | | 1 | | | 2 | | | | | | 3 | | 3 | 5 |
| 932810 | 10524 | Nephelodes minians | | | | | | | | | | | 1 | | | | | 1 | | | | | | 5 | | 5 | 6 |

Appendix I. Raw light trap data from two Rhode Island barrier dune systems (Napatree Point Conservation Area and Goosewing Beach Preserve) recorded in 2020.

# = counts of individuals red = species that have not been previously found by the author in MA or RI barrier dune systems

| | | Town | Westerly (Watch Hill) | | | | | | | | | | | | | | | Little Compton | | | | | | | | |
|---|---|---|---|---|---|---|---|---|---|---|---|---|---|---|---|---|---|---|---|---|---|---|---|---|---|---|
| | | System | Napatree (N) | | | | | | | | | | | | | | | Goosewing (GW) | | | | | | | | |
| | | Month | June | June | June | June | June | July | July | Aug | Aug | Aug | Sept | Sept | Sept | Oct | Oct | total | June | July | July | Aug | Sept | Sept | Oct | total | total |
| | | day | 3 | 17 | 17 | 30 | 30 | 19 | 19 | 6 | 6 | 24 | 16 | 16 | 24 | 6 | 6 | | 23 | 14 | 27 | 26 | 13 | 24 | 6 | | N + GW |
| Zook. # | MONA # | Station | N1 | N2 | N3 | N2 | N4 | N2 | N4 | N2 | N5 | N5 | N2 | N4 | N1 | N4 | N6 | | GW1 | GW1 | GW1 | GW1 | GW1 | GW1 | GW1 | GW1 | |
| 932882 | 10300 | Lacanobia grandis | | 6 | 2 | | | | | | | | | | | | | 8 | 1 | | | | | | | 1 | 9 |
| 932931 | 10434 | Dargida rubripennis | | | | | | | | | 1 | | | | | | | 1 | | | 1 | | | | | 1 | 2 |
| 932933 | 10436 | Mythimna oxygala | 1 | 3 | 5 | | | | | | | | | | | | | 9 | | | | 1 | 2 | | | 3 | 12 |
| 932935 | 10438 | Mythimna unipuncta | | | 1 | | 1 | 1 | 8 | | 2 | 17 | 20 | 69 | | 1 | 2 | 122 | | 2 | 34 | 6 | 5 | 22 | 4 | 73 | 195 |
| 932937 | 10439 | Leucania extincta | 18 | 20 | 41 | | | | | 9 | 5 | | | | | | | 93 | | | 21 | | | | | 21 | 114 |
| 932938.1 | 10440.1 | Leucania amygdalina | 1 | 1 | | | | | | 2 | 3 | | | | | | | 7 | | | | | 1 | | | 1 | 8 |
| 932943 | 10444 | Leucania phragmitidicola | | 1 | 5 | | | | | | | | 2 | | | | | 8 | 16 | | 1 | 6 | 9 | 5 | | 37 | 45 |
| 932944 | 10445 | Leucania linda | | 2 | 2 | | | | | | | | | | | | | 4 | | | | | | | | | 4 |
| 932946 | 10446.1 | Leucania lapidaria | | | | | | | | | | | | | | | | | | 1 | | | | | | 1 | 1 |
| 932960 | 10456 | Leucania adjuta | 2 | | | | | | | | | | 1 | 1 | | | | 4 | | | 5 | | | | | 5 | 9 |
| 932965 | 10461 | Leucania ursula | | | | | | | | | | | | | | | | | | | 1 | | | | | 1 | 1 |
| 933044 | 10397 | Lacinipolia renigera | 1 | 7 | 3 | | | | | | | | 4 | 7 | | 1 | | 23 | 15 | 2 | | 3 | 13 | 27 | 3 | 63 | 86 |
| 933052 | 10405 | Lacinipolia lorea | | | | | | | | | | | | | | | | | | | 1 | | | | | 1 | 1 |
| 933067 | 10413 | Lacinipolia explicata | | | | | | | | | | | | | | | | | | | | | | 1 | | 1 | 1 |
| 933136 | 10585 | Orthodes majuscula | | 5 | | | | | | | 3 | 8 | 1 | | | | | 17 | | | | 1 | 1 | | | 2 | 19 |
| 933138 | 10587 | Orthodes cynica | | 95 | | 3 | 7 | | | | | | | | | | | 105 | 1 | | | | | | | 1 | 106 |
| 933146 | 10288 | Orthodes detracta | | 1 | | | | | | | | | | | | | | 1 | | | | | | | | | 1 |
| 933193 | 10627 | Tricholita signata | | | | | | | | | | | 3 | 3 | | | | 6 | | | | | 12 | 24 | | 36 | 42 |
| 933211 | 10915 | Peridroma saucia | | 1 | | | | | 1 | | 1 | | | | | | | 3 | | | | 1 | | | | 1 | 4 |
| 933212 | 10911 | Anicla infecta | | | | | | | | | | | 2 | | | | | 2 | | | | | 1 | 3 | | 4 | 6 |
| 933329 | 10715 | Euxoa scandens | | | | 1 | 1 | | | 6 | 1 | | | | | | | 9 | | | 9 | | | | | 9 | 18 |
| 933346 | 10727 | Euxoa pleuritica | | | | | | | | | | | | | | | | | | | | 1 | | | | 1 | 1 |
| 933461 | 10838 | Euxoa detersa | | | | | | | | | | | 14 | 18 | 20 | 7 | 16 | 75 | | | | | 14 | 315 | 35 | 364 | 439 |
| 933495 | 10680 | Feltia geniculata | | | | | | | | | | | | | 9 | | | 9 | | | | | 18 | 3 | | 21 | 30 |
| 933498 | 10670 | Feltia jaculifera | | | | | | | | | | | 2 | 1 | | | | 3 | | | | | | 2 | | 2 | 5 |
| 933501 | 10674 | Feltia subgothica | | | | | | | | | | | | | | | | | | | | 2 | 4 | 1 | | 7 | 7 |
| 933503 | 10676 | Feltia herilis | | | | | | | | | | 2 | | 4 | | 1 | | 7 | | | | 1 | 3 | 6 | 3 | 13 | 20 |
| 933504 | 10664 | Feltia subterranea | | | | | | | | | 1 | | | | | | 1 | 2 | | | | | | | | | 2 |
| 933506 | 10641 | Agrotis vetusta | | | | | | | | | | | 32 | 8 | 21 | | 3 | 64 | | | | | 12 | 93 | 6 | 111 | 175 |
| 933515 | 10650 | Agrotis gladiaria | | | | | | | | | | | | | | | | | | | | | | 1 | | 1 | 1 |
| 933528 | 10663 | Agrotis ipsilon | 2 | | | | | | | | | 10 | 1 | 3 | | | | 16 | | | | | 1 | 2 | | 3 | 19 |
| 933529 | 10891 | Ochropleura implecta | | | | | | | | | 1 | 2 | | | | | | 3 | | | 1 | | | 1 | | 2 | 5 |
| 933551 | 11003.1 | Noctua pronuba | | 7 | 5 | | | | | | | 4 | 4 | 2 | 1 | | | 23 | 13 | | | | 14 | 11 | | 38 | 61 |
| **933554** | **10926** | **Spaelotis clandestina** | | | | | | | | | | | | | | | | | | | | | | 2 | | 2 | 2 |
| 933572 | 10944 | Xestia smithii | | | | | | | | | | | | | | | | | | | | | | 3 | | 3 | 3 |
| 933588 | 10942 | Xestia c-nigrum | | | | | | | | | | | | | | | | | | | | | | 1 | | 1 | 1 |
| 933589 | 10942.1 | Xestia dolosa | | 2 | 1 | | | | | | | 3 | | 2 | | | | 8 | | | | | 2 | 1 | 2 | 5 | 13 |
| 933626 | 10955 | Agnorisma badinodis | | | | | | | | | | | | | | 10 | 1 | 11 | | | | | | 1 | 2 | 3 | 14 |
| 933649 | 11006 | Protolampra brunneicollis | | | | | | 1 | | | | | 5 | | | | | 6 | 1 | | | | 1 | 1 | | 3 | 9 |
| 933663.1 | 11024.1 | Abagrotis benjamini | | | | | | 7 | 3 | 1 | 1 | | 5 | 3 | 9 | | | 29 | | 3 | 5 | | 3 | 2 | | 13 | 42 |
| 933680 | 11029 | Abagrotis alternata | | | | | | | | | | | | | 1 | | | 1 | | | | | | | | | 1 |
| 933686 | 11043.2 | Abagrotis magnicupida | | | | | | 1 | 1 | | | | 3 | | | | | 5 | | 1 | 1 | | 1 | 6 | 1 | 10 | 15 |
| | | | | | | | | | | | | | | | | | | | | | | | | | | | |
| | | Total # species | **14** | **56** | **24** | **23** | **23** | **19** | **22** | **18** | **47** | **35** | **35** | **24** | **11** | **13** | **12** | **161** | **31** | **26** | **66** | **31** | **35** | **43** | **15** | **145** | **216** |
| | | Total # individuals | **38** | **297** | **180** | **73** | **54** | **34** | **39** | **57** | **195** | **128** | **186** | **209** | **71** | **45** | **35** | **1641** | **144** | **39** | **321** | **95** | **166** | **593** | **64** | **1422** | **3063** |
| | | Number samples | | | | | | | | | | | | | | | | **15** | | | | | | | | **7** | **22** |

Appendix II. List of microlepidoptera recorded from light trap samples at Napatree Point Conservation Area and Goosewing Beach Preserve during 2020.
"1" = occurance, not count

| | month | June | Sept | June | June | July | Aug | Sept | June | June | July | Sept | Oct | Aug | Aug | Oct | total | June | July | July | Aug | Sept | Sept | Oct | total | total occurences |
|---|---|---|---|---|---|---|---|---|---|---|---|---|---|---|---|---|---|---|---|---|---|---|---|---|---|---|
| | day | 3 | 24 | 17 | 30 | 19 | 6 | 16 | 17 | 30 | 19 | 16 | 6 | 6 | 24 | 6 | occ. | 23 | 14 | 27 | 26 | 13 | 24 | 6 | occ. | N + GW |
| | station | N1 | N1 | N2 | N2 | N2 | N2 | N2 | N3 | N4 | N4 | N4 | N4 | N5 | N5 | N6 | N1-6 | GW1 | GW1 | GW1 | GW1 | GW1 | GW1 | GW1 | GW1 | |
| | **OECHOPHORIDAE** | | | | | | | | | | | | | | | | | | | | | | | | | |
| 856 | *Agonopterix hyperella* | | | | | | | | | | | | | | 1 | | 1 | | | | | | | | | 1 |
| 955 | *Psilocorsis querciella* | | | | | | | | | | | | | | | | | 1 | | | | | | | 1 | 1 |
| 1014 | *Antaeotricha leucillana* | | | | | | | | | | | | | 1 | | | 1 | | | 1 | | | | | 1 | 2 |
| | **BLASTOBASIDAE** | | | | | | | | | | | | | | | | | | | | | | | | | |
| | *Blastobasis sp.* | | | | | | | | | | | | | | | | | | | 1 | | | | | 1 | 1 |
| | **COLEOPHORIDAE** | | | | | | | | | | | | | | | | | | | | | | | | | |
| 1257 | *Coleophora (atromarginata ?)* | | | | | | | | | 1 | | | | | | | 1 | | | | | | | | | 1 |
| | **GELECHIIDAE** | | | | | | | | | | | | | | | | | | | | | | | | | |
| 1986 | *Gnorimoschema gallaesolidaginis* | | | | | | | | | | | | 1 | | | | 1 | | | | 1 | 1 | | | 2 | 3 |
| 2060 | *Chionodes arenella* | | | | | | | 1 | | | | | | | | | 1 | | | | | | 1 | | 1 | 2 |
| 2187 | *Aroga compositella* | | 1 | | 1 | | 1 | | | | | | | | | | 3 | | | | | | | | | 3 |
| 2301 | *Dichomeris serrativittella* | | | | | | | | | | | | | | | | | 1 | | | | | | | 1 | 1 |
| | **PLUTELLIDAE** | | | | | | | | | | | | | | | | | | | | | | | | | |
| 2366 | *Plutella xylostella* | | | | | | | | 1 | 1 | | | | | | | 2 | | | | | | | | | 2 |
| | **ATTEVIDAE** | | | | | | | | | | | | | | | | | | | | | | | | | |
| 2401 | *Atteva aurea* | | | | | | | | | | | | | | 1 | | 1 | | | | | | | | | 1 |
| | **TORTRICIDAE** | | | | | | | | | | | | | | | | | | | | | | | | | |
| 2701 | *Episimus argutana* | | | | | | | | | | | | | | | | | | | | 1 | | | | 1 | 1 |
| 2770 | *Orthotaenia undulana* | | | 1 | | | | | 1 | | | | | | | | 2 | | | | | | | | | 2 |
| 2823 | *Olethreutes fasciatana* | | | | | | | | | 1 | | | | | | | 1 | 1 | | | | | | | 1 | 2 |
| 2860 | *Hedya separatana* | | | | | | | | | | | | | | | | | | | 1 | | | | | 1 | 1 |
| 2867 | *Rhyacionia buoliana* | | | | 1 | | | | | | | | | | | | 1 | | | | | | | | | 1 |
| 2927 | *Eucosma ochrocephala* | | | | | | | | | | | | | | | | | | | | 1 | | | | 1 | 1 |
| 2928 | *Eucosma raracana* | | | | | | | | | | | | | | | | | | | | 1 | | | | 1 | 1 |
| 3009 | *Pelochrista robinsonana* | | | | | 1 | | | 1 | | | | | | | | 2 | 1 | | | | | | | 1 | 3 |
| 3038 | *Pelochrista argentialbana* | 1 | | | | | | | 1 | 1 | | | | | | | 3 | | | | | | | | | 3 |
| 3074 | *Eucopina tocullionana* | | | | | | | | | | | | | 1 | | | 1 | | | | | | | | | 1 |
| 3116 | *Pelochrista dorsisignatana* | | | | | | | | | | | | | | | 1 | 1 | | | | | | 1 | | 1 | 2 |
| 3116.2 | *Pelochrista oraria* | | | 1 | | | | 1 | | | | 1 | | | | | 3 | | | | | 1 | 1 | | 2 | 5 |
| 3120 | *Pelochrista derelicta* | | | | | | | | | | | | | 1 | | | 1 | | | 1 | | | | | 1 | 2 |
| 3142 | *Pelochrista cataclystiana* | | | | | | | | | | | | | | | | | | | 1 | | | | | 1 | 1 |
| 3173 | *Epiblema abruptana* | | | | | | | | | | | | | | 1 | | 1 | | | | | | | | | 1 |
| 3186a | *nr. Epiblema scudderiana* | | | | | | 1 | | | | | | | | | | 1 | | | 1 | | | | | 1 | 2 |
| 3208 | *Notocelia rosaecolana* | | | | | | | | 1 | 1 | | | | | | | 2 | 1 | | | | | | | 1 | 3 |
| 3211 | *Notocelia culminana* | | | | | | | | | | | 1 | | | | | 1 | | | | | | | | | 1 |
| 3486 | *Cydia toreuta* | | | 1 | 1 | | | | | | | | | | | | 2 | | | | | | | | | 2 |
| 3633 | *Choristoneura parallela* | | | | | | | | | | | | | | | | | | 1 | 1 | | | | | 2 | 2 |
| 3635 | *Choristoneura rosaceana* | | | | 1 | | | | | | | | | | | | 1 | | 1 | 1 | 1 | | | | 3 | 4 |
| 3648 | *Archips argyrospila* | | | | 1 | | | | | | | | | | | | 1 | | 1 | | | | | | 1 | 2 |
| 3658 | *Archips purpurana* | | | | | | | | | | | | | | | | | | 1 | 1 | | | | | 2 | 2 |
| 3688 | *Clepsis peritana* | | | | | | | | 1 | 1 | | | | 1 | | | 3 | | | | | | | | | 3 |
| 3695 | *Sparganothis sulfureana* | | | | | | | | | | | | | | | | | | | | | | 1 | | 1 | 1 |
| 3720 | *Cenopis reticulatana* | | | | | | | | | | | | | 1 | | | 1 | | | | | | | | | 1 |
| 3732 | *Platynota flavedana* | | | | | | | | | | | | | | | | | 1 | | | | | | | 1 | 1 |
| 3740 | *Platynota idaeusalis* | | | | | | | | | 1 | | | | | | | 1 | | | | | | | | | 1 |

Appendix II. List of microlepidoptera recorded from light trap samples at Napatree Point Conservation Area and Goosewing Beach Preserve during 2020.

"1" = occurance, not count

| | month | June | Sept | June | June | July | Aug | Sept | June | June | July | Sept | Oct | Aug | Aug | Oct | total | June | July | July | Aug | Sept | Sept | Oct | total | total |
|---|---|---|---|---|---|---|---|---|---|---|---|---|---|---|---|---|---|---|---|---|---|---|---|---|---|---|
| | day | 3 | 24 | 17 | 30 | 19 | 6 | 16 | 17 | 30 | 19 | 16 | 6 | 6 | 24 | 6 | occ. | 23 | 14 | 27 | 26 | 13 | 24 | 6 | occ. | occurences |
| | station | N1 | N1 | N2 | N2 | N2 | N2 | N2 | N3 | N4 | N4 | N4 | N4 | N5 | N5 | N6 | N1-6 | GW1 | GW1 | GW1 | GW1 | GW1 | GW1 | GW1 | GW1 | N + GW |
| 3812 | *Phtheochroa baracana* | | | | | | | | | | | | | | | | | | | 1 | | | | | 1 | 1 |
| | **ZYGAENIDAE** | | | | | | | | | | | | | | | | | | | | | | | | | |
| 4624 | *Harrisina americana* | | | | 1 | | | | | | | | | | | | 1 | | | | | | | | | 1 |
| | **MEGAJOPYGIDAE** | | | | | | | | | | | | | | | | | | | | | | | | | |
| 4644 | *Megalopyge crispata* | | | | | | | | | | | | | | | | | 1 | 1 | | | | | | 2 | 2 |
| | **LIMACODIDAE** | | | | | | | | | | | | | | | | | | | | | | | | | |
| 4665 | *Lithacodes fasciola* | | | 1 | | | | | | | | | | | | | 1 | 1 | 1 | | | | | | 2 | 3 |
| 4671 | *Prolimacodes badia* | | | | | 1 | | | | | | | | | | | 1 | | 1 | | | | | | 1 | 2 |
| 4697 | *Euclea delphinii* | | | 1 | | | | | | 1 | | | | | | | 2 | 1 | | | | | | | 1 | 3 |
| | **CRAMBIDAE** | | | | | | | | | | | | | | | | | | | | | | | | | |
| 4751 | *Elophila gyralis* | | | 1 | | | | | | | | | | | | | 1 | | | | | | | | | 1 |
| 4761 | *Paraponyx badiusalis* | | | | | | | | | | | | | | | | | | | | 1 | | | | 1 | 1 |
| 4794 | *Eustixia pupula* | | | | | | | | 1 | | | | | 1 | | | 2 | | | 1 | | | | | 1 | 3 |
| 4870 | *Glaphyria sequistrialis* | | | | | | | | | | | | | 1 | | | 1 | | | | | | | | | 1 |
| 4951 | *Perispasta caeculalis* | | | | | | | | | | | | | 1 | | | 1 | | | 1 | 1 | | | | 2 | 3 |
| 4953 | *Anania tertialis* | | | 1 | | | | | | | | | | 1 | 1 | | 3 | | | | | | | | | 3 |
| 4975 | *Achyra rantalis* | | | | | | | | | | | 1 | | | | | 1 | | | | | | | | | 1 |
| 4986.1 | *Sitochroa palealis* | | | | | | | | | 1 | 1 | | | | | | 2 | | | | | | | | | 2 |
| 5040 | *Pyrausta bicoloralis* | | | | | | | | | | | | | 1 | | | 1 | | | | | | | | | 1 |
| 5079 | *Udea rubigalis* | 1 | | 1 | | | | | | 1 | 1 | | | 1 | | | 5 | | | | 1 | | | 1 | 2 | 7 |
| 5117 | *Loxostegopsis merrickalis* | | | | | | | | | | | | | | | | | | 1 | | | | | | 1 | 1 |
| 5143 | *Diacme adipaloides* | | | | | | | | | | | | | | | | | | 1 | | | | 1 | | 2 | 2 |
| 5156 | *Nomophila nearctica* | | | | | | | | | | | | | | | | | | | | | | | 1 | 1 | 1 |
| 5159 | *Desmia funeralis* | | | | | | | | | | | | | 1 | | | 1 | | | | | | | | | 1 |
| 5160 | *Desmia maculalis* | | | 1 | | | | | | | | | | 1 | | | 2 | 1 | | 1 | | | | | 2 | 4 |
| 5169 | *Hymenia perspectalis* | | | | | | | | | | | | | | | | | | | | | 1 | | 1 | 2 | 2 |
| 5170 | *Spoladea recurvalis* | | | | | | | 1 | | | | 1 | 1 | | | | 3 | | | | | | | | | 3 |
| 5176 | *Anageshna primordialis* | | | | | | | | | | | | | 1 | | | 1 | | | | | | | | | 1 |
| 5177 | *Apogeshna stenialis* | | | | 1 | | | | | | | | | | | | 1 | | | | | | | | | 1 |
| 5275 | *Herpetogramma pertextalis (?)* | | | | | | | | | | | | | | | | | | 1 | 1 | | | | | 2 | 2 |
| 5277 | *Herpetogramma thestealis* | | | | | | | | | | | | | | | | | | | 1 | | | | | 1 | 1 |
| 5280 | *Herpetogramma aeglealis* | | | | | | | | | | | | | | | | | | | 1 | | | | | 1 | 1 |
| 5298.1 | *Sclerocona acutellus* | | | 1 | | | | | | 1 | | | | 1 | | | 3 | | | | | | | | | 3 |
| 5299 | *Acentria ephemerella* | | | | | | | | | 1 | | | | | | | 1 | | | | | | | | | 1 |
| 5357 | *Crambus leachellus* | | 1 | | | | | | | | | | | | | | 1 | | | | | | | | | 1 |
| 5399 | *Agriphila ruricolellus* | | | | | | | | | | | | | | | | | | | | | 1 | 1 | | 2 | 2 |
| 5403 | *Agriphila vulgivagellus* | | 1 | | | | | | | | | 1 | | | | | 2 | | | | | | 1 | | 1 | 3 |
| 5417 | *Pediasia dorsipunctellus* | | | | | | | | | | | | | | | | | | | | | 1 | | | 1 | 1 |
| 5420 | *Microcrambus elegans* | | | | | | | | | 1 | | | | 1 | | | 2 | | | | | | | | | 2 |
| 5435 | *Fissicrambus mutabilis* | | | | | | 1 | | 1 | | | | | | | | 2 | | | | 1 | | | | 1 | 3 |
| 5443 | *Thaumatopsis floridella* | | | 1 | | | | | | | | | | 1 | | | 2 | | | | | | | | | 2 |
| 5450 | *Parapediasia decorellus (?)* | | | | 1 | | | | | | | | | | | | 1 | | | | | | | | | 1 |
| 5451 | *Parapediasia teterrellus* | | | | | | 1 | | | | | | | | | | 1 | 1 | | | | | | | 1 | 2 |
| | **PYRALIDAE** | | | | | | | | | | | | | | | | | | | | | | | | | |
| 5533 | *Hypsopygia olinalis* | | | 1 | 1 | | | | | | | | | | | | 2 | | 1 | 1 | | | | | 2 | 4 |
| 5552 | *Galassa nigrinodis* | | | | | | | | | | | | | 1 | | | 1 | | | 1 | | | | | 1 | 2 |

Appendix II. List of microlepidoptera recorded from light trap samples at Napatree Point Conservation Area and Goosewing Beach Preserve during 2020.
"1" = occurance, not count

| | month | June | Sept | June | June | July | Aug | Sept | June | June | July | Sept | Oct | Aug | Aug | Oct | total | June | July | July | Aug | Sept | Sept | Oct | total | total |
|---|---|---|---|---|---|---|---|---|---|---|---|---|---|---|---|---|---|---|---|---|---|---|---|---|---|---|
| | day | 3 | 24 | 17 | 30 | 19 | 6 | 16 | 17 | 30 | 19 | 16 | 6 | 6 | 24 | 6 | occ. | 23 | 14 | 27 | 26 | 13 | 24 | 6 | occ. | occurences |
| | station | N1 | N1 | N2 | N2 | N2 | N2 | N2 | N3 | N4 | N4 | N4 | N4 | N5 | N5 | N6 | N1-6 | GW1 | GW1 | GW1 | GW1 | GW1 | GW1 | GW1 | GW1 | N + GW |
| 5566 | *Arta statalis* | | | | | | | | | | | | | 1 | | | 1 | | | 1 | | | | | 1 | 2 |
| 5577 | *Epipaschia superatalis* | | | 1 | | | | | | 1 | | | | | | | 2 | | | | | | | | | 2 |
| 5579 | *Macalla zelleri* | | | 1 | | | | | | | 1 | | | 1 | | | 3 | 1 | 1 | 1 | | | | | 3 | 6 |
| 5629 | *Aphomia sociella* | | | | | | | | | 1 | | | | | | | 1 | | | | | | | | | 1 |
| 5691 | *Acrobasis comptoniella* | | | | | | | | | | | | | | | | | | 1 | | | | | | 1 | 1 |
| 5744 | *Etiella zinckenella* | | | 1 | | | | | 1 | | | | | | | | 2 | 1 | | 1 | | | | | 2 | 4 |
| 5747 | *Pima albiplagiatella* | 1 | | | 1 | | | | 1 | | | | | | | | 3 | 1 | | | | | | | 1 | 4 |
| 5787 | *Meroptera pravella* | | | | | | | | | | | | | | | | | 1 | | | | | | | 1 | 1 |
| 5847 | *Dioryctria disclusa* | | | | | | 1 | | | | | | | | | | 1 | | | | | | | | | 1 |
| 5896 | *Elasmopalpus lignosellus* | | | | | | | | | | | | | | | | | | | | | | | 1 | 1 | 1 |
| 5946.2 | *Phycitodes reliquellum* | | | | | | | | | | | | | | | | | | | | | | 1 | | 1 | 1 |
| 5999 | *Eulogia ochrifrontella* | | | | | | | | | | | | | 1 | | | 1 | | | | | | | | | 1 |
| 6005 | *Moodna ostrinella* | | | 1 | | | | 1 | | | | | | | | | 2 | | | | | | | | | 2 |
| 6048 | *Peoria gematella* | | | | | | | | | | | | | 1 | | | 1 | 1 | | 1 | 1 | | | | 3 | 4 |
| | # species | 3 | 3 | 16 | 10 | 2 | 5 | 4 | 10 | 15 | 3 | 5 | 2 | 22 | 4 | 1 | 66 | 16 | 13 | 22 | 10 | 5 | 8 | 4 | 57 | 94 |

*Immature Cooper's Hawk feeding on a Sora Rail.*
*Photo Credit: The Watch Hill Conservancy*

CHAPTER 17

# Camera Trap Reconnaissance of Wildlife in the Napatree Point Conservation Area: 2022 Sampling

*Emily Bodell,[1] Peter August,[1,2] Daniel Cole[1] & Janice Sassi[1]*

[1] Napatree Point Conservation Area, The Watch Hill Conservancy

[2] Department of Natural Resources Science, University of Rhode Island

## INTRODUCTION

The western end of Napatree Point consists of a variety of ecological communities—maritime shrubland, salt scrub, maritime shrub dune, high marsh, maritime herbaceous dunes, rocky shore, and maritime beach. Previous years of camera trapping have shown that the meso-mammal (midsize) fauna of this section of Napatree is surprisingly diverse. This group of animals plays an important role in the ecosystem as predators, prey, plant consumers, and seed dispersers. The overall purpose of our camera trap survey is to assess the diversity and relative abundance of midsize mammals (and occasionally birds) in the different habitats in the Napatree Point Conservation Area (NTPCA), Watch Hill, Rhode Island. This is the tenth year we have been camera trapping at Napatree, and each year we strive to monitor different habitats in the conservation area. These data will contribute to the overall ecological knowledge of the NTPCA, serve as a baseline inventory of meso-mammals of Napatree, and provide objective data on the relative abundance of potential predators of shorebirds in the area. With these baseline data on predator and prey populations, we are beginning to discern trends in the relative abundance of various species on Napatree and how abundance varies from year to year.

## METHODS

We deployed camera traps (Reconyx Rapid Fire, Browning Strike Force, and Browning Dark Ops) in four different locations in the NTPCA to monitor animal activity (Figure 1, Table 1). At each site, a camera was set approximately 1 m (3.3 feet) off the ground affixed to a stout branch or mounted on a wooden or metal pole. The SD card containing photographs was retrieved at two- to eight-week intervals. Each camera ran continuously during the period of deployment and was set to record three to five consecutive photographs every time the motion sensor beam was broken. Each photo was stamped with the date, time, temperature, and moon phase. Photos taken in the day were color, and photos taken at night were black-and-white and were illuminated by an infrared flash. One camera was set to record in video mode. For data analysis, an animal that was continuously present for a string of consecutive photos was recorded as a single occurrence (visit). When photo sequences were more than five minutes apart they were counted as separate animal visits. Our data do not reflect the density of each species (#/unit area). It is impossible to know if five visits to a camera trap site represents five individuals or one individual visiting five times. Our data do, however, provide good estimates of animal diversity and relative abundance.

Nomenclature of mammals follows ITIS (2022), Wilson and Reeder (2005), and August *et al.* (2001). The authoritative genus name for American Mink (*Neovison* and *Mustela*) is in dispute so both are given.

*Figure 1. Camera trap sampling locations in 2021-2022.*

## RESULTS AND DISCUSSION

The species photographed and their capture frequencies are given in Table 2. Human activity and domestic dog sightings were infrequent this year compared to previous years. This is likely due to the off-the-beat locations of the camera traps rather than a change in visitors.

We recorded a new species to Napatree this season, a woodchuck (*Marmota monax*) photographed at the bat detector on 17 May 2022. It is a long haul from Watch Hill to the west end of Napatree!

Coyotes were regularly photographed at the log but in fewer numbers (3 visits per 100 camera trap days) than previous years. The eight-year average is 7.8 visits per 100 days camera trapping. Abundance estimates have ranged from 0 visits per 100 days to 26 visits per 100 days over this period. Fox abundance was also low this year with only 1.8 visits per 100 days; this is well below the eight-year average of 3.2 visits per 100 days. It is tempting to speculate that fox and coyote are competitors on Napatree. When coyote visits to the log cam are high (2016, 2019, and 2021), fox numbers are low. But coyote numbers are variable and can be high or low when fox numbers are high (2015 and 2018). Another factor to consider is denning sites for fox and coyotes. In 2014, we found an active den of a pair of red fox, and we accumulated thousands of photos. If fox or coyotes are denning near the log location, we would expect a high frequency of visits as they commute through the area. Given the abundance of prey on Napatree (rabbits, mice, skunks, opossums, and birds), it is not surprising this small area supports a seemingly stable population of predators.

Mink were frequent visitors at the log with 4.8 visits per 100 days, which is almost double the eight-year average of 2.5 visits per 100 days. They appear to be stable resident(s) on Napatree and are usually photographed in a variety of habitats, albeit in low numbers.

| Site | Sampling Period | Total Days of Camera Sampling | Habitat |
|---|---|---|---|
| Bat Detector | 13 May 2022 to 22 Nov 2022 | 144 | In dense stand of bayberry, Rosa, and poison ivy |
| Log | 27 Nov 2021 to 13 May 2022 | 166 | Along a large log at the edge of a *Phragmites* patch and shrubland |
| Pines | 8 Mar 2022 to 22 Mar 2022 | 14 | Camera on large branch of a black pine with a view of a small 10' x 10' clearing |
| Salt Marsh Video | 27 Nov 2021 to 18 Jan 2022 | 47 | At edge of the salt marsh, mounted on U-pole |
| Salt Marsh Stills | 27 Nov 2021 to 15 Apr 2022 | 138 | At edge of the salt marsh, mounted on U-pole |

*Table 1: Camera trap sampling locations in 2021–2022.*

| | | Location<br>Total # Visits Recorded<br>(Visits/100 Days Sampling Effort) | | | | |
|---|---|---|---|---|---|---|
| **Species** | **Common Name** | Bat Detector | Log | Pines | Salt Marsh Video | Salt Marsh Stills |
| *Didelphis virginiana* | Virginia Opossum | 5<br>(3.5) | | | | |
| *Sylvilagus floridanus* | Eastern Cottontail Rabbit | 7<br>(4.9) | 203<br>(122.3) | 15<br>(107.1) | 18<br>(38.3) | |
| *Marmota monax* | Woodchuck | 1<br>(0.7) | | | | |
| *Peromyscus leucopus* | White-footed Mouse | 40<br>(27.8) | 3<br>(1.8) | | | |
| *Canis latrans* | Coyote | 3<br>(2.1) | 5<br>(3.0) | | | 7<br>(5.1) |
| *Canis familiaris* | Domestic Dog | | | | | 1<br>(0.7) |
| *Procyon lotor* | Raccoon | | 3<br>(1.8) | | | |
| *Vulpes vulpes* | Red Fox | 1<br>(0.7) | 3<br>(1.8) | | | |
| *Neovison (Mustela) vison* | Mink | 2<br>(1.4) | 8<br>(4.8) | 1<br>(7.1) | | |
| *Mephitus mephitus* | Striped Skunk | | 3<br>(1.8) | | | |
| *Odocoileus virginianus* | White-tailed Deer | 16<br>(11.1) | 2<br>(1.2) | | | 1<br>(0.7) |
| Unknown Mammal | | 2<br>(1.4) | 2<br>(1.2) | 3<br>(21.4) | | |
| Misc Birds | | 77<br>(53.5) | 91<br>(54.8) | | 6<br>(12.7) | 1<br>(0.7) |

*Table 2. Species recorded in photographs in the 2021/2022 sampling period. The total number of visits per species and the frequency of visits (in visits/100 camera trap days) are presented.*

Deer abundance was relatively low this year except at the bat detector site (11.1 visits per 100 days). In previous sampling, deer visits could be up to 25 visits per 100 camera trap days (e.g., the fort location). Although deer were only photographed twice at the log, one was a herd of seven individuals (see Appendix 1).

Eastern cottontail rabbits were extremely abundant, especially at the log and pines. Given the regularity with which rabbits were observed in both locations, we suspect we are capturing one or a few resident rabbits on a regular basis. We commonly photographed a pair of rabbits at the log. They appeared very social and showed no sign of aggressively chasing each other off a territory.

There was a resident deer mouse at the bat detector site. It was photographed on a regular basis dashing down a branch in front of the camera.

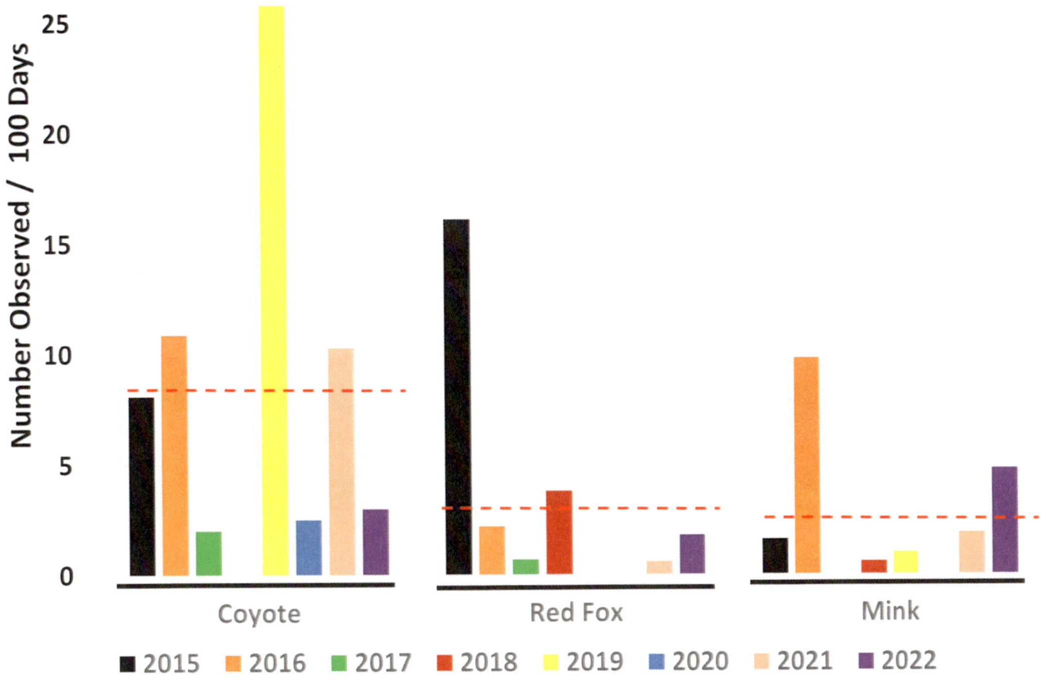


*Figure 2. Annual variation in predator abundance at the log site. The dashed red lines indicate the average annual number of observations over the eight-year period for each species.*

The complete list of mammals photographed in our camera trap monitoring is presented in Table 3.

| Notes | | | |
|---|---|---|---|
| Species | Common Name | Habitat | Frequency Observed |
| *Didelphis virginiana* | Virginia Opossum | Ubiquitous | Common |
| *Homo sapiens* | Human | Ubiquitous | Common |
| *Sylvilagus floridanus* | Eastern Cottontail Rabbit | Ubiquitous | Common |
| *Marmota monax* | Woodchuck | Shrubland, Ecotone | Uncommon |
| *Peromyscus leucopus* | White-footed Mouse | Shrublands, Open Understory | Common |
| *Microtus pennsylvanicus** | Meadow Vole | Dune Grass | Common |
| *Canis latrans* | Coyote | Ubiquitous | Common |
| *Canis familiaris* | Domestic Dog | Ubiquitous | Common |
| *Vulpes vulpes* | Red Fox | Shrublands, Ecotones | Occasional |
| *Procyon lotor* | Common Raccoon | Pines, Shrublands | Uncommon |
| *Neovison (Mustela) vison* | Mink | Shrublands, Ecotones | Common |
| *Mephitus mephitus* | Striped Skunk | Ubiquitous | Common |
| *Felis catus* | Domestic Cat | Pines | Uncommon |
| *Odocoileus virginianus* | White-tailed Deer | Shrublands, Salt Marsh, Ecotones | Common |

*Table 3. Mammal species recorded in camera trap sampling since the inception of the study (February 2013). Estimates of dominant habitat and frequency of observation based on the subjective judgement of PVA. *Red fox were photographed bringing dead meadow voles to the fox den in 2014. Vole sign (runways, clipped dune grass) are common in open dune grass habitats.*

## CONCLUSIONS

The 2021–2022 camera trapping season yielded a new species to the meso-mammal inventory of Napatree, a woodchuck (*Marmota monax*). Given that it was only observed once, he might have been a transient visitor.

The log continues to be a useful sentinel location to understand annual variation in species presence. Mink visits to the log were very high this year. It would be interesting to know if the site is part of a regular hunting circuit or if there is a den nearby. We will continue to monitor the log in future sampling.

The ecotone between the maritime shrublands and the salt marsh is heavily travelled by coyotes. We observed the same pattern with cameras set at the salt marsh osprey pole; commuting coyotes were frequently photographed at the marsh/shrub edge. Compared to the densely vegetated shrublands, low, grassy vegetation would afford easier movement.

## ACKNOWLEDGMENTS

This project was supported by The Watch Hill Conservancy and the University of Rhode Island Coastal Institute.

## PREVIOUS SUMMARIES

Each year's camera trapping results are tabulated in the *State of Napatree Report* and can be downloaded from https://thewatchhillconservancy.org/napatree/napatree-resources/.

## REFERENCES

August P., R. Kenney, and T. Husband. 2001. "Mammals." In *Vertebrates of Rhode Island*, edited by P. August, R. Enser, and L. Gould. Volume 2. The Biota of Rhode Island. Kingston, RI: Rhode Island Natural History Survey.

ITIS. 2022. Retrieved 23 January 2022 from the Integrated Taxonomic Information System (ITIS) On-line Database. https://www.itis.gov/.

Wilson D., and D. Reeder. 2005. "Mammal Species of the World." In *A Taxonomic and Geographic Reference*, edited by D. Wilson and D. Reeder. 3rd ed. Johns Hopkins University Press.

**Appendix 1**. *Sample photos from the 2021-2022 season.*

*Napatree deer herd at the log.*

*Newborn fawn at the bat detector site.*

*Mink on the move at the log.*

*Fresh powder at the log.*

*Northern Flicker at the log.*

*Red-winged Blackbird showing off at the log.*

*Big brown bat, the most common species on Napatree.*
*Photo Credit: MerlinTuttle.org*

CHAPTER 18

# Summertime Bat Activity on Napatree Point (2022)

*Peter V. August,[1,2] Janice M. Sassi[2] & Daniel T. Cole[2]*
[1] Department of Natural Resources Science, University of Rhode Island
[2] Napatree Point Conservation Area, The Watch Hill Conservancy

## INTRODUCTION

In the *2016 State of Napatree Report*, we presented the initial results of bat surveys at various locations in the Napatree Point Conservation Area (August *et al.*, 2016). Using a Wildlife Acoustics (Maynard, MA) Echo Meter Touch bat detector, we recorded bats passing in the vicinity of the instrument between the hours of 20:00 and 22:00 in three locations. We found a very typical (for this area) suite of bat species using Napatree. Bat activity was generally low with few bat passes per hour compared to high quality bat habitats in the region. However, our sampling period was only a few hours after sunset. In 2017, we conducted night-long acoustic bat surveys all summer at one location in the Maritime Shrubland habitat west of the Pines (August *et al.*, 2017, Figure 1). This continuous record of bat activity over the summer showed a number of dramatic spikes in flight activity. Notably, big brown bats (*Eptesicus fuscus*) showed a surge in activity in the last few weeks of August. Similarly, red bats (*Lasiurus borealis)* demonstrated a spike in activity in early October. We presume these patterns reflect migratory activity as bats fly to the southern United States (red bats) or nearby hibernacula (brown bats) for the winter. Similar trends have been observed in subsequent years (August *et al.*, 2021).

To validate the patterns and trends in bat activity observed in previous years, we monitored bats in 2022 at the same location using the same equipment used in previous years.

## METHODS

We sampled bat activity with a Pettersson D500X bat detector (Pettersson Elektronik AB, Uppsala, Sweden) with an external microphone attached to a 7.6 m (25-foot) cable. This is the same device used by the RI Department of Environmental Management (RI DEM) for its statewide bat monitoring program. We wanted to make sure any data we collected would be consistent with RI DEM's so they could include Napatree in their statewide monitoring network. The instrument was powered by a Power Sonic 6-volt, 20-amp-hour wet cell battery. The battery would typically last three weeks before discharging. The bat detector was run on a second wet cell battery while the first battery recharged. The bat detector was enclosed in multiple weather-proof cases to protect the electronics in inclement weather. Using a telescoping pole, the microphone was deployed 1 m (3.3 feet) above the bayberry shrub canopy in the vicinity of the Pines (Figure 1) between 18 April and 18 November 2022 for a total of 214 days. Data were logged for 146 days. Because the SD card filled or the battery drained, data were not logged on 68 days. The total height of the microphone from the ground was approximately 4 m (13.1 feet). The microphone was pointed toward the northwest. The ocean was south (behind) of the microphone, thus minimizing acoustic clutter from breaking waves. The detection distance of a D500X system is approximately 40 to 50 m (131 to 164 feet). Any bats flying above the shrubs in this region in front of the microphone would be recorded. The bat recorder was set to turn on an hour before dusk and turn off at 6:00 AM. Calls were written to a 32 GB (UDMA 7) SanDisk compact flash (CF) card. Each bat call recording was 2.9 MB and lasted 60 to 1,300 msec. The number of calls recorded in a week ranged from under ten to almost 3,700 bat passes.

At two-to-three-week intervals, the CF card was retrieved and downloaded into a desktop computer. Sonobat 4.2.0 software was used to process bat call recordings made by the D500X. The workflow for processing calls was as follows:

1. The contents of the CF card were downloaded to the computer.
2. The Sonobat D500X file attributer 2.7 utility was run to write necessary header information to each recording.
3. Calls were processed with Sonobat Batch Scrubber 5.7 to exclude from analysis any calls <20 kHz in frequency and containing mostly noise. The *medium* threshold setting for call quality was used.
4. Recordings that were retained by the batch scrubber were processed by the Sonobatch utility using the northeastern United States bat species reference database. This tool attempts to identify the species making each call based on the acoustic properties of the recording. Two categories of identifications are made—high confidence species identifications when the software's maximum likelihood classifier has a very strong match to reference data. In this study, the high confidence identifications had average likelihood probabilities of a match greater than 99 percent. If a call does not match the reference data well, a high-quality identification is not made. Only high-confidence classifications are reported here. This gave us a very conservative but high-quality estimate of bat diversity and activity patterns.
5. The Sonovet utility was used to create a data table of date, time, probable species, and acoustic parameters of each call. This table was downloaded to Excel for analysis.

It is important to note that the number of bat passes that are recorded is not equivalent to the number of bats flying through the monitoring area. It is impossible to know if multiple recordings are from one bat flying back and forth by the microphone or many bats.

*Figure 1. Bat sampling location (yellow symbol). The inset shows the microphone extending above the canopy of the shrubs.*

*Aerial Photo: USFWS, 2017*

## RESULTS

In the 2022 bat sampling season, 48,630 recordings were obtained by the bat detector. Of these, the software retained 25,379 recordings (52 percent of total) as possible bat passes. The remainder were noise or too poor a quality for analysis. Of the 25,379 recordings, 12,162 were classified as high-quality bat vocalizations (Figure 2). Thus, 25 percent of all sound recordings made in 2022 were high-quality bat passes.

The greatest activity occurred on 4 August 2022, when 1,431 bat passes were recorded. A total of 12,162 bat passes were recorded in 146 days of sampling in 2022 (Table 1, Figure 3). Big brown bats were by far the most common species on Napatree this year. In fact, big brown bats were more abundant this year compared to the previous years. Tree bats (silver-haired, hoary, and red bats) were also very common in 2022. No little brown bats were recorded with high confidence. The lack of activity by little brown bats is consistent with their declining population in this region as a result of white-nose syndrome. A small number of tricolored bats were recorded in 2022.

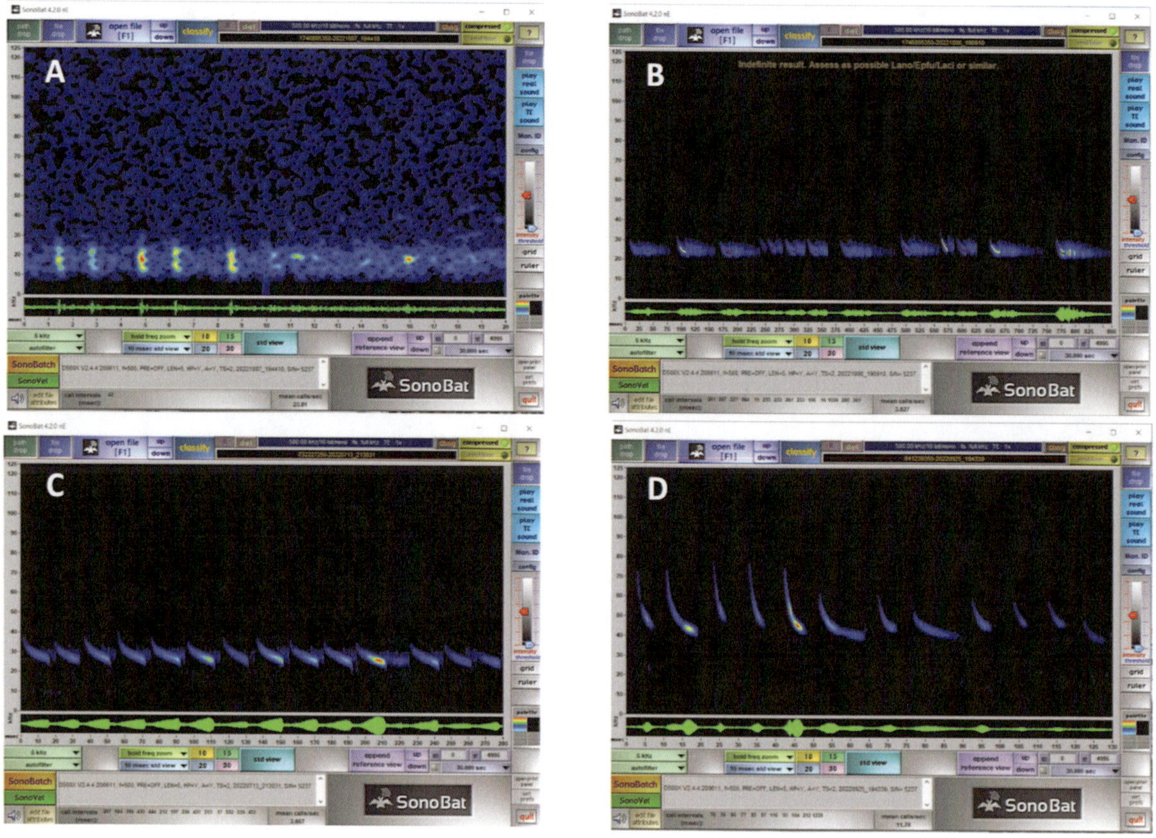


*Figure 2. Sample sonograms of recordings from the bat detector. Graphs show time (msec) on the X axis and sound frequency (kHz) on the Y axis. The strength of the sound is shown in the green pulse below the call. A, Noise, not considered for analysis. B, Indefinite classification, too little information to make a determination. C, Big brown bat echolocation call. D, Red bat echolocation call.*

| Species | | Total Number of Passes (Passes/Sampling Day) | | | | |
|---|---|---|---|---|---|---|
| Latin Name | Common Name | 2022 | 2021 | 2019 | 2018 | 2017 |
| *Eptesicus fuscus* | Big Brown Bat | 6,461 (44.2) | 5,129 (25.8) | 3,859 (19.8) | 917 (4.6) | 1,921 (10.9) |
| *Lasiurus borealis* | Eastern Red Bat | 1,649 (11.3) | 607 (3.1) | 61 (0.3) | 165 (0.8) | 995 (5.7) |
| *Lasionycteris noctivagans* | Silver-haired Bat | 2,814 (19.3) | 1,470 (7.4) | 569 (2.9) | 651 (3.2) | 875 (5.0) |
| *Lasiurus cinereus* | Hoary Bat | 1,163 (8.0) | 601 (3.0) | 283 (1.5) | 108 (0.5) | 430 (2.5) |
| *Perimyotis subflavus* | Tricolored Bat | 54 (0.4) | 30 (0.2) | 0 (0.0) | 12 (0.1) | 17 (0.1) |

*Table 1. Number of passes for bat species detected in the Napatree Point Conservation Area in 2022. Because the number of sampling days differed between years, the number of bat passes per sampling day (146 days in 2022) is presented so years can be compared. Species with fewer than ten passes over the whole season are not shown in this table.*

The different species showed very different patterns of activity over the summer (Figure 3). There was a large surge in activity in August by big brown bats. This would correspond to the time when females would have weened their young and began their flight to nearby (within 30 km) winter hibernacula. It probably also represents activity by young-of-the-year big brown bats. This surge in activity by big brown bats in mid- to late-summer has been observed in previous years on Napatree (August *et al.*, 2019).

Silver-haired bats were abundant this year and were very active from mid-August to early September. They mirrored the activity pattern of big brown bats. This pulse of activity occurred earlier this year than in previous years. Red bats had a spectacular peak of activity in mid-September. Red bats and hoary bats were the last bats flying on Napatree in the fall. This has been the pattern the past few years. Hoary bats were very active from mid-September to November. Red bats, hoary bats, and silver-haired bats do not hibernate locally; instead they migrate to southern states in the winter.

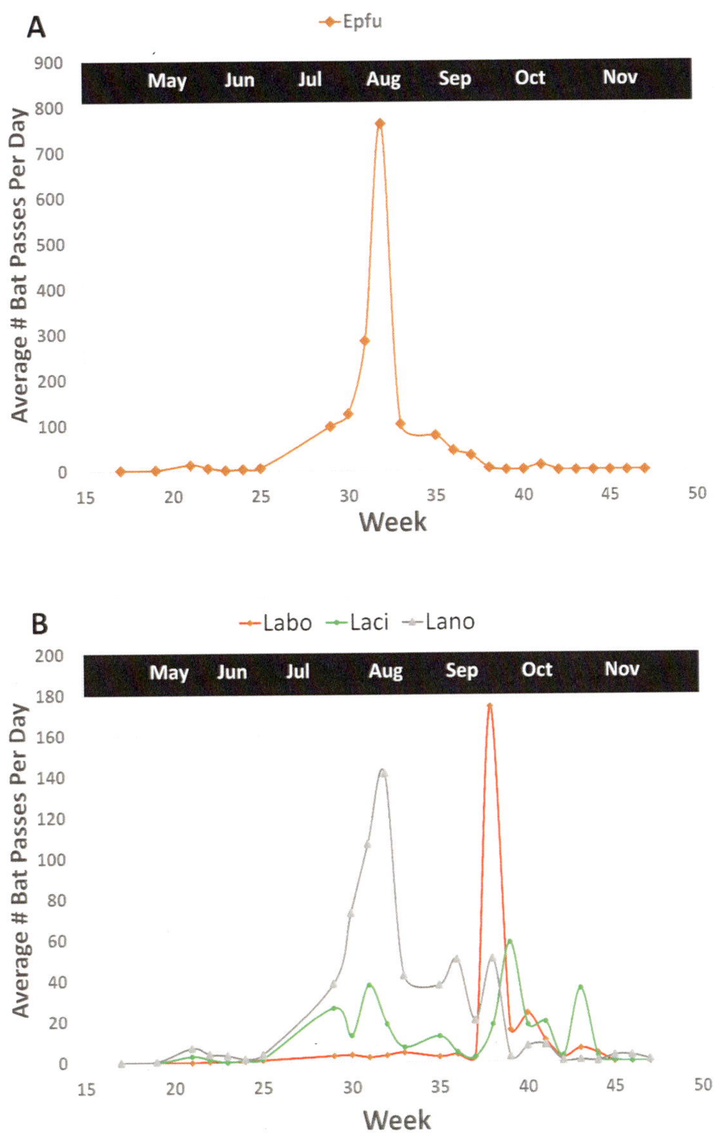


*Figure 3. Seasonal activity of bats in the Napatree Point Conservation Area. The Y axis shows the average number of bat passes per day on a weekly basis. The X axis shows the week of the year. A, big brown bats, Eptesicus fuscus; B, the tree bats: red bats, Lasiurus borealis; hoary bats, Lasiurus cinereus; and silver-haired bats Lasionycteris noctivagans.*

We analyzed patterns of nocturnal bat activity over the course of the summer. We classified nights into three temporal categories: evening (7:00 PM to 11:00 PM); night (11:00 PM to 2:00 AM); and morning (2:00 AM to 6:00 AM). For each species for which we had adequate sample sizes (>1,000 passes), we calculated the percent total bat passes each month in each of the three temporal categories (Figure 4).

Big brown bats showed a dramatic shift in being less active in the middle of the night and more active in early evenings later in the season. It would be interesting to learn if this pattern of activity is related to weaning of pups mid/late-summer and adults commuting back to winter hibernacula.

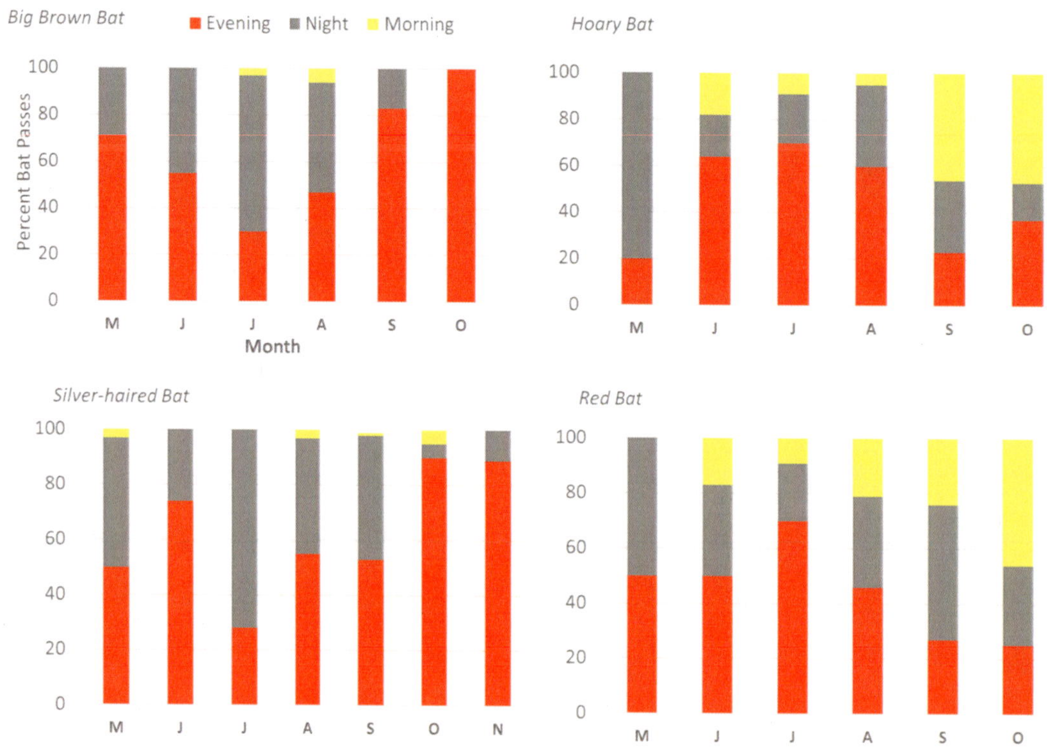


*Figure 4. Seasonal shifts in nocturnal activity patterns of bats in the Napatree Point Conservation Area. Three nocturnal time periods were established: evenings (7:00 PM to 11:00 PM); nights (11:00 PM to 2:00 AM); mornings (2:00 AM to 6:00 AM). Histograms show the percent of bat passes each month that fall in the three time periods.*

Conversely, red bats and hoary bats showed a marked increase in morning activity; evening and night activity decreased as summer progressed. It is tempting to speculate on the implications of this observation in view of the long-distance migration red bats and hoary bats have in the fall. Compared to the other species of tree bats, silver-haired bats tended to fly more in the early evening as the summer progressed. The patterns we observed in 2022 were quite similar to those observed in 2021 (August *et al.*, 2021).

## CONCLUSIONS

The end-of-summer surge in activity by tree bats is consistent with the hypothesis that these species follow the coast in their southerly fall migration (Cryan, 2003; Gorman *et al.*, 2021; Johnson *et al.*, 2011) to the mid-Atlantic and southeastern United States.

The peak of activity for big brown bats in 2022 was in the month of August. This is the same pattern we have observed in previous years. Charles Brown of the RI DEM (pers. comm.) has reported seeing the mid- to late-August surge in big brown bat activity in other monitoring stations in Rhode Island.

Napatree does not appear to be high quality bat foraging habitat during the summer, although bats are continuously present albeit in modest numbers (August *et al.*, 2021). We have searched the ruins of Fort Mansfield for roosting bats in all seasons. None have been found; however, Charles Brown reports bats roosting inside the interior chambers at other coastal forts in Rhode Island.

Napatree supports large numbers of migratory bats on their way to local hibernacula (e.g., big brown bats) or the southeastern United States (e.g., red, hoary, and silver-haired bats). The sudden surge in numbers of red bats, silver-haired bats, and big brown bats is reminiscent of the staging aggregations of swallows and other migratory birds that are seen on Napatree in the late summer months. For most species of birds and bats, the Atlantic Ocean presents a major migratory navigational barrier. Thus, Napatree Point, with the ocean on one side and Little Narragansett Bay on the other, concentrates individuals as they fly along the coast, with most individuals eventually crossing over to Connecticut to continue their journey south.

## ACKNOWLEDGMENTS

This project was supported by the University of Rhode Island Coastal Institute and The Watch Hill Conservancy. The Pettersson D500X bat detectors were purchased using a grant-in-aid of research from the Coastal Institute and an equipment grant in support of monitoring by The Watch Hill Conservancy. Special thanks to Charles Brown, the mammologist for the RI Department of Environmental Management. We are most grateful for the advice and support Charlie provided us when we began these surveys. Peter Paton offered very helpful guidance on the physical geography of Napatree Point and the role it plays in directing the migratory patterns of birds in the region.

## REFERENCES

August, P., E. Bodell, R. Kleinert, C. Fox, A. Peterson, and L. Craver-Rogers. 2016. "Bats of Napatree: A Preliminary Assessment." In *The State of Napatree Report: 2016*, edited by J. Sassi. The Watch Hill Conservancy. https://thewatchhillconservancy.org/napatree/napatree-resources/.

August P., A. Murphy, and J. Sassi. 2017. "Summertime Bat Activity on Napatree Point." In *The State of Napatree Report: 2017*, edited by J. Sassi. The Watch Hill Conservancy. https://thewatchhillconservancy.org/napatree/napatree-resources/.

August P., D. Cole, and J. Sassi. 2021. "Summertime Bat Activity on Napatree Point (2021)." In *The State of Napatree Report: 2021*, edited by J. Sassi. The Watch Hill Conservancy. https://thewatchhillconservancy.org/napatree/napatree-resources/.

Cryan, P. M. 2003. "Seasonal Distribution of Migratory Tree Bats (*Lasiurus* and *Lasionycteris*) in North America." *Journal of Mammalogy*, 84: 579–593.

Gorman, K. M., E. L. Barr, L. Ries, T. Nocera, and W. Ford. 2021. "Bat Activity Patterns Relative to Temporal and Weather Effects in a Temperate Coastal Environment." *Global Ecology and Conservation* 30.

Johnson, J. B., J. Gates, and N. Zegre. 2011. "Monitoring Seasonal Bat Activity on a Coastal Barrier Island in Maryland, USA." *Environmental Monitoring and Assessment*, 173:685–699.

*Nekton monitoring team at work.*
*Photo Credit: The Watch Hill Conservancy*

CHAPTER 19

# An Ecological Reconnaissance of the Napatree Lagoon: Fish and Water Quality in 2018


*Nicole Rohr,*[1] *Peter August*[2] *& Janice Sassi*[3]

[1] Coastal Institute, University of Rhode Island

[2] Department of Natural Resources Science, University of Rhode Island

[3] Napatree Point Conservation Area, The Watch Hill Conservancy


## INTRODUCTION

The 4 ha (9.9-acre) lagoon on the western end of Napatree Point is critical habitat for shorebirds, horseshoe crabs, and presumably marine organisms. Until 2014, to our knowledge, there had been no systematic study of the biology of the Napatree Lagoon. Over the past five years, we have investigated the physical and biological characteristics of the lagoon and surrounding area. Our studies, when taken together, begin to form a picture of habitat structure and species interactions within the lagoon itself, and how the lagoon serves the larger Napatree ecosystem. In the first year, we mapped the bathymetry of the lagoon (Rohr *et al.*, 2014); in the second year, we began a biological inventory of the lagoon with a focus on fish and algae (Rohr *et al.*, 2015); in the third year, we continued the biological inventory to examine annual variability (Rohr *et al.*, 2016); in the fourth year, we continued the biological inventory as well as added subtidal crab surveys and collected water samples at high tide to analyze for nitrogen and phosphorous content (Rohr *et al.*, 2017). Due to consistency of results in fish abundance and species composition from 2015 through 2017, the 2018 survey was limited to July only. The goal of the study was to determine what fish species occurred in the lagoon and the spatial variability in species composition and mean total length in July. Furthermore, we measured water quality parameters in the lagoon over the summers of 2015 to 2018. See Green *et al.* (2017) on algae surveys and Croteau *et al.* (2018) on crab surveys of the lagoon that were done concurrently with the nekton surveys.

These data will add to the baseline of the ecology of the Napatree Lagoon and allow us to monitor changes in the fish community and water quality of the lagoon, which have larger ecological ramifications for shorebirds that feed and nest along the shores. The lagoon is sensitive to changes in tidal flushing as large storms change the breach in the lagoon. When future storms hit Napatree, and if the breach is altered, we will be able to determine the impacts of increased or decreased tidal flushing of this system.

## METHODS

On 24 July 2018, a subtidal seine survey was conducted, and physical water quality parameters were recorded at three sites in the Napatree Lagoon (hereafter referred to as NW, SE, and SW; see Figure 1) during low tide. Seining was not conducted at the NE site as in 2015 to 2017 because shifting sediments have accumulated in that quadrant of the lagoon, resulting in water shallower than just a couple of inches deep for a distance of well over 40 feet. At each of the three monitored sites, a 6.1 m (20-foot) 0.3 cm (0.1-inch) mesh Douglas Net Minnow seine was pulled from the interior of the lagoon perpendicular to the shore for 12.2 m (40 feet), ending on the beach. The maximum depth at the start of each seine pass was never more than 1 m (3.3 feet, waist deep). All fish were identified to the species level and the total length (cm) was recorded. In seine hauls where large numbers of fish were collected, a minimum of thirty individuals of each species were measured, and all remaining fish were counted and recorded as adult (>2 cm, 0.8-inch) or small juvenile (<2 cm, 0.8-inch). In addition to fish, all crabs were identified to the species level and carapace width measured (cm), except hermit crabs which were identified to the genus level and counted only. All snails were identified to the species level and counted. All organisms were released alive once data recording was completed.

*Figure 1. Sampling regions in the Napatree Lagoon for fish, algae, and water quality monitoring.*

Physical water quality parameters were recorded at each sampling site from the bottom of the water column at the start of the seine haul 12.2 m (40 feet) from shore using a YSI Pro2030 DO/Salinity water quality meter equipped with a YSI 2003 Polargraphic DO sensor. The instrument was maintained and calibrated by the URI Watershed Hydrology Laboratory. The following parameters were measured during water quality assessments: dissolved oxygen (mg/L; hereafter DO), salinity (parts per thousand, PPT), temperature (degrees Celsius), and depth (cm). Water quality data were logged into an iPad in the field using a data entry form developed with Fulcrum software. Data records were immediately uploaded into a cloud-based database.

Comparison of means was done using full factorial ANOVAs, and JMP 14 software was used for all statistical analyses.

## RESULTS

### Fish and Invertebrates

A total of five different fish species were captured in 2018, with the most prevalent species being *Fundulus majalis* (striped killifish), *F. heteroclitus* (mummichogs), and *Menidia menidia* (Atlantic silversides; Table 1). The other two species were *Brevoortia tyrannus* (Atlantic menhaden) and *Microgadus tomcod* (Atlantic tomcod). This species abundance is a decrease from all other years: nine total species were found in 2016, seven total species in 2017, and six total species in 2015. The three most prevalent species remained the same in all years (SoN 2015, 2016, 2017). A total of three invertebrate species were found in Napatree Lagoon in 2018: *Carcinus maenas* (European green crab), *Tritia obsolete* (Eastern mud snail), and *Pagurus* spp. (hermit crabs; Table 1). These three species were present all three years, and very low numbers of other invertebrates were found intermittently in other years. Other invertebrates are regularly seen in the lagoon, often at high levels, but are fast moving and not easily captured in the seine net.

In 2018, because only one seine survey was conducted at each site in only one month, statistical analyses could not be conducted on this year's data. A total of 850 fish were captured, measured, and released at all sites in July 2018 (Table 2). Of the three most common fish, *F. heteroclitus* had a mean length of 20.94 ± 0.92 cm (8.24 ± 0.36 inches), *F. majalis* had a mean length of 4.21 ± 2.36 cm (1.65 ± 0.93 inches), and *M. menidia* had a mean length of 3.83 ± 1.49 cm (1.51 ± 0.59 inches) (Table 2).

When compared with previous years' July results, species abundance did not significantly vary in Napatree Lagoon among years or sites (years, p=0.36; sites, p=0.29; interaction, p=0.55; Figure 2). In general, total fish abundance (mean number of individuals ± SD) across all sites was, from lowest to highest, 142.25 ± 189.95 in 2016; 165.00 ± 233.99 in 2015; 266.70 ± 112.48 in 2018; and 355.33 ± 325.21 in 2017. These fish abundances were largely driven by at least one site having the majority of the fish captured while other site(s) had few fish. For example, in 2017, there were 717 fish captured at the SW site, but none at the NE site. In 2018, the SW and NW sites had similar fish abundances (SW, 325; NW, 388), but only 137 were captured at the SE site. While there was no significant difference among sites in fish abundance, largely due to extremely variable numbers at each site between years, there was a general trend of fewer fish on the eastern side of the lagoon (NE, 141.67 ± 244.51; SE, 62.75 ± 60.65) than the western side (SW, 405.50 ± 269.81; NW, 210.00 ± 149.82). It is worth noting that one or zero fish were captured at the NE site in all years except for in 2016, when 424 fish were captured, almost all of which were less than 1.0 cm in length. The abundance at the SE site also gradually increased from 2015 to 2018.

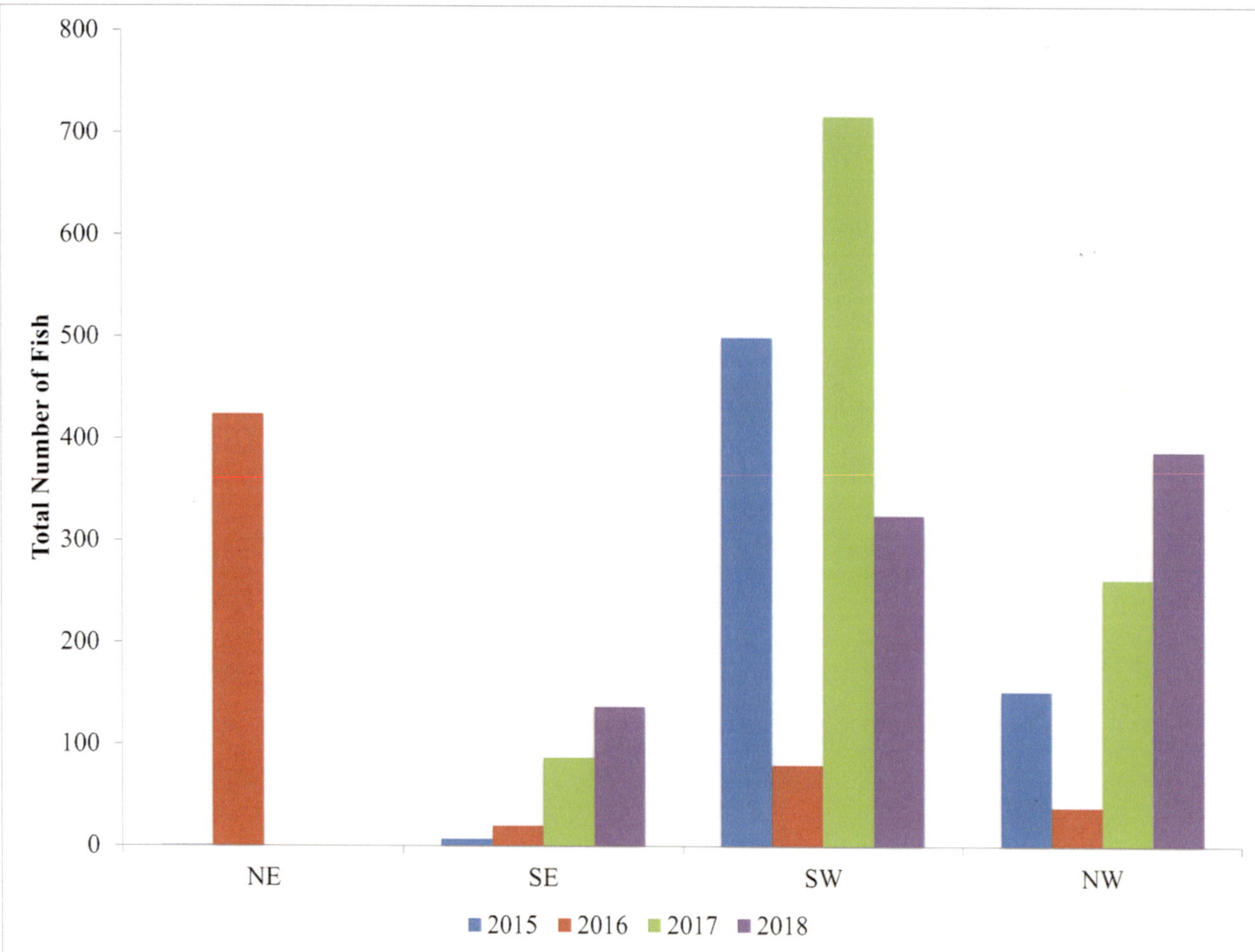


*Figure 2. Numbers of fish captured at each sampling location during July in each year from 2015 through 2018.*

The mean total length (cm + SD) of fish varied significantly among sites ($p<0.0001$), but not among years ($p=0.96$), and there was an interaction effect ($p<0.0001$; Figure 3). The mean total length was smallest at the NE and SW sites (1.99 ± 0.16 cm (0.78 ± 0.06 inches) and 2.04 ± 0.02 cm (0.80 ± 0.01 inches), respectively), significantly larger at the NW site (2.65 ± 0.04 cm (1.04 ± 0.02 inches), and largest at the SE site (4.09 ± 0.12cm (1.61 ± 0.05 inches) (Figure 3, three most common species). In 2018, mean total length over all sites for fish >2 cm was 2.94 ± 0.92 cm (1.16 ± 0.36 inches) for *F. heteroclitus*, 4.21 ± 2.36 cm (1.66 ± 0.93 inches) for *F. majalis*, and 3.83 ± 1.49 cm (1.51 ± 0.59 inches) for *M. menidia*.

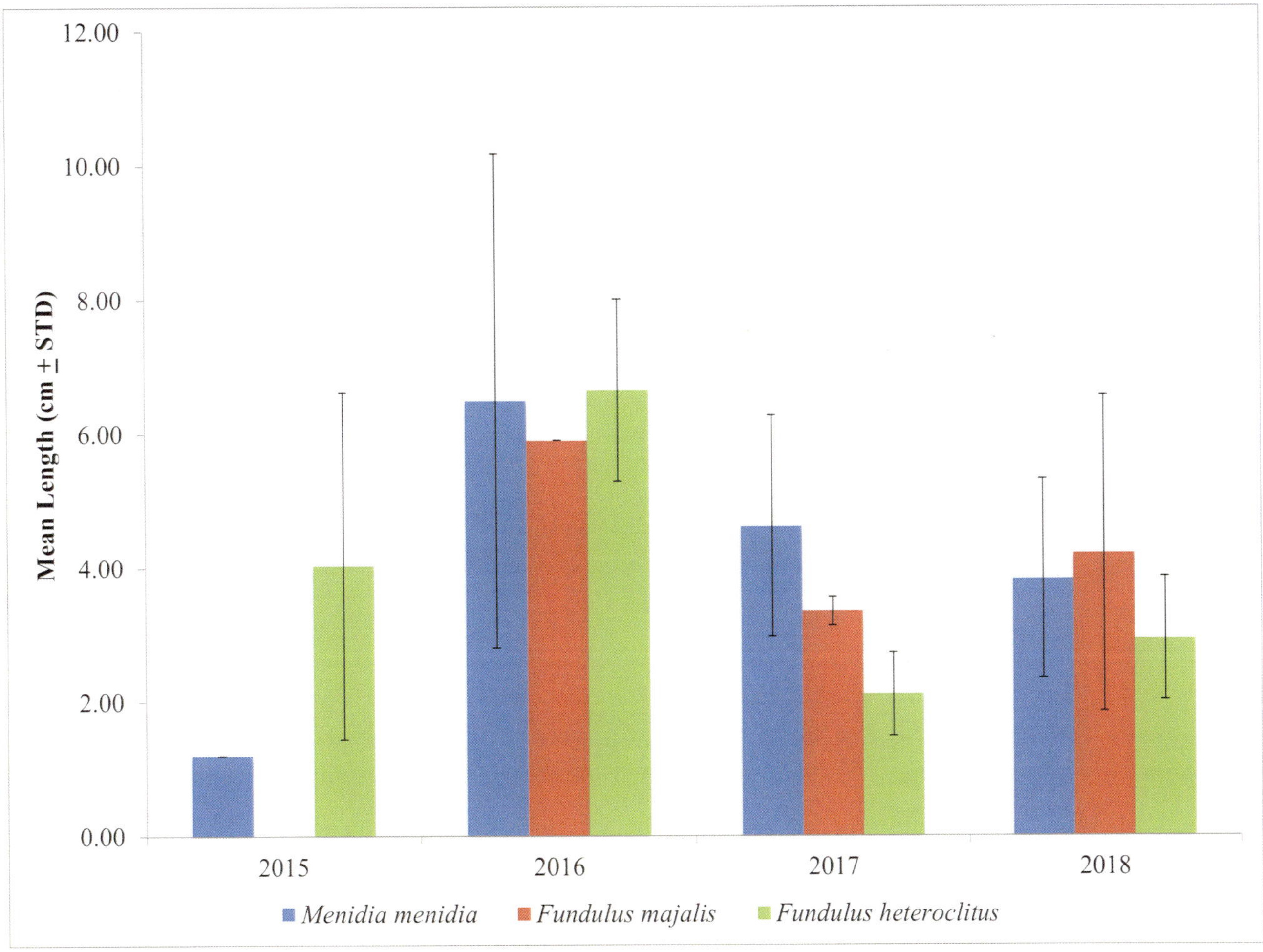


*Figure 3. Mean (± SD in cm) total length of the three most common fish species >2 cm across all sites in July from 2015 through 2018.*

## Water Quality

July DO levels significantly varied among sites (p=0.01), but not among years (p=0.55), and there was no interaction effect (p=0.45). In general, over all years, the eastern side of the lagoon had significantly lower DO levels (in mg/L) than the western side (SE, 3.05 ± 1.83; NE, 2.18 ± 1.83; SW, 10.32 ± 1.83; NW, 11.63 ± 1.83).

July temperature levels did not significantly vary among sites or years (p=0.37 and p=0.43, respectively) with no interaction effect (p=0.71). When combined over all sites and years, mean temperature was 25.44 ± 2.37°C. July salinity levels also did not significantly vary among sites or years (p=0.73 and p=0.43, respectively) with no interaction effect (p=0.42). When combined over all sites and years, mean salinity was 28.98 ± 2.97 PPT (July 2018 physical characteristic measurements in Table 3).

| W – Present in 2015<br>X – Present in 2016<br>Y – Present in 2017<br>Z – Present in 2018 | July | | | |
|---|---|---|---|---|
| FISH SPECIES | NE | SE | NW | SW |
| *Apeltes quadracus* (Fourspine Stickleback) | | Y | | |
| *Brevoortia tyrannus* (Atlantic Menhaden) | | | Z | |
| *Cyprinodon variegatus* (Sheepshead Minnow) | | | W | |
| *Fundulus* spp. (Striped Killifish/Mummichogs) | W | W | W | W |
| *Fundulus heteroclitus* (Mummichogs) | X | XYZ | XYZ | XYZ |
| *Fundulus majalis* (Striped Killifish) | | YZ | YZ | XZ |
| *Menidia menidia* (Atlantic Silverside) | | WXYZ | XYZ | XZ |
| *Microgadus tomcod* (Atlantic Tomcod) | | | Z | Z |
| INVERTEBRATE SPECIES | | | | |
| *Carcinus maenas* (European Green Crab) | WX | XYZ | WXY | WY |
| *Hemigrapsus sanguineus* (Asian Shore Crab) | | | Y | |
| *Tritia obsoleta* (Eastern Mud Snail) | WXY | WXY | WZ | WXYZ |
| *Pagurus* spp. (Hermit Crab) | WX | XZ | Z | XZ |

*Table 1. Fish and invertebrate species observed during subtidal seine surveys conducted at low tide in Napatree Lagoon, Rhode Island, in July from 2015 to 2018. W indicates 2015 sample results, X indicates 2016 results, Y indicates 2017 results, and Z indicates 2018 results.*

| July 2018 – Fish Abundance and Size | | | | | | |
|---|---|---|---|---|---|---|
| | NE | SE | SW | NW | Mean | SD |
| Number of Fish | | | | | | |
| | | 137 | 325 | 388 | *266.7* | *112.5* |
| Size, (Mean (SD) cm) | | | | | | |
| Mummichogs | | 3.06 (0.85) | 2.54 (1.01) | 3.32 (0.71) | *2.94* | *0.92* |
| Striped Killifish | | 2.27 (1.07) | 4.73 (0.49) | 8.50 (--) | *4.21* | *2.36* |
| Atlantic Silverside | | 4.33 (1.45) | 2.78 (0.59) | 3.88 (1.68) | *3.83* | *1.49* |
| *Mean* | | 3.34 | 2.66 | 3.58 | | |
| *SD* | | 1.16 | 1.01 | 1.29 | | |

*Table 2. Fish abundance and size for the Napatree Lagoon during July 2018. Means and standard deviations (SD) are provided for species and site samples.*

| July 2018 – Water Quality Measurements | | | | | | |
|---|---|---|---|---|---|---|
| | NE | SE | SW | NW | Mean | SD |
| Bottom DO (mg/L) | 2.4 | 4.2 | 4.7 | 5.5 | 4.20 | 1.31 |
| Temperature (°C) | 25.7 | 25.0 | 25.2 | 25.3 | 25.30 | 0.29 |
| Salinity (PPT) | 29.6 | 29.6 | 29.0 | 26.9 | 28.78 | 1.28 |

*Table 3. Water quality parameters for the Napatree Lagoon during July 2018. Means and standard deviations (SD) are provided for site samples.*

## OTHER ITEMS OF NOTE

### Localized Fish Kill

The lagoon experienced at least one fish kill on 17 September 2018. The dead young menhaden covered the entire lagoon shore from the tip of the spit, around most of the saltmarsh, and almost to the pines on the north and west shore of the lagoon. Likely, the menhaden were chased into the lagoon by schooling stripers or other predatory fish and were then trapped in the lagoon at low tide. Large piles of decaying algae were seen covering the tidal delta near the lagoon channel, and the lagoon typically has low DO in September, resulting in the menhaden dying once they became trapped in the lagoon. The Rhode Island Department of Environmental Management was notified of the fish kill and relayed that no other events had been reported in the area.

A

B

C

D

*Figures 5A–D. Fish kill of young menhaden in Napatree Lagoon, 17 September 2018.*
*Photos: Kevin Rogers*

### Renaming of the Eastern Mud Snail

As a result of recent phylogenetic studies, the Eastern mud snail has been renamed *Tritia obsoleta* (previously *Ilyanasa obsoleta*) (Galindo *et al.*, 2016).

## CONCLUSIONS

The fish fauna of the lagoon is typical for this habitat; *Menidia menidia* (silversides), *Fundulus heteroclitus* (mummichogs), and *Fundulus majalis* (striped killifish) tend to be most common in shallow lagoons in Rhode Island. The abundance and total length of fish varied over years and sites, but clear patterns are not discernable because of the high variability. Future sampling efforts should consider hauling multiple seines per site per month to allow for more advanced statistical analyses that better examine spatial and temporal patterns in combination; however, this should be balanced against the small ecosystem being sampled and perhaps limited to one western and one eastern site.

The abundance of fish did not vary among years 2015, 2016, 2017, and 2018, and the same species were found each year with the exception of single individuals of rarer species collected in previous years.

Consistent with previous years, the lowest DO levels were in the eastern lagoon, and higher DO levels were on the western side. These DO levels are inversely related with algal abundance: algae washes into the lagoon with the incoming tide, settles out of the water column on the eastern shore, and decomposes, a process that consumes oxygen from the surrounding water (Green *et al.*, 2017). July 2018 DO plummeted to near-zero DO levels, but the milky, sulfurous plume that has been seen in the past was not present (Rohr *et al.*, 2015; Rohr *et al.*, 2017).

While the physical characteristics of the lagoon remained largely unchanged in 2014, 2015, and 2016, there were noticeable changes in 2017 that became more pronounced in 2018. The NE site is now a settling location for sand coming in through the inlet from Little Narragansett Bay, and the depth all along this quadrant is only a few inches deep for more than 40 feet into the center of the lagoon from the shore. This resulted in the elimination of the NE site from our sampling locations. In addition, the sediment composition appears to be changing with the SE site shifting from sand to mud, and mud of the NW site becoming firmer. These are qualitative observations, and researchers may consider doing another survey of the physical properties (Rohr *et al.*, 2014).

## ACKNOWLEDGMENTS

This project was supported by a grant-in-aid from the University of Rhode Island Coastal Institute. Thank you to the numerous students, colleagues, and Napatree Naturalists who helped in conducting surveys each month. Seine netting was conducted under a valid RI Department of Environmental Management scientific collecting permit (# 277, Type 1, Issued 5/1/2018) issued to Nicole Rohr.

## REFERENCES

Croteau, J., B. Oakley, and N. Rohr. 2017. "Crab Abundance in the Napatree Lagoon." In *The State of Napatree Report: 2017*, edited by J. Sassi. The Watch Hill Conservancy. https://thewatchhillconservancy.org/napatree/napatree-resources/.

Galindo, LA. *et al.* 2016. "The Phylogeny and Systematics of the Nassariidae Revisited (Gastropoda, Buccinoidea)." *Molecular Phylogenetics and Evolution*, 99:337–353.

Green-Gavrielidis, L., E. Ernst, J. Guttandin, and Carol Thornber. 2017. "Monitoring Seaweed Abundance and Species Composition at Napatree Lagoon: 2017." In *The State of Napatree Report: 2017*, edited by J. Sassi. The Watch Hill Conservancy. https://thewatchhillconservancy.org/napatree/napatree-resources/.

Rohr, N., J. Turenne, M. Bradley, P. August, and J. Sassi. 2014. "An Ecological Reconnaissance of the Napatree Lagoon: Physical Properties." In *The State of Napatree Report: 2014*, edited by J. Sassi. The Watch Hill Conservancy. https://thewatchhill-conservancy.org/napatree/napatree-resources/.

Rohr, N., P. August, and J. Sassi. 2015. "An Ecological Reconnaissance of the Napatree Lagoon: Fish and Water Quality." In *The State of Napatree Report: 2015*, edited by J. Sassi. The Watch Hill Conservancy. https://thewatchhillconservancy.org/napatree/napatree-resources/.

Rohr, N., P. August, C. Fox, E. Bodell, and J. Sassi. 2016. "An Ecological Reconnaissance of the Napatree Lagoon: Fish and Water Quality." In *The State of Napatree Report: 2016*, edited by J. Sassi. The Watch Hill Conservancy. https://thewatchhillcon-servancy.org/napatree/napatree-resources/.

Rohr, N., P. August, A. Murphy, and J. Sassi. 2017. "An Ecological Reconnaissance of the Napatree Lagoon: Fish and Water Quality." In *The State of Napatree Report: 2017*, edited by J. Sassi. The Watch Hill Conservancy. https://thewatchhillconservancy.org/napatree/napatree-resources/.

**Appendix 1.** *Photographs of some of the fish species captured. A, Striped killifish (Fundulus majalis); B, Atlantic silversides (Menidia menidia); C, Silversides in the seine; D, Atlantic menhaden (Brevoortia tyrannus).*

*Photo Credit: The Watch Hill Conservancy*

*Mating pair of horseshoe crabs.*
*Photo Credit: The Watch Hill Conservancy*

CHAPTER 20

# Project Limulus on Napatree Point: Horseshoe Crab Surveys in 2022

*Jennifer Croteau*
Napatree Point Conservation Area, The Watch Hill Conservancy

## INTRODUCTION

Horseshoe crabs (*Limulus polyphemus*) are an ecologically and medically important species that have existed for over 400 million years on the muddy bottoms of bays and estuaries, feeding on clams, small crustaceans, and worms. Horseshoe crabs deposit eggs beneath the wet sand along shorelines during a critical time of shorebird migration and nesting, thus providing a necessary nutrient source to many threatened and endangered shorebirds. Additionally, their "blue blood" is used to help detect contaminants in intravenous drugs.

Horseshoe crabs live along the Atlantic coast of the United States and spawn in the intertidal zones of protected, sandy beaches. Often, this means coming into the shallow waters of coves and bays, such as Little Narragansett Bay, which borders Napatree Point to the north. Napatree provides excellent breeding habitat that allows horseshoe crabs to spawn without being flipped over by large waves. Horseshoe crabs come to shore during the extreme high tides nearest to the full and new moons during May, June, and into July. It is during these spawning periods that horseshoe crabs are the easiest to count.

The Watch Hill Conservancy (WHC) and the Watch Hill Fire District (WHFD) have partnered with Sacred Heart University's Project *Limulus* (PL) to monitor the horseshoe crabs that use Napatree Point. Beginning in 1998 in collaboration with the US Fish and Wildlife Service (USFWS), PL relies on public volunteers to contribute valuable data to this scientific research. This information helps PL to estimate and monitor the population size, identify the most important spawning and nursery areas, and help answer other research questions that may be critical for the conservation of this species.

## METHODS

Traditionally, there are four separate parts to this project: spawning surveys, tagging adults, reporting previously tagged adults (recaptures), and juvenile surveys. Since 2020, WHC has decided to focus mainly on spawning surveys.

### Spawning Surveys

Spawning surveys occur during the high tides surrounding the full and new moons during May and June, when spawning is most likely to occur. The surveys are conducted twice a day, morning and night, the day of the new/full moon as well as two days before and two days after. Napatree's survey route runs along the northern shore from the rock jetty adjacent to Watch Hill Harbor westward into the lagoon and ending at the middle Osprey nesting pole. All horseshoe crabs within 3 m of the shoreline are counted, sexed (male or female), and recorded whether they are paired or alone. In 2016, we broke the survey route into five zones to assess which stretches of beach were preferred by horseshoe crabs and have continued this through 2022. The routes were determined by easily identified landmarks, not distance.

### Tagging

The tagging of adults can occur at any time, but typically occurs during the return trip after the spawning survey. Sex, shell (carapace) width, and shell condition are recorded when animals are tagged. Tags are attached by creating a small hole in the shell of the horseshoe crab and then inserting a small, plastic tag into the hole. These round white tags are imprinted with a unique USFWS Federal ID number and a telephone number to call to report locating a tagged animal.

### Reporting Recaptures

During horseshoe crab surveys, previously tagged individuals are frequently located. They are carefully captured long enough to record the tag number along with the sex, shell width, and condition. These recapture data are integral in estimating population size and determining whether an animal returns to the same beaches to spawn. Data from tagged horseshoe crabs located outside survey times are also submitted to PL.

## RESULTS

In twenty-one surveys, Napatree Naturalists counted 6,729 spawning horseshoe crabs with the most found as pairs (Figure 1). Single males outnumbered single females (1,354 single males, 86 single females). The area most used by horseshoe crabs was section C, the middle section of the beach (Figures 2 and 3). The wave energy and bottom substrate are not consistent throughout the sections. Horseshoe crabs prefer more sandy, calmer water because it is easier to navigate and dig into the shoreline. The conditions of the northern side change daily, depending on the weather and tidal cycle. It is very common to have entire sections covered in mounds of seaweed one day and clear the next. No horseshoe crabs were tagged during the 2022 season due to a tag shortage.

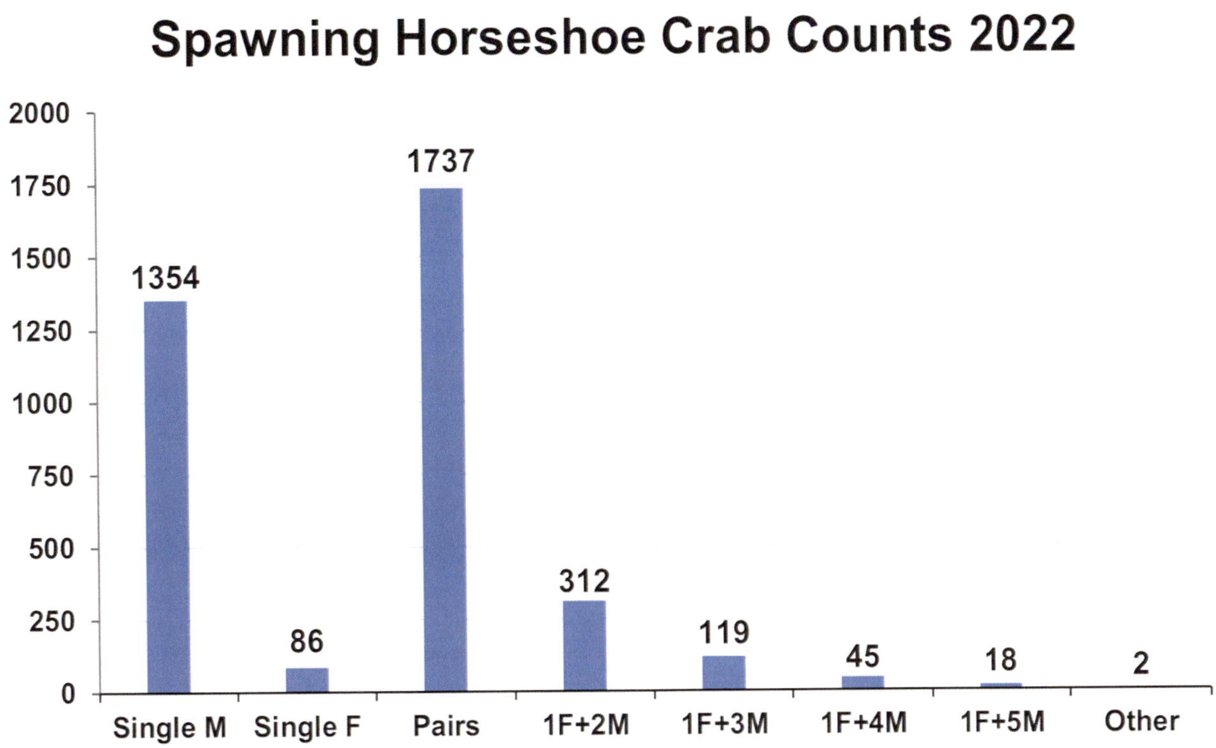


*Figure 1. Total number of horseshoe crabs by class (tallies) for the 2022 season. 1F+2M indicates one female and two males, 1F+3M represents one female and three males, and so on. The numbers presented here are the number of times each configuration was observed. For example, 1,737 pairs of horseshoe crabs were found, and this consisted of 3,474 individual crabs.*

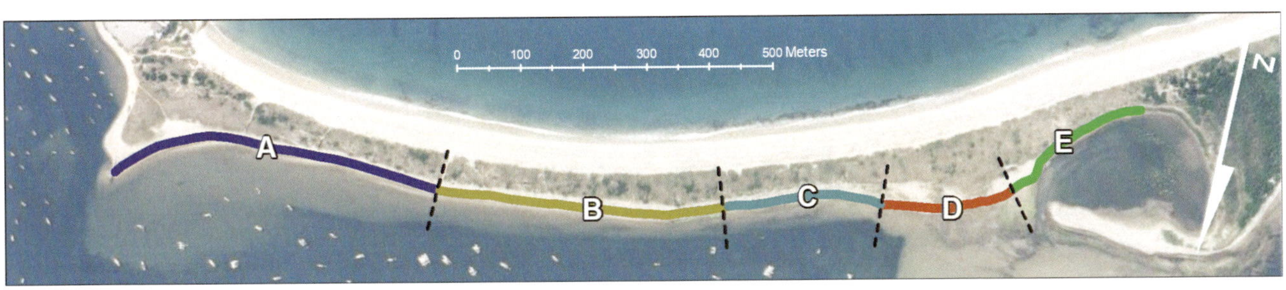


*Figure 2. A map depicting the five zones of the Napatree spawning horseshoe crab survey.*

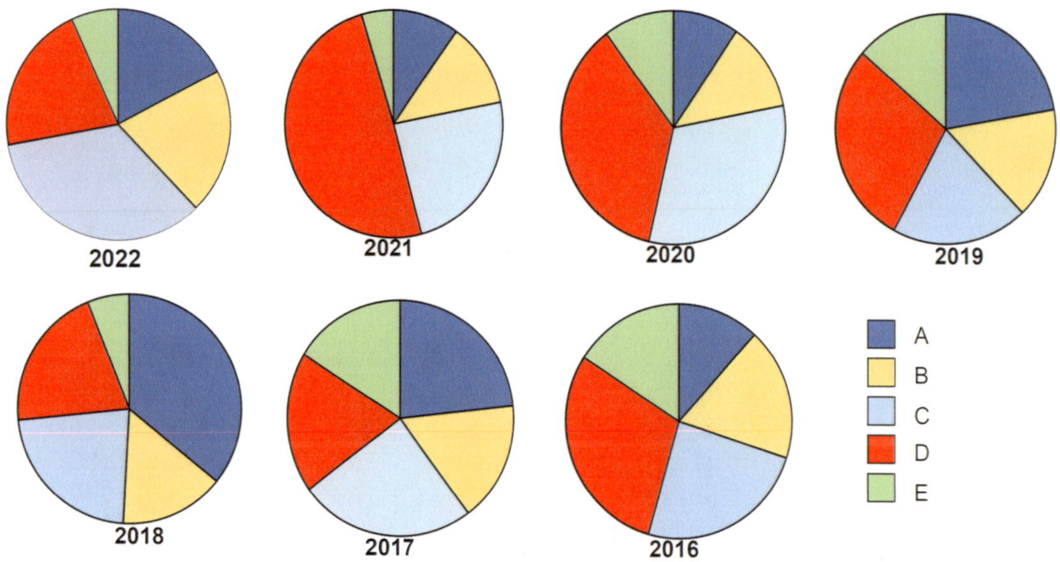


*Figure 3. Pie charts showing the percent total horseshoe crabs counted per beach section for each year since 2016.*

The 2022 results showed a significant increase in the total number of horseshoe crabs counted per survey compared to past seasons (Table 1). More horseshoe crabs were observed in 2022 than any other season (Figure 4).

| Year | # Surveys | # HSC | # HSC/ Survey |
|---|---|---|---|
| 2013 | 13 | 1,685 | 129.6 |
| 2014 | 23 | 2,070 | 90.0 |
| 2015 | 16 | 1,837 | 114.8 |
| 2016 | 20 | 4,759 | 238.0 |
| 2017 | 25 | 4,840 | 193.6 |
| 2018 | 20 | 2,110 | 105.5 |
| 2019 | 25 | 3,050 | 122.0 |
| 2020 | 16 | 3,402 | 212.6 |
| 2021 | 14 | 2,899 | 207.1 |
| 2022 | 21 | 6,729 | 320.4 |

*Table 1. Number of surveys conducted each season, the total number of horseshoe crabs (HSC) counted each season, and the average number of HSC counter per survey.*

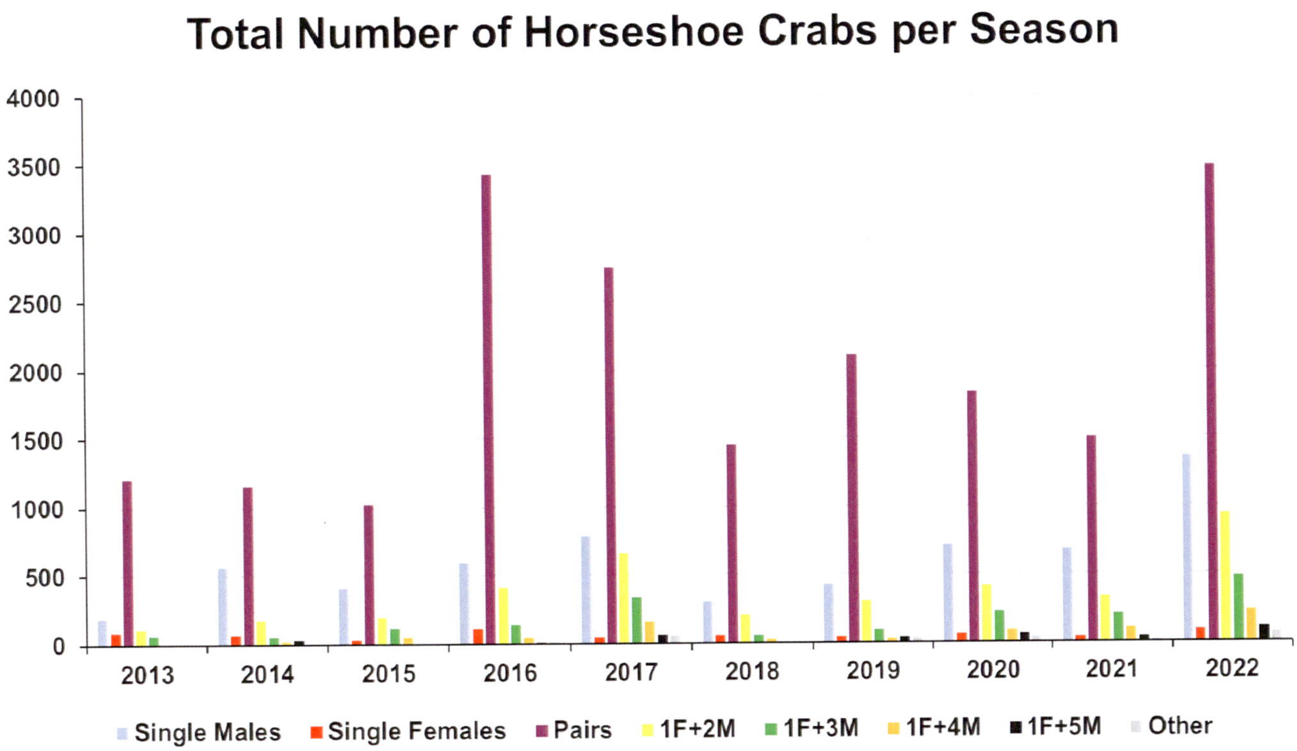


*Figure 4. The total number of horseshoe crabs of each class per season for the 2013–2022 seasons.*

## CONCLUSIONS

Horseshoe crabs are an important part of the Napatree Point ecosystem and should continue to be monitored. Their eggs provide a source of fat and protein to shorebirds that stop at Napatree during spring migration, as well as for many aquatic predators, including fish. Horseshoe crab shells also provide habitat for several species of ocean-dwelling invertebrates and seaweeds, including common whelk, blue mussels, and barnacles. Since horseshoe crabs have been harvested for bait and medical research, the population has declined sharply in recent years, making it even more important that this species is protected on Napatree Point. The 2022 survey results were outstanding, and we hope to see this trend continue in 2023.

In 2023, we should continue to make a special effort to maximize the number of evening surveys that are completed and continue to recruit public volunteers to assist with the spawning surveys.

Horseshoe crabs play an important role in the education and outreach activities in the Napatree Point Conservation Area. The Investigators summer education program for children has lessons on horseshoe crab ecology (see chapter 3 by Brown *et al.*). Visitors to Napatree frequently ask Naturalists about the horseshoe crabs they find, thus creating an opportunity to educate the public about this ecologically important species. Finally, the volunteer-based horseshoe crab surveys in the summer are a valuable way to engage the public in scientific data collection to support conservation and resource management. During 2022, over sixty volunteers helped with surveys, including two girl scout troops and college students. Along with Piping Plovers, Oystercatchers, and Ospreys, horseshoe crabs are one of the iconic species occurring on Napatree Point.

The University of Rhode Island communications division created a beautiful video documentary of one of our night-time surveys. A link to the video can be found at https://youtu.be/Vurtmpy-Aw9Q?si=c8v8wb1_tcEKQZG7.

## ACKNOWLEDGMENTS

The following individuals assisted in horseshoe crab surveys in 2022: Napatree Naturalists Emily Bodell, Emily Watling, Noah Barreto, Lucianna Faraone Coccia, Marco Greco-Byrne, Finn Harty, and Maddie Smith. A special thank you to all the public volunteers who came out this season as well.

*Crab team processing a sample.*
*Photo Credit: The Watch Hill Conservancy*

CHAPTER 21

# Crab Abundance and Interactions with the Fringing Saltmarsh in the Napatree Lagoon*

*Jennifer Croteau,*[1] *Bryan A. Oakley*[1] *& Nicole Rohr*[2]
[1] Department of Environmental Earth Science, Eastern Connecticut State University
[2] Coastal Institute, University of Rhode Island

## INTRODUCTION

The lagoon, adjacent tidal flats, and fringing saltmarsh on the northwest end of the Napatree Barrier (Figure 1) are an important part of the overall Napatree ecosystem for a variety of organisms ranging from fish to horseshoe crabs to shorebirds. The western shoreline of the lagoon consists of a fringing saltmarsh containing saltmeadow cordgrass (*Spartina patens*) and smooth cordgrass (*Spartina alterniflora*) interspersed with other marsh vegetation, which is typical of a Rhode Island salt marsh. Anecdotal observations at this site suggest that the fringing marsh has been decreasing in width as the marsh erodes, potentially jeopardizing an important component of the lagoon ecosystem.

Marsh loss is not unique to the Napatree Lagoon or Rhode Island; it is a growing concern throughout New England, and there are multiple hypotheses as to why marshes are in decline. Among these possible causes are accelerating rates of sea level rise coupled with the lack of vertical marsh accretion, wave erosion, and marsh drowning (Cole-Ekberg *et al.*, 2017). One additional hypothesis suggests that crab activity—notably burrowing—can destabilize and weaken marsh sediment, making it more susceptible to erosion (Salgado-Kent *et al.*, 2006; Raposa *et al.*, 2015; Cole-Ekberg *et al.*, 2017). Previous studies at other New England marshes have examined the marsh burrowing activities and impacts of European green crabs (*Carcinus maenas*), fiddler crabs (*Uca* spp.) and purple marsh crabs (*Sesarma reticulatum*). Results show that each of these crab species relies on the marsh habitat, and their burrows contribute to marsh erosion.

Benthic trapping surveys were continued as part of a suite of lagoon monitoring activities in 2018 to determine if crab burrowing is a possible contributor to the erosion of the fringing marsh at the Napatree

* Originally published in 2017

Lagoon (Figure 1). Time-lapse cameras were also deployed to observe crab behavior. Burrow counts and observations of the marsh edge morphology were conducted along the marsh edge during June.

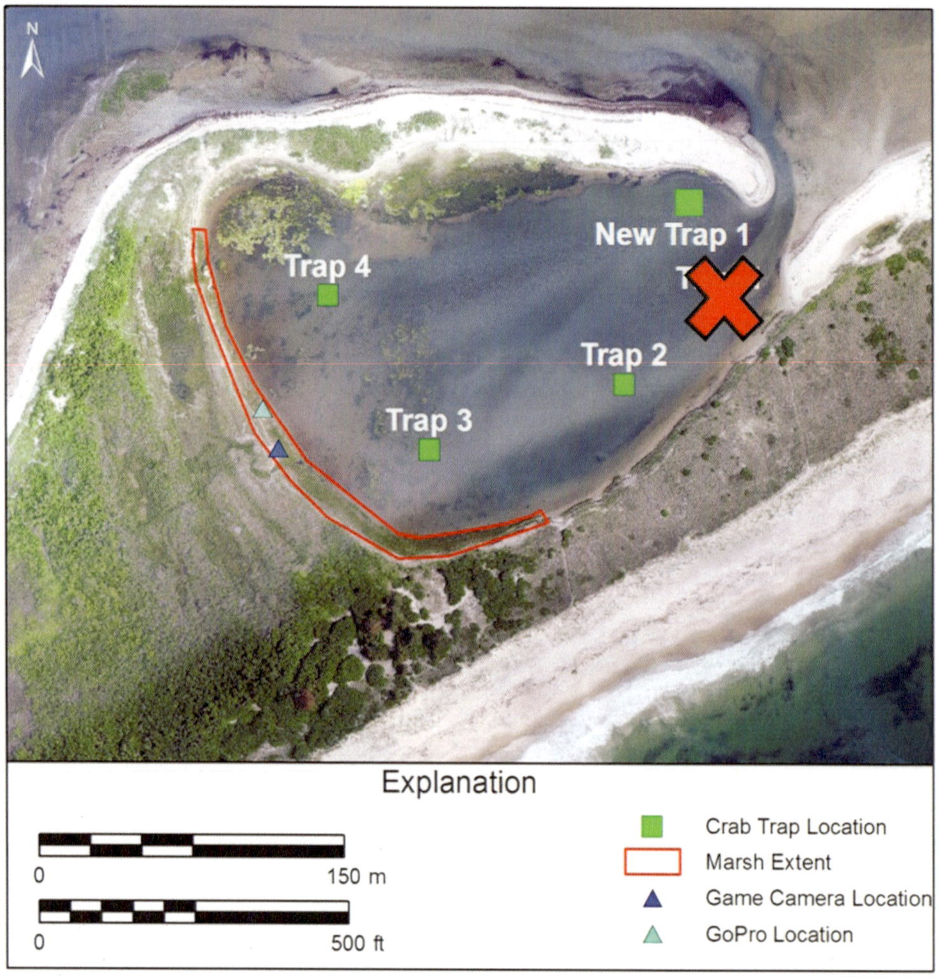


*Figure 1. Napatree Lagoon showing the extent of the fringing marsh as well as the positions of cameras and crab traps. Note the location of trap 1 had to be shifted north because of shoaling at the inlet. The red X marks the trap 1 position in the 2017 surveys. Basemap, June 2017 orthophotograph (URI Environmental Data Center).*

## METHODS

### Benthic Trap Survey

Four traps were placed in a designated "corner" of the lagoon (Figure 1), and the locations roughly correspond to the fish and algal sampling quadrants (Rohr *et al.*, 2018). Trap 1 had to be moved from its 2017 location because of shoaling related to expansion of the flood-tidal delta. Ketcham crab traps (0.3 x 0.3 x 0.6 m (1 x 1 x 2 feet); Figure 2A), baited with either whole mackerel, menhaden, or salmon, were deployed in the lagoon monthly from April through October. The dates of deployment were 14/15 April; 16/17 May; 13/14 June; 24/25 July; 21/22 August; 28/29 September; and 27/28 October. Trap deployment occurred within two hours of low tide and were left overnight. Traps were recovered at roughly the same time the second day, allowing for two high- and two low-tide cycles per trapping period. Water quality parameters were measured, typically during trap recovery, using an YSI Pro2030 DO/Salinity water quality meter equipped with an YSI 2003 Polargraphic DO sensor. The instrument was maintained and calibrated by the URI Watershed Hydrology Laboratory. Measured parameters included temperature (°C), dissolved oxygen (mg/L), conductivity (mscm), salinity (ppt), and barometric pressure (mmHg). The positions of the traps (latitude/longitude) and water depth were also measured at each station.

All trapped crabs were identified, sexed, and the carapace width was measured (mm) (Figure 2*B*). Other species captured were American eel (*Anguilla rostrata*), striped killifish (*Fundulus majalis*), and common mummichog (*Fundulus heteroclitus*). All organisms were released back into the lagoon after being recorded.

A

B

*Figure 2. A, Trap photograph taken prior to the May 2017 survey. Note the trap is vertical in this photograph; the deployed position is horizontal (Photo Credit: The Watch Hill Conservancy). B, Measuring the carapace width of a green crab using calipers.*

## Marsh Edge Morphology

Burrow counts were done during June 2018 along the marsh edge at each transect starting point (Figure 3). Twelve steel stakes were placed along the marsh edge by Dr. Peter August to track possible marsh erosion or advancement. A measuring tape (meters) was extended in a straight line from the stake to high marsh vegetation. The length of each transect is dependent on the marsh platform's width in that area. To expand the study site, five additional transects were added outside the range of the stakes for a total of seventeen transects. A 0.5 $m^2$ (5.4 square feet) quadrat was laid to the east and west of each transect point.

Visible burrows were counted and measured by depth (cm), and visible flaps were measured by width (cm), depth (cm), and height (cm). The average marsh edge height (cm) was also recorded. In this study, a *flap* is defined as an area of the marsh edge where sediment has been undercut by an undetermined cause. The condition and present vegetation were also described.

*Figure 3. The locations of the marsh edge stakes and transects discussed in the text.*

### Time-lapse Cameras

Two cameras were used to monitor crab activity in conjunction with the crab surveys in May, June, and July. A GoPro Hero 3 and GoPro Hero 5 were deployed to collect oblique images of the marsh platform and the marsh edge. The cameras were placed in various locations corresponding with transects. The GoPros were also deployed on the adjacent tidal flat to monitor crab activity along the marsh edge. The GoPro was limited by battery life and recorded one photograph every minute for approximately 1.5 hours.

## RESULTS

### Crab Traps

All of the crabs captured in the trap survey were European green crabs (*Carcinus maenas*). There was considerable temporal variation in crab abundance over the summer and fall, with peak abundances of 450 total crabs in October (Table 1). More males than females were captured each month, with the exception of October. More than twice as many males than females in April through September, and then proportions inversed so that there were nearly twice as many females than males in October (Figure 4). No gravid females were found during any survey. There were also no clear spatial patterns in crab abundance (Figure 5); however, no crabs were captured in trap 1 between April and July. Mean carapace widths ranged from 31–57 mm (1.2–2.2 inches) (Table 2) with a mean of 36 mm (1.4 inches). The DO levels (Table 3, Figure 6) varied from a low of 2.4 mg/L in July (trap 1) to 13.4 mg/L in June (trap 3). Trap 1 consistently had lower DO concentrations than traps 2, 3, and 4 from May through August 2018, and all four traps had similar DO concentrations in April, September, and October 2018.

## Marsh Edge Morphology

The marsh edge has three general forms of morphology: areas with portions of the marsh edge undercut, producing a flap of sediment, and burrows; areas without burrows or flaps; and areas of low relief relative to the adjacent tidal flat without burrows or undercut areas (flaps). The height of the marsh edge at the transects (Figure 3) ranged from 2–20 cm (0.8–7.9 inches) (mean = 7 cm, 2.7 inches). A total of twenty-one burrows were observed and with some transects having up to a density of 5 $m^{-1}$, but the average density of burrows per transect was 1 $m^{-1}$. Transects 7 and 12 had the highest density of burrows, which was 5 $m^{-1}$ and 4 $m^{-1}$, respectively. The average burrow depth (measured as the horizontal distance in from the marsh edge) was 8 cm (3.1 inches). A total of thirteen flaps were observed and the density of flaps per transect was less than 1 $m^{-1}$. The average flap width was 15 cm (5.9 inches), depth was 7 cm (2.7 inches), and height was 3 cm (1.2 inches).

## Time-lapse Cameras

Crab activity was captured by the cameras during the May and July surveys. Fiddler crabs were observed moving atop the marsh platform in both deployments, and in July two green crabs were observed burrowing into the marsh edge.

| Month | Trap 1: Number of Crabs | Trap 2: Number of Crabs | Trap 3: Number of Crabs | Trap 4: Number of Crabs |
|---|---|---|---|---|
| April | 0 | 3 | 39 | 22 |
| May | 0 | 0 | 11 | 16 |
| June | 0 | 65 | 12 | 68 |
| July | 0 | 1 | 37 | 6 |
| August | 15 | 21 | 87 | 8 |
| September | 17 | 95 | 96 | 25 |
| October | 102 | 110 | 125 | 113 |

*Table 1. Total number of crabs per trap.*

| Month | Trap 1: Mean Carapace Width – mm (SD) | Trap 2: Mean Carapace Width – mm (SD) | Trap 3: Mean Carapace Width – mm (SD) | Trap 4: Mean Carapace Width – mm (SD) |
|---|---|---|---|---|
| April | nc | 31.76 (±18.13) | 45.93 (±7.98) | 45.61 (±11.43) |
| May | nc | nc | 39.90 (±9.17) | 41.53 (±7.98) |
| June | nc | 42.10 (±8.88) | 49.03 (±15.13) | 45.90 (±12.15) |
| July | nc | nc | 47.00 (±8.42) | 50.90 (±6.35) |
| August | 44.99 (±8.32) | 46.52 (±8.41) | 49.80 (±7.05) | 57.48 (±9.96) |
| September | 44.94 (±6.50) | 42.97 (±9.44) | 46.32 (±6.70) | 48.80 (±6.61) |
| October | 48.69 (±8.01) | 46.71 (±10.59) | 49.66 (±8.88) | 50.43 (±6.93) |

*Table 2. Mean carapace width per trap in millimeters (standard deviation in parentheses); "nc" indicates no captures at the station for the sample.*

| Month | Trap 1: DO (mg/L) | Trap 2: DO (mg/L) | Trap 3: DO (mg/L) | Trap 4: DO (mg/L) |
|---|---|---|---|---|
| April | 11.4 | 13.1 | 11.6 | 10.6 |
| May | 3.02 | 5.95 | 8.46 | 8.43 |
| June | 2.9 | 7.7 | 13.4 | 7.45 |
| July | 2.4 | 4.2 | 4.7 | 5.5 |
| August | 3.2 | 6.9 | 6.6 | 5.6 |
| September | 7.2 | 6.5 | 7.3 | 8.8 |
| October | 5.8 | 6.3 | 6.3 | 6.4 |

*Table 3. Dissolved oxygen levels at each sampling station.*

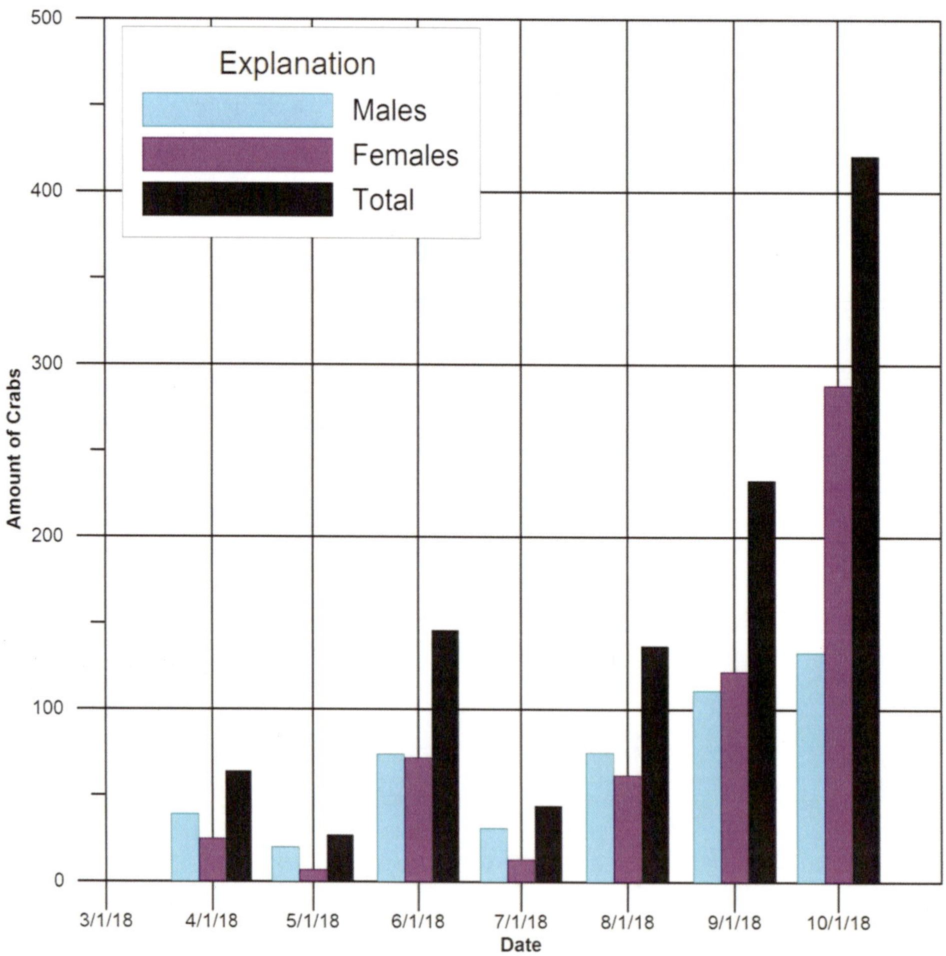


*Figure 4. Total crabs captured per month, by gender, for each of the monthly surveys.*

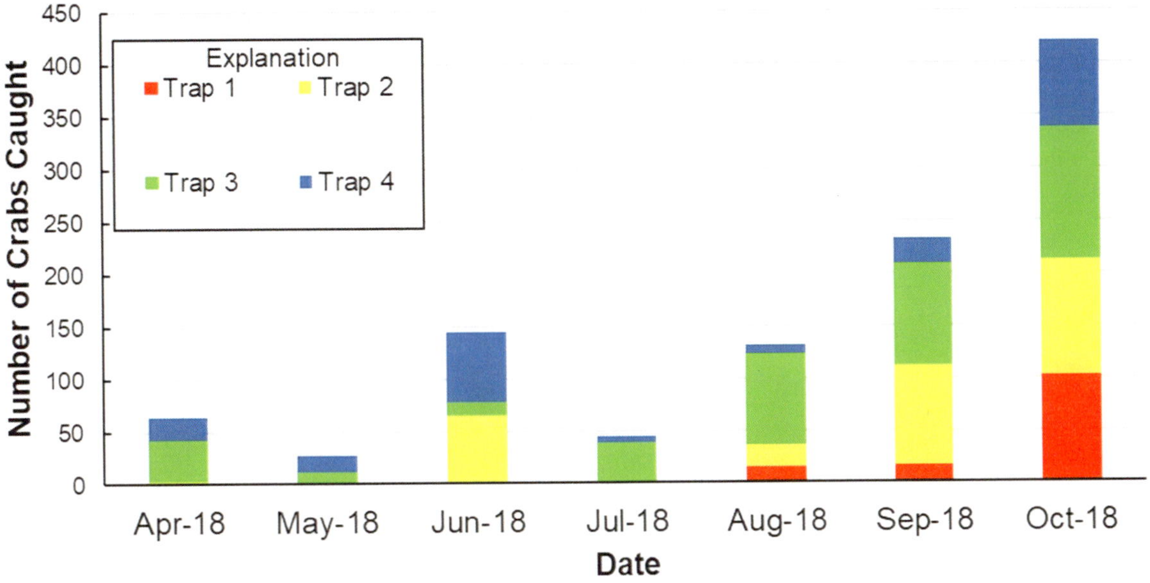


*Figure 5. Number of crabs captured at each trap location for each of the monthly surveys.*

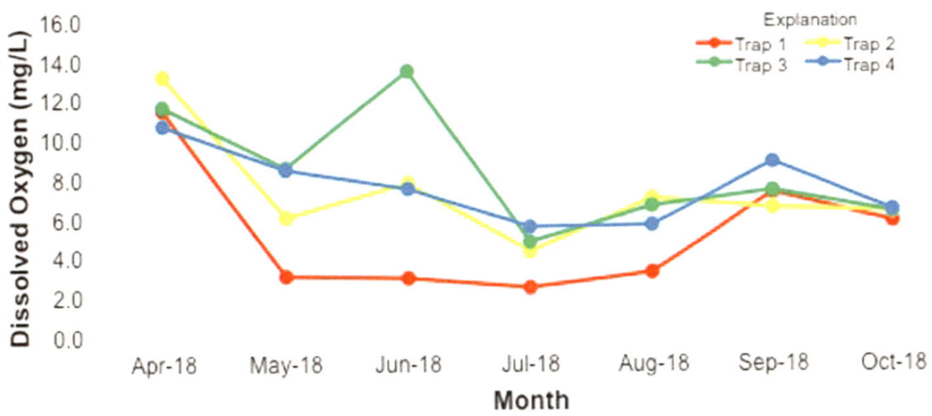


*Figure 6. The dissolved oxygen concentration (mg/L) at each trap location during each monthly survey.*

## DISCUSSION

### Crab Traps

While the second summer of sampling did not reveal any significant patterns in crab abundance or size, it did reveal that there is a population of green crabs in the Napatree Lagoon during the sampling period and that these crabs may be burrowing in the marsh edge. The number of crabs captured in the lagoon decreased relative to the 2017 surveys (Croteau *et al.*, 2017). The crab population across all trap locations varied from April to July but increased from August to October. There were no crabs found in trap 1 from April to July (Figure 6). This is mostly due to the location of trap 1 within the lagoon. Wind generated current within the lagoon coupled with the morphology of the flood-tidal delta and lagoon spit causes a considerable amount of macroalgae to congregate in this location and the measured DO to be consistently lower as the substrate is often covered by about a meter (3.3 feet) of squishy, mucky material. Crabs may either have trouble detecting the bait and/or navigating to its location due to this. However, the condition of this trap remained the same throughout the entire survey season, and it is not clear as to why there were crabs captured here later in the season.

### Marsh Edge Morphology

There appears to be a low density of burrows and flaps along the marsh edge. The flaps are wide and can extend deep underneath the marsh platform. These flaps represent marsh edge undercutting from crab burrowing and/or wave action; at present it is not known which process is more important regarding the formation of the undercut areas. Regardless, undercutting the marsh edge reduces the overall stability of the marsh platform. The overall marsh edge height is often below mean higher high water (i.e., inundated at some high tides and susceptible to wave damage).

### Time-lapse Cameras

Camera deployments occurred during months coinciding with lower numbers of crabs captured during trapping surveys. The cameras were not deployed during October, which had the highest survey results. However, the results of the cameras do prove crabs interact with, and potentially degrade, the marsh edge and platform.

## ACKNOWLEDGMENTS

Fieldwork was assisted by Janice Sassi, Laura Craver-Rogers, Melissa Cote (The Watch Hill Conservancy); Madie Varney, Jeff Fontaine and Bryce Masse, Alyson Augustein, Tom Zimmerman and Katrina Bell (Eastern Connecticut State University undergraduate students); and Drew Hyatt (Eastern Connecticut State University faculty). This work was supported by grants from The Watch Hill Conservancy, The Mystic Garden Club, the University of Rhode Island Coastal Institute, and the Eastern Connecticut State University Exemplary Program Fund. Pete August provided constructive edits of this document.

## REFERENCES


Cole-Ekberg, M. L., K. Raposa, W. Ferguson, K. Ruddock, and E. Watson. 2017. "Development and Application of a Method to Identify Salt Marsh Vulnerability to Sea Level Rise." *Estuaries and Coasts*, 40:694–710.

Croteau, J., B. Oakley, P. August, and N. Rohr. 2017. "Crab Abundance in the Napatree Lagoon." In *The State of Napatree Report: 2017*, edited by J. Sassi. The Watch Hill Conservancy. https://thewatchhillconservancy.org/napatree/napatree-resources/.

Environmental Data Center. 2017. Aerial Imagery of Napatree Point Rhode Island from 21 June 2017, Digital Raster Dataset. University of Rhode Island, unpublished work.

Gittman, R., A. Popwich, J. Bruno, and C. Peterson. 2014. "Marshes with and without Sills Protect Estuarine Shorelines from Erosion Better Than Bulkheads during a Category 1 Hurricane." *Ocean and Coastal Management*, 120:94–102.

Green-Gavrielidis, L., E. Ernst, J. Valentin Guttandin, and C. Thornber. 2017. "Monitoring Seaweed Abundance and Species Composition at Napatree Lagoon: 2017." In *The State of Napatree Report: 2017*, edited by J. Sassi. The Watch Hill Conservancy. https://thewatchhillconservancy.org/napatree/napatree-resources/.

Raposa, K. B., and K. Wasson. 2016. "Assessing Tidal Marsh Resilience to Sea Level Rise at Broad Geographic Scales with Multi-Metric Indices." *Biology Conservation*, 204:263–275.

Raposa, K. B., R. Weber, M. Cole-Ekberg, and W. Ferguson. 2017. "Vegetation Dynamics in Rhode Island Salt Marshes during a Period of Accelerating Sea Level Rise and Extreme Sea Level Events." *Estuaries and Coasts*, 40:640–650.

Rohr, N., P. August, A. Murphy, and J. Sassi. 2017 "An Ecological Reconnaissance of the Napatree Lagoon: Fish and Water Quality." In *The State of Napatree Report: 2017*, edited by J. Sassi. The Watch Hill Conservancy. https://thewatchhillconservancy.org/napatree/napatree-resources/.

Salgado-Kent, C. P., and K. McGuinness. 2006. "A Comparison of Methods for Estimating Relative Abundance of Graspid Crabs." *Wetlands and Ecology Management*, 14:1–9.

*RI DEM biologists conducting a seine net fish survey from shore.*
*Photo Credit: RI DEM*

CHAPTER 22

# The Juvenile Fish Assemblage of Little Narragansett Bay*

*Katherine Rodrigue,*[1] *Conor McManus,*[1] *Peter August,*[2,4] *Alan Desbonnet*[3] *& Janice Sassi*[4]
[1] Rhode Island Department of Environmental Management, Division of Marine Fisheries
[2] Department of Natural Resources Science, University of Rhode Island
[3] Rhode Island Sea Grant, University of Rhode Island
[4] Napatree Point Conservation Area, The Watch Hill Conservancy

## INTRODUCTION

The watershed of the Wood and Pawcatuck Rivers in Rhode Island encompasses 300 square miles, equal to approximately one-third the land area of the state. The watershed drains into Little Narragansett Bay. Fortunately, the Wood-Pawcatuck watershed is quite pristine. Under the leadership of the Wood-Pawcatuck Watershed Association (wpwa.org), seven of the rivers in the watershed have been granted the National Park Service's Wild and Scenic River System designation, one of only seventeen river systems in the country to be so recognized (http://wpwildrivers.org/). The Wood-Pawcatuck Wild and Scenic system is only one of three partnership programs in the nation that has received *both* "Wild" and "Scenic" designations. Little Narragansett Bay, the receiving waters for the Wood-Pawcatuck watershed, is a coastscape of special distinction (Figure 1). Barn Island Wildlife Management Area is an undeveloped 1,000-acre natural area of upland forest and salt marsh. To the south is Sandy Point, a nature reserve and breeding habitat for many species of endangered birds. The Napatree Point Conservation Area has been recognized as a Globally Important Bird Area by the National Audubon Society and is arguably one of the most pristine barrier spit ecosystems in southern New England. Offshore from Napatree is a complex system of reefs that extend from Watch Hill Lighthouse to Fishers Island. These reefs comprise essential fish habitat for fourteen species of commercially important fish. Clean water quality flowing into Little Narragansett Bay from the Wood-Pawcatuck watershed is a fundamental component of the health of this ecosystem. Inasmuch as this coastscape is home to a number of rare or endangered fish-eating birds, for example Least Terns (*Sternula antillarum*), Common Terns (*Sterna hirundo*), Roseate Terns (*Sterna dougalli*), American Osprey (*Pandion haliaetus*), and Black Skimmers (*Rynchops niger*), the small baitfish community of the bay is the

* Originally published in 2019

base of the food chain for these species, especially Sand Lance, Mummichog, Silversides, and Killifish (Cornell Lab of Ornithology, 2019).

The Rhode Island Department of Environmental Management (RI DEM) Division of Marine Fisheries surveys juvenile fish in Narragansett Bay, the coastal ponds, and the lower Pawcatuck River in Little Narragansett Bay to assess the recruitment of recreationally important finfish stocks in Rhode Island. Sampling in Little Narragansett Bay began in 2011. This report presents the species of fish that have been captured in the RI DEM seine surveys in Little Narragansett Bay from 2011 through 2018. A thorough presentation of the monitoring and analysis of these results can be found in RI DEM's annual performance report for Project Number F-61-R, *Assessment of Recreationally Important Finfish Stocks in Rhode Island Waters* (http://www.dem.ri.gov/programs/marine-fisheries/surveys-pubs/index.php).

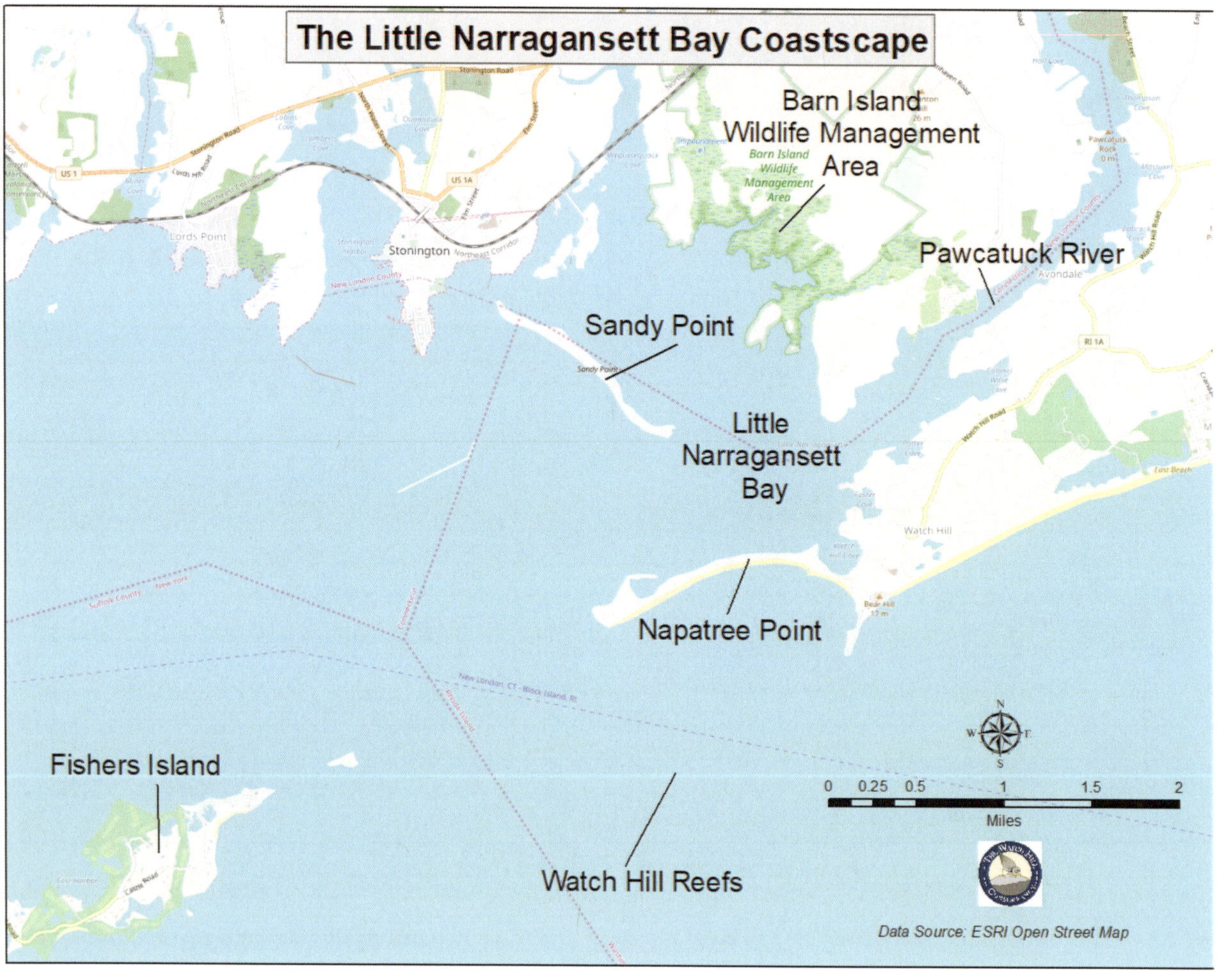


*Figure 1. Little Narragansett Bay coastscape.*

## METHODS

The three sampling locations at the mouth of the Pawcatuck River in Little Narragansett Bay are shown in Figure 2. Each site is sampled once a month from May through October. Station PR-1 is a protected beach in a small cove surrounded by large boulders. The bottom substrate is fine sand. PR-2 is located on the north (protected) side of Sandy Point in the middle of the bay. The bottom substrate is coarse sand.

PR-3 was originally located in the southern part of the bay on the protected bayside of Napatree Point. After it was initially sampled in May 2011, the station was relocated because it was extremely shallow and in a high wave energy area. PR-3 is currently located in the northern section of Little Narragansett Bay at the mouth of the river near Colonel Willie Cove. The station is on a *Spartina* spp. covered bank at the head of the cove. The bottom substrate is cobble. This station was selected to best characterize the species assemblage in the Lower Pawcatuck River, as the majority of the shoreline consists of marsh grass covered banks. The station has been sampled in all six months since 2012.

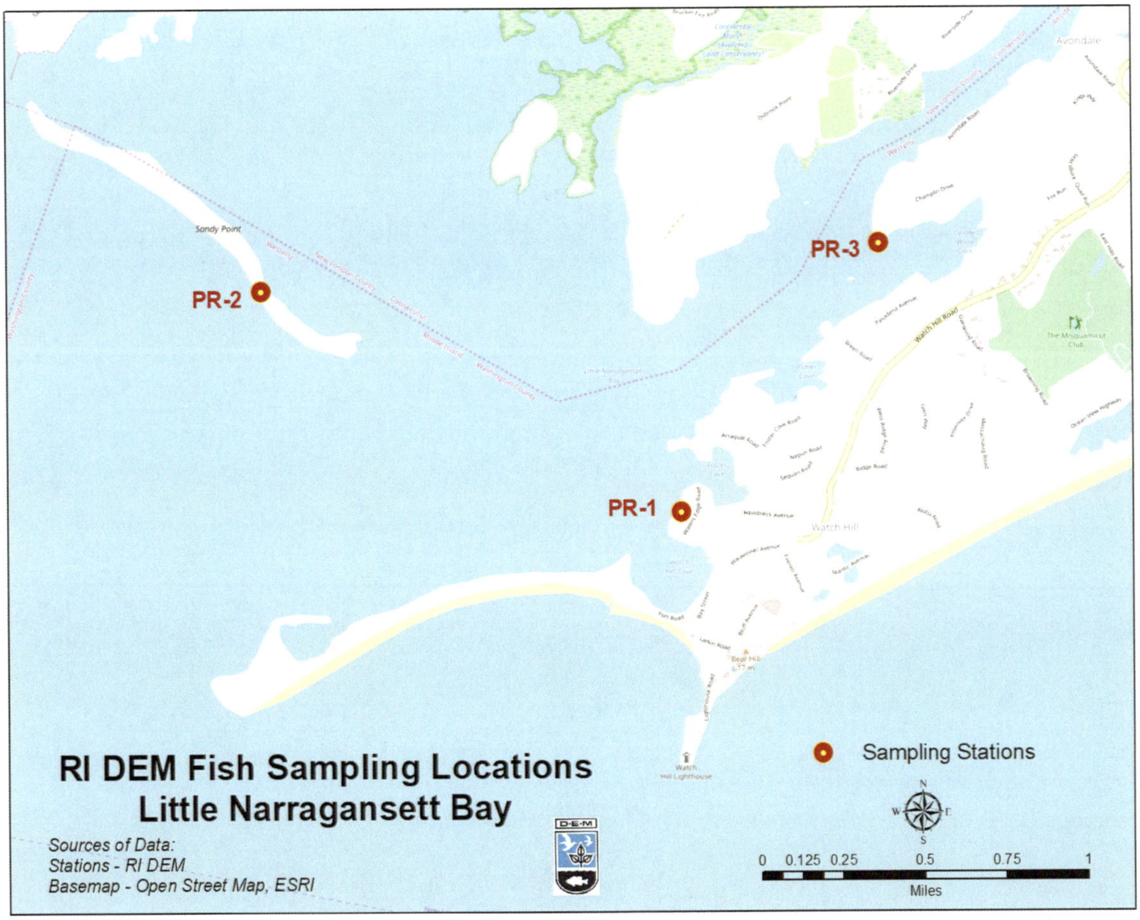


*Figure 2. Map of the RI DEM juvenile fish seine survey sampling locations.*

Seining is done on an outgoing tide. To collect animals, investigators use a 40 x 1.8 m beach seine [39.62 m long (130 feet), 1.83 m deep (6 feet) deep] with 6.4 mm mesh (0.25 inches) (Taylor *et al.*, 2016). The seine has a bag at its midpoint, a weighted footrope, and floats on the head rope. The beach seine is set in a semi-circle, away from the shoreline and back again using an outboard powered 4.9 m (16-foot) Polarkraft aluminum boat. The net is then hauled toward the beach by hand, and the bag is emptied into a large water-filled tote. All animals collected are identified to species, measured, enumerated, and sub-samples taken when appropriate. Water quality parameters (e.g., temperature, salinity, and dissolved oxygen) are measured at each station.

## RESULTS

The species of fish (and crabs) and the number captured at each of the three locations over the eight years of sampling are presented in Table 1. Briefly, winter flounder, tautog, silversides, and killifish have been the most prevalent species over the sites and time period analyzed (Table 1). Sites PR-1 and PR-3 had greater catch than PR-2. However, the species composition with PR-1 and PR-3 have differences. For example, winter flounder is more prevalent at the more southerly PR-1 site and tautog is more abundant at the more riverine PR-3 site.

| Species | Sampling Station # Captured | | | |
|---|---|---|---|---|
| Common Name, Scientific Name | PR-1 | PR-2 | PR-3 | Total |
| Alewife, *Alosa pseudoharengus* | | 10 | 28 | 38 |
| Anchovy Bay, *Anchoa mitchilli* | 3 | 5 | 5 | 13 |
| Bass Striped, *Morone saxatilis* | | | 9 | 9 |
| Blue Crab Female, *Calinectes sapidius* | 3 | 5 | 9 | 17 |
| Blue Crab Male, *Calinectes sapidius* | 10 | 6 | 10 | 26 |
| Blue Crab, *Callinectes sapidius* | 11 | 4 | 13 | 28 |
| Bluefish, *Pomatomus saltatrix* | 17 | 24 | 28 | 69 |
| Cod Atlantic, *Gadus morhua* | 2 | | | 2 |
| Cunner, *Tautogolabrus adspersus* | 12 | 5 | 35 | 52 |
| Eel American, *Anguilla rostrata* | | | 3 | 3 |
| Flounder Smallmouth, *Etropus microstomus* | 17 | 1 | | 18 |
| Flounder Summer, *Paralichthys dentatus* | 3 | 2 | | 5 |
| Flounder Twospot, *Bothus bobinsi* | 1 | | | 1 |
| Flounder Winter, *Pseudopleuronectes americanus* | 381 | 73 | 43 | 497 |
| Goby Naked, *Gobiosoma bosc* | 1 | | 2 | 3 |
| Grubby, *Myoxocephalus aenaeus* | 37 | 14 | 12 | 63 |
| Gunnel Rock, *Pholis gunnellus* | 6 | 1 | 1 | 8 |
| Herring Atlantic, *Clupea harengus* | | 1 | | 1 |
| Herring Blueback, *Alosa aestivalis* | | 7 | 1 | 8 |
| Horseshoe Crab Female, *Limulus polyphemus* | 3 | 2 | | 5 |
| Horseshoe Crab Male, *Limulus polyphemus* | 2 | 1 | | 3 |
| Horseshoe Crab, *Limulus polyphemus* | 4 | 1 | | 5 |
| Jack Crevalle, *Caranx hippos* | | 1 | | 1 |
| Killifish Striped, *Fundulus majalis* | 72 | 79 | 11 | 162 |
| Kingfish Northern, *Menticirrhus saxatilis* | 12 | 2 | | 14 |
| Lizardfish Inshore, *Synodus foetens* | 29 | | | 29 |
| Menhaden Atlantic, *Brevoortia tyrannus* | 1 | 11 | 15 | 27 |
| Minnow Sheepshead, *Cyprinodon variegatus* | 13 | | 3 | 16 |
| Mullet White, *Mugil curema* | 3 | | | 3 |
| Mummichog, *Fundulus heteroclitus* | 13 | 9 | 65 | 87 |
| Needlefish Atlantic, *Strongylura marina* | 1 | | 2 | 3 |
| Perch White, *Morone americana* | | | 2 | 2 |
| Permit, *Trachinotus falcatus* | | 1 | | 1 |

| | | | | |
|---|---|---|---|---|
| Pipefish Northern, *Syngnathus fuscus* | 20 | 14 | 13 | 47 |
| Puffer Northern, *Sphoeroides maculatus* | 13 | | 5 | 18 |
| Rainwater Killifish, *Lucania parva* | | 2 | 16 | 18 |
| Sand Lance American, *Ammodytes americanus* | | 3 | | 3 |
| Scup, *Stenotomus chrysops* | 14 | 5 | 10 | 29 |
| Sea Bass Black, *Centropristis striata* | 5 | 4 | 1 | 10 |
| Seahorse Lined, *Hippocampus erectus* | 1 | | | 1 |
| Searobin Northern, *Prionotus carolinus* | 3 | 1 | | 4 |
| Searobin Striped, *Prionotus evolans* | 21 | 2 | | 23 |
| Sennet Northern, *Sphyraena borealis* | 1 | | | 1 |
| Shad American, *Alosa sapidissima* | | | 3 | 3 |
| Silverside Atlantic, *Menidia menidia* | 171 | 159 | 202 | 532 |
| Snakefish, *Trachinocephalus myops* | 4 | | | 4 |
| Spot, *Leiostomus xanthurus* | 1 | | | 1 |
| Stickleback Fourspine, *Apeltes quadracus* | 12 | 19 | 67 | 98 |
| Tautog, *Tautoga onitis* | 26 | 37 | 230 | 293 |
| Toadfish Oyster, *Opsanus tau* | 1 | | 4 | 5 |
| Tomcod Atlantic, *Microgadus tomcod* | 14 | 7 | 35 | 56 |
| Weakfish, *Cynoscion regalis* | | | 2 | 2 |
| Windowpane, *Scophthalmus aquosus* | 3 | | | 3 |
| **GRAND TOTAL** | 967 | 518 | 885 | 2370 |

*Table 1. Total catch by species in RI DEM seine net survey of Little Narragansett Bay. Counts are presented by station and represent the total fish caught from 2011 through 2018.*

There is a clear pattern in abundance over the months of the sampling season (Figure 3). Fish abundance peaked in the month of August and declined into the late-summer and fall months. There was no clear pattern of changes in fish abundance across years (Figure 4). If all species were caught in equal numbers each of the eight years, the average percent total caught in any one year would be 12.5 percent. The average percent total caught among all years in Little Narragansett Bay is 13 percent, which suggests there is not a pattern of change across all species. It will be interesting to continue monitoring annual changes in fish abundance. Visual inspection of Figure 4 suggests that fish abundance declined between 2011 and 2014. However, 2015 was an abnormally high year in fish abundance, and fish numbers have been gradually increasing since 2016.

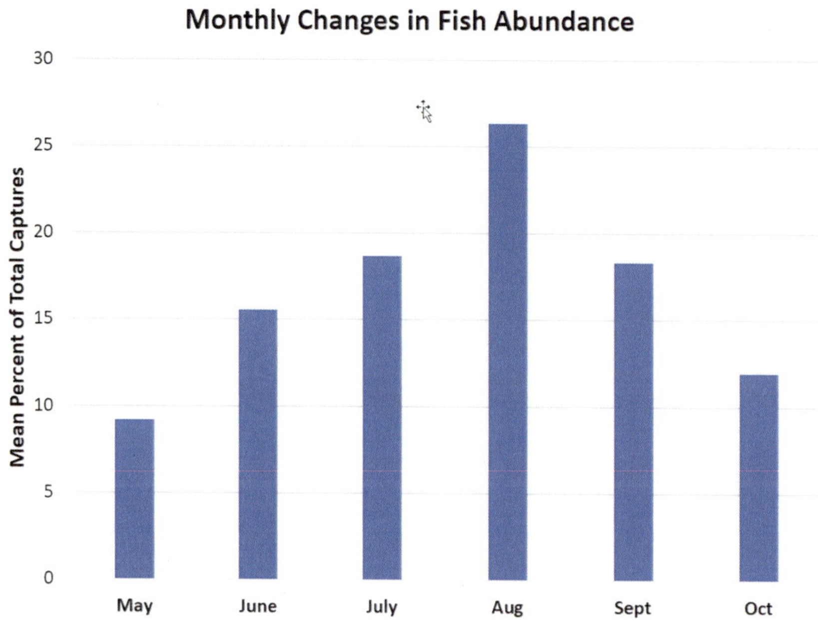


*Figure 3. Monthly changes in fish abundance. For the ten species for which more than fifty individuals were captured (see Table 1), the mean of the monthly percentage of all captures per species is presented. For example, in May on average, 9 percent of the years captures occurred in that month. All years and all stations are combined.*

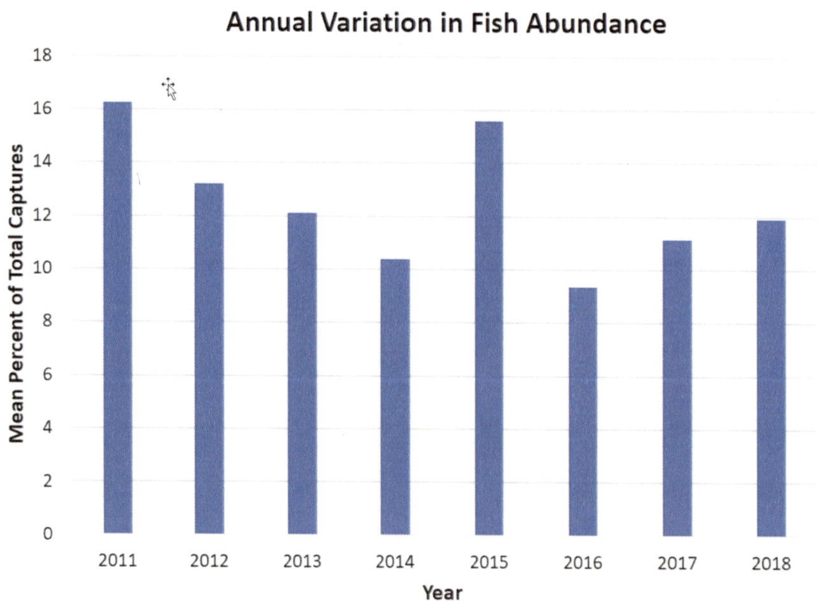


*Figure 4. Annual changes in fish abundance. For the ten species for which more than fifty individuals were captured (see Table 1), the mean of the annual percentage of all captures per species is presented. For example, in 2011, on average, 16 percent of all the captures for these ten species occurred in that year. All months and all stations are combined.*

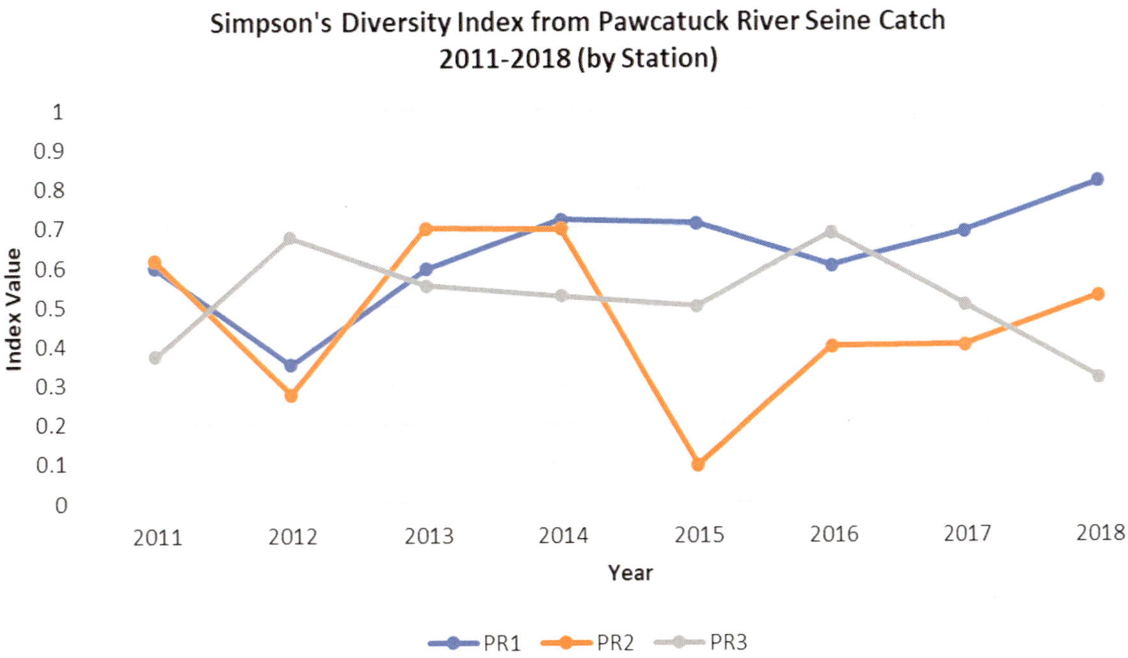


*Figure 5. Annual changes in fish diversity at each of the three sampling stations.*

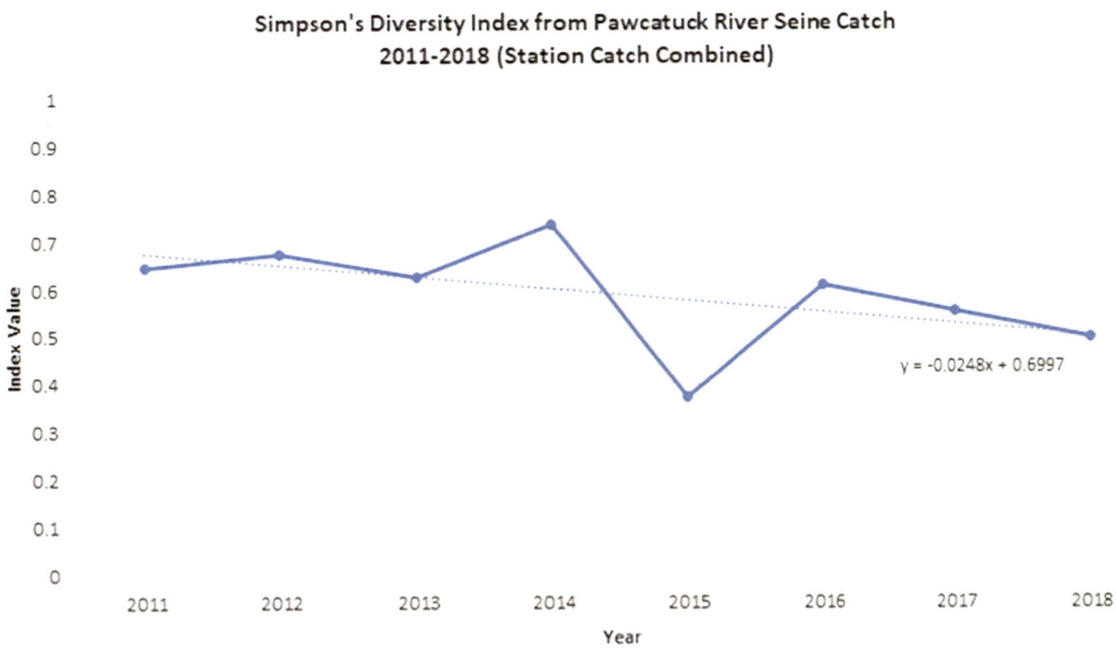


*Figure 6. Overall changes in fish diversity with stations combined.*

There are interesting changes in the diversity of fish caught in Little Narragansett Bay as indicated by Simpson's Index (Figures 5 and 6). A Simpson's Index of 1.0 reflects high species diversity: many different species all equitably represented. A Simpson's Index approaching zero indicates few species captured or when a single species dominates the catch. Site PR-2 dropped in diversity in 2012 and 2015 (Figure 5). It is not completely clear why this might be so. Events that resulted in changes to the bottom morphology occurred in 2012 (Superstorm Sandy) and 2015 (dredging of the western end of the navigation channel between Sandy Point and Barn Island); however, both events occurred after the summer seining was complete. One explanation for the low diversity at this site in 2015 may be due to a higher-than-usual proportion of silversides in the catch compared to other species. Finally, there is a slight decreasing trend in fish diversity in Little Narragansett Bay when the data from all stations are pooled (Figure 6). Continued monitoring will help us understand if this is a true decline or simply natural variation in the data.

## CONCLUSIONS

The RI DEM seine survey data allow for a comparison of how previous sampling efforts on Napatree compare to the larger Little Narragansett Bay region. The fish fauna of the bay is typical for a coastal estuary. The dominant species observed in seining surveys of the Napatree Lagoon (Rohr *et al.*, 2018) are similar to the results observed in the eight years of RI DEM seining surveys in Little Narragansett Bay. Silversides, killifish, and mummichog were common to both monitoring programs. Tautog and winter flounder were common in the RI DEM samples but uncommon in the lagoon. Winter flounder were especially abundant at site PR-1 between Watch Hill Cove and Foster Cove. PR-1 and PR-3 had similar overall densities of fish. The Sandy Point site (PR-2) had fewer numbers of fish captured in the seines. Small baitfish that are prey for seabirds are quite abundant in the samples. Important recreational and commercial species, such as striped bass, bluefish, tautog, winter flounder, and blue crab, are common to Little Narragansett Bay.

Juvenile fish increase in abundance as the summer progresses but begin to decline in September. A clear trend in annual changes in fish abundance is not evident. Comparisons of the Little Narragansett Bay seine data relative to the coastal ponds and Narragansett Bay sampling sites can be found in the annual *Assessment of Recreationally Important Finfish Stocks in Rhode Island Waters, Annual Performance Reports*, which are available from http://www.dem.ri.gov/programs/marine-fisheries/pubs.php.

## ACKNOWLEDGMENTS

This work is supported by the US Fish and Wildlife Sportfish Restoration Program grant awarded to RI DEM Division of Marine Fisheries.

## REFERENCES

Cornell Lab of Ornithology. 2019. "Birds of North America." https://birdsna.org/Species-Account/bna/species/leater1/foodhabits.

Simpson, E. 1949. "Measurement of Diversity." *Nature*, 163:688–688.

Taylor, D., J. McNamee, J. Lake, C. Gervasi, and D. Palance. 2016. "Juvenile Winter Flounder (*Pseudopleuronectes americanus*) and Summer Flounder (*Paralichthys dentatus*) Utilization of Southern New England Nurseries: Comparisons among Estuarine, Tidal River, and Coastal Lagoon Shallow-Water Habitats." *Estuaries and Coasts*, 39:1505–1525.

Photo Credit: Janice Sassi, The Watch Hill Conservancy

SECTION FIVE

# PLANTS

*Accumulation of the green seaweed Ulva in June 2017 at Napatree Lagoon.*
*Photo Credit: Lindsay Green-Gavrielidis*

CHAPTER 23

# Monitoring Seaweed Abundance and Species Composition at Napatree Lagoon: 2015–2018

*Lindsay Green-Gavrielidis, Alec Mauk & Carol Thornber*
College of Environment and Life Sciences, University of Rhode Island

## INTRODUCTION

Seaweeds are critical to the health of coastal ecosystems. There are three major groups of seaweeds, known as the greens (Chlorophyta), reds (Rhodophyta), and browns (Phaeophyceae). In addition to serving as a primary food source, seaweeds also create a three-dimensional structure that provides a habitat to countless marine organisms. Seaweed communities are sensitive to changes in the nitrogen concentration of the coastal environment. Nitrogen is generally the limiting nutrient in coastal systems (Howarth and Marino, 2006), and elevated levels can lead to a disruption in the balance of natural ecosystems.

Napatree Lagoon is connected to Little Narragansett Bay through an inlet near the northeast corner. Little Narragansett Bay is largely influenced by the Pawcatuck River watershed and is one of the most heavily nitrogen-loaded estuaries on the Atlantic coast (Fulweiler and Nixon, 2005). In shallow, low wave energy environments influenced by nitrogen loading, seaweed blooms can form (Valiela *et al.*, 1997). These dense mats of drifting seaweeds can have negative consequences on the environment and can cause the decline of seagrass (Valiela *et al.*, 1997) and overall community diversity (Worm and Lotze, 2006). Furthermore, once seaweed blooms begin to decay, oxygen levels are depleted, which can lead to fish kills. In 2003, a fish kill occurred in Greenwich Bay (Warwick, Rhode Island) as the result of a phytoplankton bloom encouraged by nitrogen loading (Rhode Island Department of Environmental Management, 2003).

In 2015, researchers at the University of Rhode Island, along with undergraduate research fellows, began conducting monthly surveys at four sites within Napatree Lagoon (northeast, northwest, southeast, and southwest; hereafter referred to as NE, NW, SE, and SW; see Rohr *et al.*, 2018, Figure 1). The objective of these surveys was to determine the species composition and biomass of seaweeds at each Napatree Lagoon location from May through September 2015. We continued these surveys in 2016, 2017, and 2018 to determine how the seaweed community changes over time.

## METHODS

Intertidal and subtidal monthly surveys were conducted from May through September at four sites in Napatree Lagoon; in 2018, surveys were only conducted in June. Subtidal surveys were not conducted at the NW site due to the sinking soil conditions. Following our pre-established protocols, on each sampling date in the intertidal, a 10 m (32.8-foot) transect line was laid down parallel to shore at the water's edge. At every other meter along each transect line (n=5 samples/transect), a 0.25 $m^2$ (2.7 square feet) quadrat was placed on the substrata, and the percent cover of all live seaweed species was recorded. We then collected all the seaweed biomass in each quadrat and returned it to the lab for processing where we determined the biomass of seaweed and counted all associated invertebrates. The most common invertebrate was the common mud snail, *Tritia* (formerly *Ilyanassa*) *obsoleta*, which was often present in very large densities (Figure 1).

*Figure 1. Tritia obsoleta, the common mud snail, in a 0.25 $m^2$ (2.7 square feet) quadrat at Napatree Lagoon. Photo Credit: Lindsay Green-Gavrielidis*

The subtidal zone at each site was surveyed on the same day, by walking a 10 m (32.8 feet) transect perpendicular to the shoreline, starting at the water's edge. At every 3.3 m (10.8 feet) for a total of three samples per site, a 0.35 m (1.15-foot) wide net was dragged along 0.5 m (1.6 feet) of seafloor to collect all seaweed and invertebrates (Figure 1). Although most biomass was typically near the seafloor, any seaweed floating in the water column above the net drag area was also collected. The depth of the water was recorded to calculate the volume of water sampled at each point. All collected material was brought back to the laboratory for processing. In the lab, we identified all species of seaweeds, determined the biomass, and counted all invertebrates (Guidone and Thornber, 2013).

During the 2017 sampling season, monthly water samples were collected via kayak using a Niskin bottle deployed 1 m (3.3 feet) below the surface by Peter August, Nicole Rohr, and Janice Sassi. The sampling site was closest to the NE sampling location. Water samples were transported to the laboratory and stored at -80°C (-112° Fahrenheit) until analysis. Prior to analysis, samples were filtered to remove particulates. The concentration of ammonium, phosphate, nitrite, and nitrate was determined in triplicate for each sample at the Marine Science Research Facility located at the University of Rhode Island's Graduate School of Oceanography.

## RESULTS

Overall, there was wide variability in the seaweed composition in Napatree Lagoon between sites, months, and years. In 2017, we collected a fewer number of species than in previous years (Tables 1 and 2) and saw a noticeable increase in the occurrence of the introduced seaweeds *Grateloupia turuturu* and *Dasysiphonia japonica*. We also documented *Agardhiella subulata* in the intertidal surveys for the first time in 2017 and *Phyllophora* spp. and *Audionella* spp. in the subtidal for the first time. Seaweed biomass in 2017 was primarily composed of *Ulva* blades and tubes, which is similar to the previous two years of surveys. In June 2018, the seaweed biomass in both the intertidal and subtidal was generally low compared to previous years.

We also recorded differences in the biomass of seaweeds amongst years and sites. In the intertidal zone, we found that seaweed abundance in May and August was generally lower in 2017 compared to 2015 and 2016 (Figure 2). Comparing between all years in June, the pattern was highly site specific (Figure 3). At the NE, SW, and SE sites, we recorded higher seaweed biomass in 2015 and 2016 than in 2017; the lowest biomass at each site was observed in 2018 (Figure 3). In July at the NE site we recorded consistently high biomass across all years sampled (2015–2017; Figure 2). In the subtidal zone, the differences between the years were less dramatic during May, July, August, and September (Figure 4). During June, we found dramatic differences between the subtidal biomass of seaweeds collected at the SE site (Figure 5). In June 2017, we collected the highest biomass ever recorded at Napatree Lagoon at this site (>12,000 $g/m^3$), which was primarily composed of *Ulva* tubes (Figure 4). In 2018, the biomass at each of the three sites was considerably lower than previous samplings (2015–2018; Figure 5).

Nutrient values in the Napatree Lagoon are presented in Table 3. Nitrite at Napatree in 2017 was very low, ranging from (mean ± 1 standard error, n=3) 0.13 µM (±0.01) to 0.53 µM (±0.08), and was therefore not examined in relation to seaweed biomass. Ammonium concentrations increased from May to June 2017, followed by a sharp reduction in July 2017 (Figure 6). In August, ammonium concentrations had begun to increase, and this pattern continued in September when ammonium concentrations peaked at 9.37 µM (±0.03). Phosphate concentrations remained below 1 µM in all months except in July when concentrations peaked at 1.52 µM (±0.02). Nitrate followed a similar pattern to ammonium with increasing concentrations from May to June (peak 2.39 µM ±0.04), a sharp reduction in July, rebound in August, and increasing concentrations in September (Figure 6). Interestingly, seaweed biomass peaked in June 2017, when ammonium and nitrate concentrations were 8.46 µM (±0.05) and 2.39 µM (±004), respectively. A similar increase in seaweed biomass was not observed in September 2017 when ammonium concentrations peaked.

| | May | | | | June | | | | July | | | | August | | | | September | | | |
|---|---|---|---|---|---|---|---|---|---|---|---|---|---|---|---|---|---|---|---|---|
| | NE | SE | NW | SW | NE | SE | NW | SW | NE | SE | NW | SW | NE | SE | NW | SW | NE | SE | NW | SW |
| *Agardhiella subulata*[R] | | | | | | | | | | | | | | 7 | 7 | | | | | |
| *Aglaothamnion* spp.[R] | | | | | | | | 5 | | | | | 5 | | | | | | | |
| *Ascophyllum nodosum*[P] | 5 | | | | | | | | 5,6 | | | | | | | | | | | |
| *Callithamnion* spp.[R] | | | | | | | 5 | | | | | | | 5 | | | | | | |
| *Ceramium* spp.[R] | 5 | 6 | | 5 | 7 | 7 | 7 | 7 | 7 | 7 | | | 5,7 | 5,7 | 5 | 6 | 6 | 6 | 6 | |
| *Chaetomorpha* spp.[C] | 5 | 6 | | 6 | | | | | 5,6 | 6 | | 6 | 5,6 | 5,6 | | 5 | 5,6 | 6 | 5,6 | 6 |
| *Champia parvula*[R] | | | | | | | | | | | | | | 5 | | | | 6 | 6 | |
| *Chondrus crispus*[R] | 5,6, 7 | | | | 5,6 | 5,6 | 5,6 | 5 | 5,6 | 6,7 | 5 | | 5,6 | | | 7 | 5,6 | | | |
| *Chordaria flagelliformis*[P] | | | | | | | | | | | | | 5 | | | | | | | |
| *Cladophora* spp.[C] | | | | | 7,8 | 7,8 | 8 | 7 | 7 | | | | 7 | 5,6,7 | 6,7 | 7 | | | 6 | 6 |
| *Codium fragile* ssp. *fragile*[C] | | | | | 5 | | | | | | | | | | | | 6 | 6 | | |
| *Cystoclonium purpureum*[R] | | | | | 7 | | | | | | | | | 7 | 7 | | | | 6 | |
| *Dasysiphonia japonica*[R] | 5,7 | 6,7 | 6,7 | 6,7 | 6,7 | 6,7,8 | 6,7,8 | 6 | 5,6,7 | 6,7 | 7 | | 5,7 | | | 7 | 6 | 6 | 6 | |
| *Desmarestia aculeata*[P] | | | | | | | 6 | | 5 | 5 | 5 | | 5 | | | | | | | |
| *Ectocarpus* spp.[P] | 5 | 5,6 | 6 | 5 | 8 | | | | | | | | | 6 | | | 5 | | | |
| *Fucus vesiculosus*[P] | | | | | 5,6 | | | | 5 | 7 | | | 5 | | | | 6 | | | |
| *Gracilaria*/ *Agarophyton* spp.[R] | 6 | | | | | 6 | | | 5 | 6 | 6 | | 5,7 | 5,6,7 | | | 6 | | | |
| *Grateloupia turuturu*[R] | | 6 | | | | | | | 6 | | | | 6 | 7 | 7 | | | | | |
| *Grinnellia americana*[R] | | | | | | | | | 7 | | 7 | | 7 | 5 | 7 | | | | | |
| *Halosiphon tomentosus*[P] | 5 | | | | | | | | | | | | | | | | | | | |
| *Palmaria palmata*[R] | | | | | | | | | 5 | 6 | | | | | | | | | | |
| *Petalonia*/ *Punctaria* spp.[P] | 5,6 | 6 | | 5,6 | 6 | 6 | 6 | 6 | 5,6 | 5,6 | | | | | | | 6 | | | |
| *Phyllophora* spp.[R] | | | | | 7 | 7 | 7 | 7 | 5 | 5 | | | | | | | | | | |
| *Polyides rotunda*[R] | 7 | | | | 5 | | | | 5,6 | | | | 5 | | | | 5,6 | | | |
| *Polysiphonia* spp.[R] | 5,6,7 | 6,7 | 6 | 7 | 6,7 | 6,7 | 6,7 | | 5,6,7 | 6,7 | 6,7 | 6 | 7 | 7 | | 7 | 6 | 6 | 6 | |
| *Saccharina latissima*[P] | | | | | 5 | | | | 5 | | | | 5 | | | | | 6 | | |
| *Scytosiphon* spp.[P] | 5 | 5 | | | | | | | 6 | | | | | | | | | | | |
| *Spermothamnion repens*[R] | | | | | | | | | | | | | | | | | 5 | | | |
| *Ulva* blades[C] | 5,6,7 | 6,7 | 5,6,7 | 5,6,7 | 5,6,8 | 5,6,7,8 | 5,6, 7,8 | 5,6,7,8 | 5,6,7 | 5,6,7 | 5,6,7 | 5,6,7 | 5,6,7 | 5,6, 7 | 5,6 | 5 | 5,6 | 5 | 6 | 5 |
| *Ulva* tubes[C] | 5,6,7 | 5,6,7 | 5,6,7 | 5,6,7 | 5,6,7,8 | 5,6,7,8 | 5,6,7,8 | 5,6, 8 | 5,6,7 | 5,6,7 | 5,6,7 | 5,6,7 | 5,6,7 | 5,6,7 | 5,6,7 | 5,6,7 | 5,6 | 5 | 6 | 5 |

*Table 1. Taxa observed during intertidal field surveys at Napatree Lagoon in 2015 (5), 2016 (6), 2017 (7), and 2018 (8). Abbreviations next to taxon names indicate the seaweed group C: Chlorophyta, R: Rhodophyta, and P: Phaeophyceae. Note: 2018 sampling only occurred in June.*

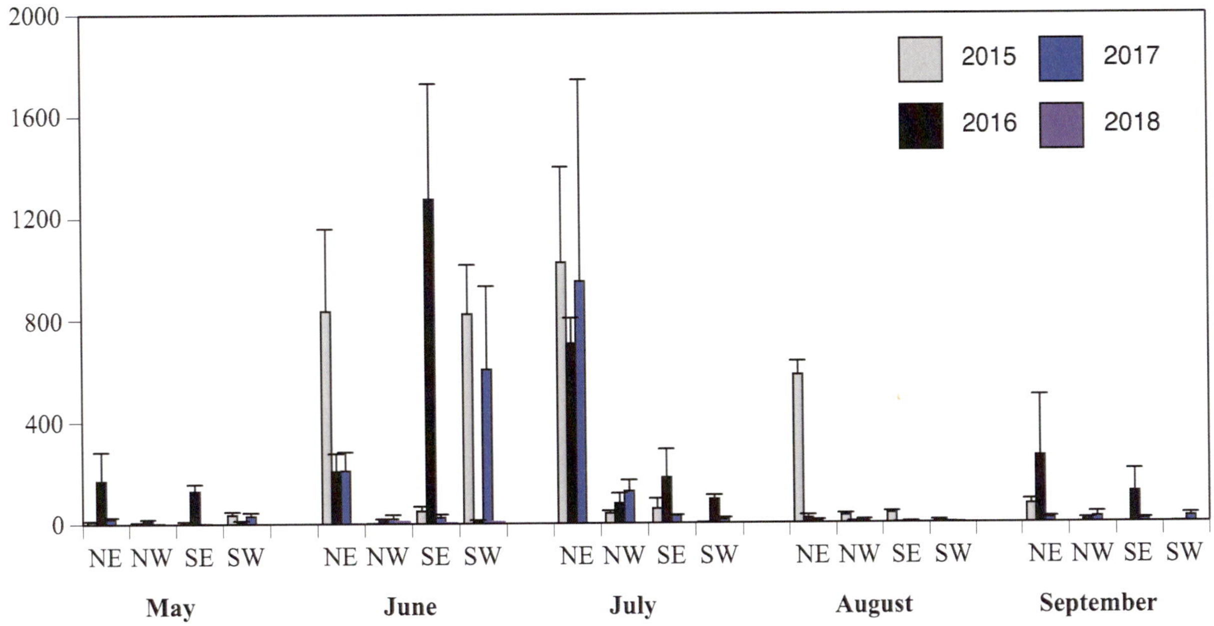


*Figure 2. Mean intertidal seaweed wet weight (g/m²) at all four study sites in Napatree Lagoon over the period of May through September 2015, 2016, and 2017 and in June 2018 (mean ± 1 SE).*

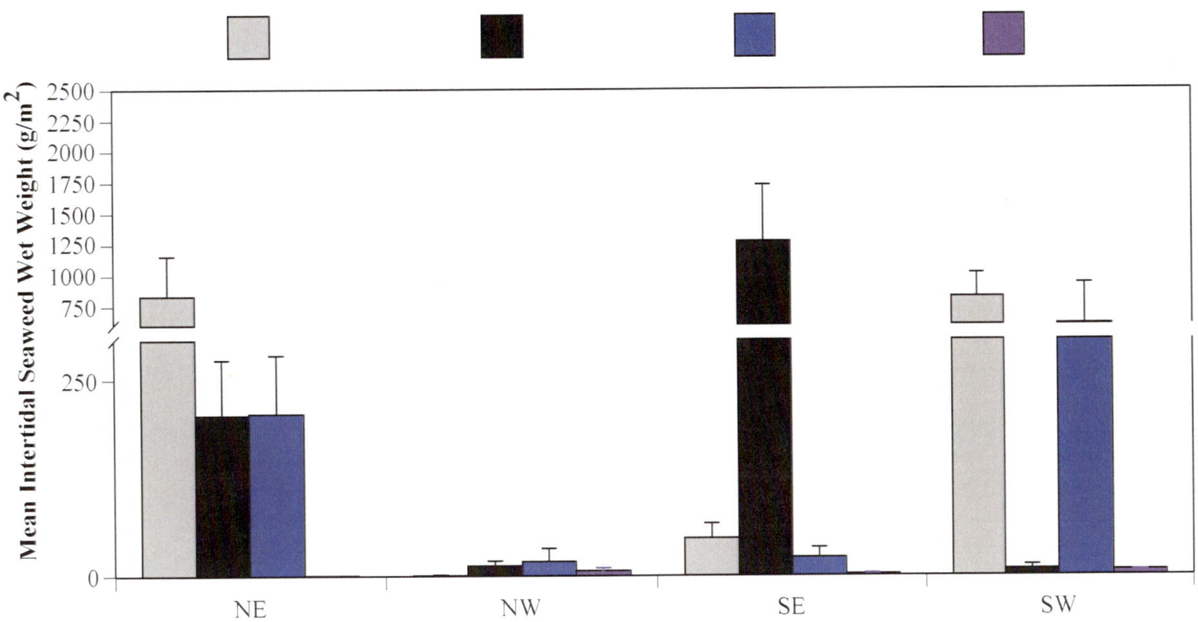


*Figure 3. Mean intertidal seaweed wet weight (g/m²) at all four study sites in Napatree Lagoon in June from 2015 through 2018 (mean ± 1 SE).*

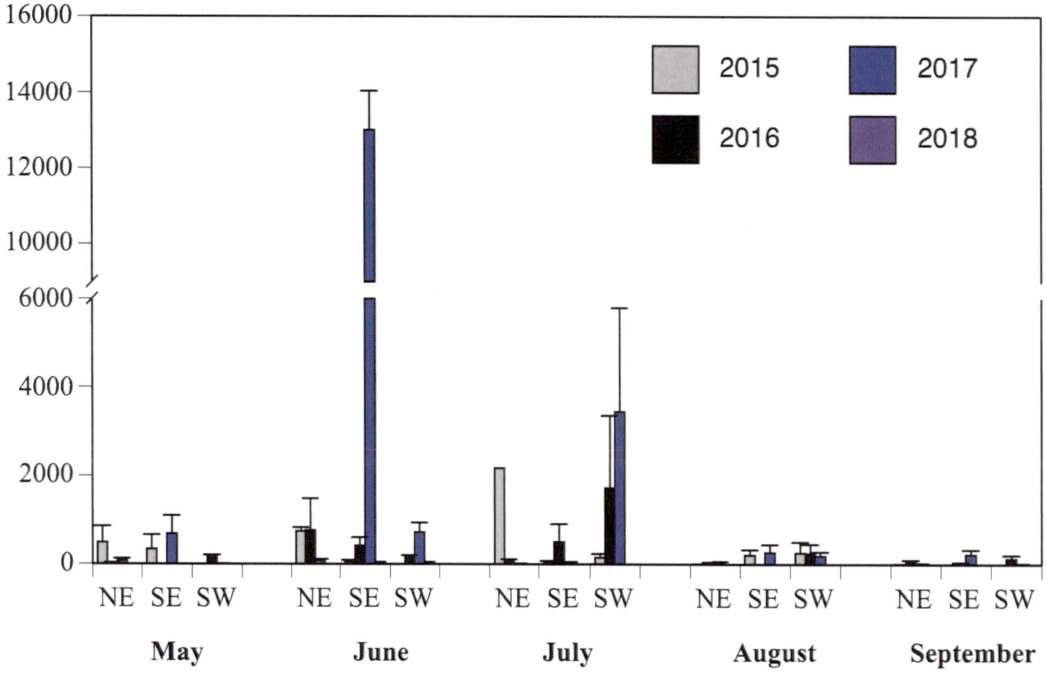


*Figure 4. Mean subtidal seaweed wet weight (g/m³) at three study sites in Napatree Lagoon over the period of May through September 2015, 2016, and 2017 and in June 2018 (mean ± 1 SE).*

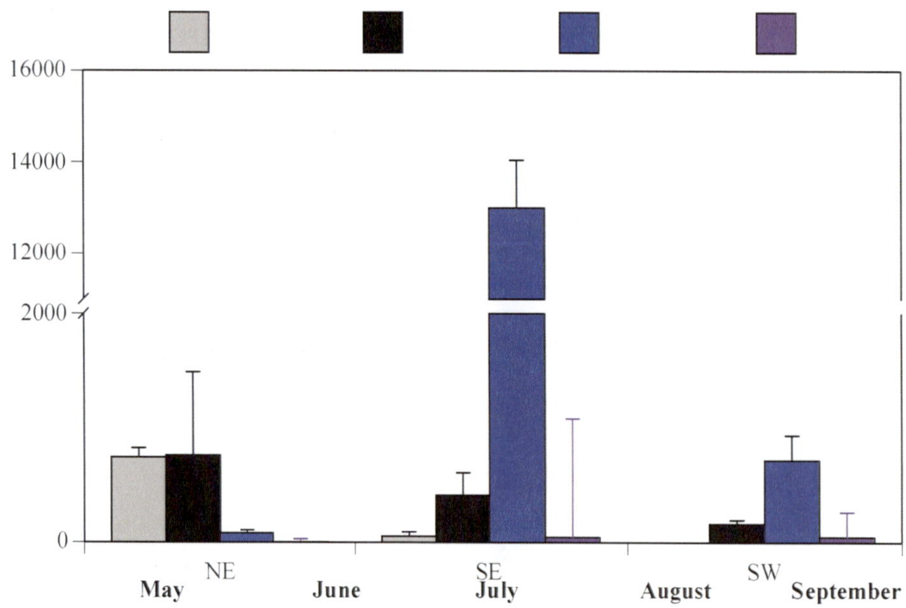


*Figure 5. Mean subtidal seaweed wet weight (g/m³) at three study sites in Napatree Lagoon in June from 2015 through 2018 (mean ± 1 SE).*

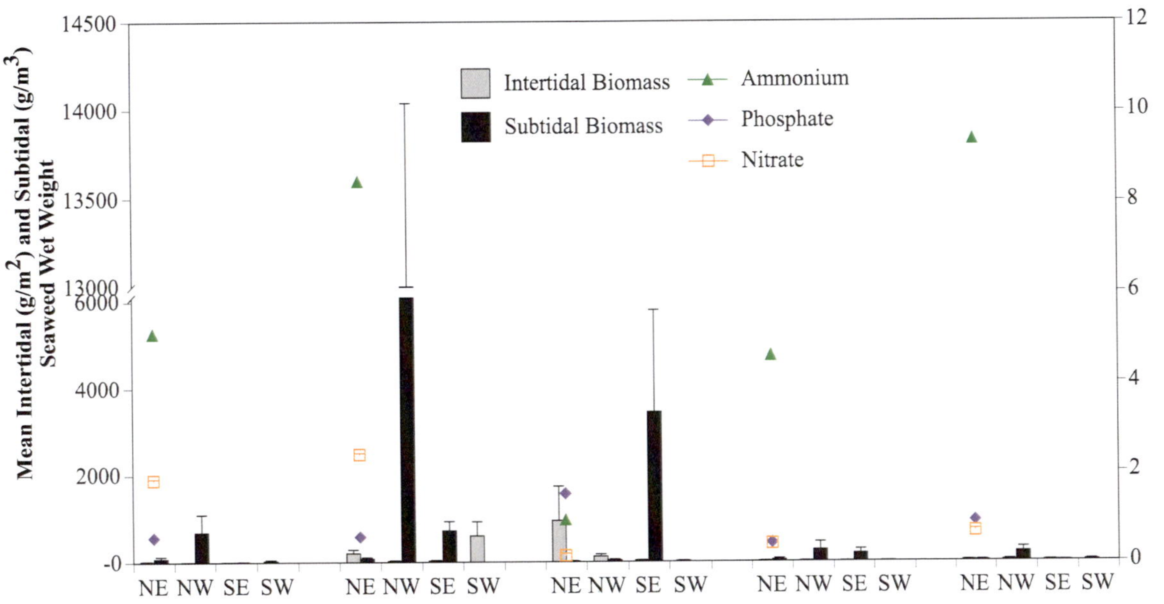


*Figure 6. Mean intertidal (g/m²) and subtidal seaweed wet weight (g/m³) at four study sites in Napatree Lagoon (subtidal not sampled at SW) from May through September 2017 plotted with the mean ammonium (µM), phosphate (µM), and nitrate (µM) concentrations in the water column (mean ± 1 SE). Note: sampling of water occurred on a different date, closest to the NE site.*

| Date | Ammonium (µM) | Phosphate (µM) | Nitrite (µM) | Nitrate (µM) |
|---|---|---|---|---|
| 15 Feb | 3.29 (0.12) | 0.49 (0.01) | 0.14 (0.01) | 4.7 (0.08) |
| 1 May | 5.08 (0.02) | 0.54 (0.02) | 0.19 (0.01) | 1.82 (0.04) |
| 12 June | 8.46 (0.05) | 0.56 (0.01) | 0.12 (0.02) | 2.39 (0.04) |
| 13 July | 0.93 (0.08) | 1.52 (0.02) | 0.53 (0.08) | 0.13 (0.07) |
| 9 Aug | 4.58 (0.03) | 0.42 (0.01) | 0.16 (0.08) | 0.48 (0.01) |
| 10 Sept | 9.37 (0.03) | 0.91 (0.01) | 0.25 (0.07) | 0.66 (0.07) |
| 9 Oct | 7.06 (0.02) | 0.90 (0.02) | 0.13 (0.01) | 1.03 (0.04) |

*Table 3. Nutrient levels recorded in the Napatree Lagoon in 2017. Values presented are the mean and standard error (in parentheses) of three replicate samples.*

## CONCLUSIONS

We found high variability in species composition and abundance within Napatree Lagoon between sites, sampling months (May–September), and sampling years (2015–2018). Species of seaweeds from all three major groups were found throughout each sampling year. The highest intertidal biomass of seaweed was recorded from June to July, and subtidal biomass was highest in June of 2017, while it peaked during July in 2015 and 2016. Both intertidal and subtidal seaweed biomass were low at all sites during sampling in June 2018. In particular, we observed a large reduction in the intertidal and subtidal seaweed biomass at the NE site in 2018. We hypothesize that the movement of the inlet to Napatree Lagoon between the fall of 2017 and spring of 2018 prevented drift seaweed from accumulating in the NE sampling site of the lagoon in 2018.

Over the course of sampling from 2015 through 2018, we have documented thirty-three different species of seaweed in Napatree Lagoon. This is higher than the amount of seaweed species present in other coastal ponds in Rhode Island during the summer months (Thornber, *unpublished data*). The green bloom-forming *Ulva* blades and tubes were the most commonly seen seaweeds at Napatree Lagoon, both in the intertidal and subtidal surveys (Tables 1 and 2). We documented the invasive red alga *Grateloupia turuturu* in Napatree Lagoon for the first time in 2016, although this alga has been present in Rhode Island since the 1990s (Villalard-Bohnsack and Harlin, 1997). In 2017, the introduced red seaweed *Dasysiphonia japonica* was collected more frequently than in previous years.

During the sampling season in 2017, monthly water samples were collected by Pete August, Nicole Rohr, and Janice Sassi to determine the relationship of dissolved inorganic nitrogen (DIN: ammonium, nitrate) and phosphate and seaweed biomass in Napatree Lagoon. Dissolved inorganic nitrogen concentrations increased during the late spring and early summer (from May to June), along with seaweed biomass. In July 2017, there was a sharp reduction in dissolved inorganic nitrogen, likely because the seaweed bloom documented in June 2017 used up all DIN in the water column. Concentrations of DIN increased from July to September, but the biomass of seaweed was relatively low during this time. Generally, algal blooms crash in the late summer due to warm water temperature; therefore, it is likely that seaweeds were limited by temperature during this time period and unable to capitalize on the surrounding nutrients.

Seaweed blooms are generally formed in shallow, low-wave energy environments as a result of nutrient loading (Valiela *et al.*, 1997). The decay of seaweed mats can lead to low oxygen concentrations in the surrounding water column (Gray *et al.*, 2002), which can negatively impact animal life. Between 2016 and 2017, the inlet to Napatree Lagoon moved due to natural processes. Where in previous years, seaweed had been deposited at the NE site (located at the mouth of the inlet), in 2017 we observed more seaweed deposited in the middle of the channel, outside of our sampling area. We hypothesize that this led to lower intertidal abundances at the NE site in May, June, and August of 2017. Interestingly, we've collected consistently high biomass in the intertidal of the NE site across all sampling years. Lastly, we collected the highest subtidal biomass of seaweeds ever recorded at Napatree Lagoon during June 2017 at the SE site. Collected biomass (>12,000 $g/m^3$) was higher than any blooms previously recorded in Narragansett Bay (Thornber *et al.*, 2017), although the spatial scale of this bloom was relatively small. The large variability of seaweed abundance and species composition at Napatree highlight the importance of long-term monitoring of seaweed populations in addition to measuring environmental variables, such as nitrogen concentrations and dissolved oxygen concentration, to determine the causes and consequences of mass accumulations of seaweeds.

## ACKNOWLEDGMENTS

We would like to acknowledge the following individuals for their extensive help with fieldwork and sample analysis in 2015 (Ivy Burns, Annie Filteau, and Fiona MacKechnie), 2016 (Ivy Burns, Fiona MacKechnie, Marguerite Kinsella, and Hannah Madison), and 2017 (Evan Ernst and Jeremy Valentin Guttandin).

## REFERENCES

Fulweiler, R., and S. Nixon. 2005. "Export of Nitrogen, Phosphorus, and Suspended Solids from a Southern New England Watershed to Little Narragansett Bay." *Biogeochemistry*, 76:567–593.

Gray, J., R. Wu, and Y. Or. 2002. "Effects of Hypoxia and Organic Enrichment on the Coastal Marine Environment." *Marine Ecology Progress Series*, 238:249–279.

Guidone, M., and C. Thornber. 2013. "Examination of *Ulva* Bloom Species Richness and Relative Abundance Reveals Two Cryptically Co-occurring Bloom Species in Narragansett Bay, Rhode Island." *Harmful Algae*, 24:1–9.

Howarth, R., and R. Marino. 2006. "Nitrogen as the Limiting Nutrient for Eutrophication in Coastal Marine Ecosystems: Evolving Views over Three Decades." *Limnology and Oceanography*, 51:364–376.

Rhode Island Department of Environmental Management. 2003. *The Greenwich Bay Fish Kill - August 2003: Causes, Impacts, and Responses*. Rhode Island Department of Environmental Management.

Thornber, C., M. Guidone, C. Deaucutis, L. Green, C. Ramsay, and M. Palmisciano. 2017. "Spatial and Temporal Variability in Macroalgal Blooms in a Eutrophied Coastal Estuary." *Harmful Algae*, 68:82–96.

Valiela I., J. McClelland, J. Hauxwell, *et al.* 1997. "Macroalgal Blooms in Shallow Estuaries: Controls and Ecophysiological and Ecosystem Consequences." *Limnology Oceanography*, 42:1105–1118.

Villalard-Bohnsack, M., and M. Harlin. 1997. "The Appearance of *Grateloupia doryphora* (Halymeniaceae, Rhodophyta) on the Northeast Coast of North America." *Phycologia* 36:324–328.

Worm, B., and H. Lotze. 2006. "Effects of Eutrophication, Grazing, and Algal Blooms on Rocky Shores." *Limnology and Oceanography*, 51:569–579.

*Seaside goldenrod (Solidago sempervirens) in flower—an Important plant species for fall migrants.*
*Photo Credit: The Watch Hill Conservancy*

CHAPTER 24

# The Plants of the Napatree Point Conservation Area*

*Keith Killingbeck,[1] Lisa Gould,[2] Hope Leeson,[2] Melissa Cote,[3] Janice Sassi[3] & Peter August[4]*
[1] Department of Biological Sciences, University of Rhode Island
[2] The Rhode Island Natural History Survey
[3] Napatree Point Conservation Area, The Watch Hill Conservancy
[4] Department of Natural Resources Science, University of Rhode Island

## INTRODUCTION

One of the fundamental stewardship goals for Napatree Point Conservation Area (NTPCA) is to document plants and animals that occur there. Thanks to long-term bird surveys conducted by Reynold Larsen, we have an excellent understanding of Napatree's avifauna (see list in chapter 11). Rothwell and Ginsberg (2017, see chapter 15) have developed a comprehensive survey of the bees of Napatree. Camera trapping for the past six years has provided a solid inventory of the meso-mammals of Napatree (see chapter 17). This paper is a comprehensive inventory of the different species of plants known to occur on Napatree. "The Plants of Napatree" is a dynamic document; we will update the list as new discoveries are made.

## METHODS

We created the inventory of the plants of Napatree from previous surveys, new field surveys, and notes from historical surveys. We limited our data gathering to sources and authors of known credibility and followed the nomenclature of Haines (2011). Table 1 lists the sources we used.

* Originally published in 2018

| Dataset | Description | Survey Period |
|---|---|---|
| Lisa Lofland Gould Field Notes | Species records from past surveys of the NTPCA | 1994–2002 |
| Hope Leeson Field Notes | Species records from past surveys of NTPCA, plant restoration projects on Napatree, and invasive species monitoring surveys | 2014–2018 |
| Keith Killingbeck Field Notes | Species records from past surveys of NTPCA | 2014–2018 |
| Baseline Documentation Report for NTPCA | Report written by Scott Ruhren, The Watch Hill Conservancy (2014) | 2014 |
| Rhode Island Natural History Survey Ecological Reconnaissance | Biological survey of NTPCA, K Puryear, 2005 | 2005 |

*Table 1. Sources of data used to compile the plants of Napatree list.*

## RESULTS

The most current list of the plants of Napatree can be found on the NTPCA data portal. The species list as of January 2019 can be found in Appendix 1. An overview of the plant inventory is presented in Table 2.

| Parameter | Number |
|---|---|
| Number of species | 133 |
| Number of genera | 101 |
| Number of families | 44 |
| Non-native species | 48 |

*Table 2. Categories of plant species in the NTPCA. Species considered to be "non-native" were those listed by Haines (2011) as being strictly non-native (i.e., we excluded species listed as having both native and non-native origins).*

## CONCLUSIONS

Our current goal is to create a comprehensive list of flora observed on Napatree Point Conservation Area. We are now photographing as many species as possible and will develop an illustrated guide to the plants of Napatree.


## ACKNOWLEDGMENTS

Special thanks to The Watch Hill Conservancy for its ongoing commitment to the stewardship of the Napatree Point Conservation Area. The plant surveys and restoration work were done with grant support from the Coastal Resources Management Council (CRMC) Coastal and Estuary Habitat Restoration Trust Fund and the URI Coastal Institute.

## REFERENCES:


Haines, A. 2011. *New England Wild Flower Society's Flora Novae Angliae: A Manual for the Identification of Native and Naturalized Higher Vascular Plants of New England*. New Haven, CT: Yale University Press.

Leeson, H., J. Sassi, and P. August. 2017. "Native Vegetation Restoration and Invasive Plant Control in the Napatree Point Conservation Area: 2017." In *The State of Napatree Report: 2017*, edited by J. Sassi. The Watch Hill Conservancy. https://thewatchhillconservancy.org/napatree/napatree-resources/.

Puryear, K. 2005. *Ecological and Land Management Survey of Napatree Beach and Napatree Point, Westerly Rhode Island*. Kingston, RI: Rhode Island Natural History Survey.

Rothwell, A., and H. Ginsberg. 2017. "The Bee Fauna of Coastal Napatree Point and Two Inland Sites in Southern Rhode Island." *Northeastern Naturalist*, 26:446–464.

Ruhren, S. 2014. *Baseline Documentation Report, Napatree Point Conservation Area*. The Watch Hill Conservancy.


***Appendix 1***. *The Plants of the Napatree Point Conservation Area, January 2019*

| Family | Scientific Name | Common Name |
|---|---|---|
| Adoxaceae | *Viburnum dentatum* var. *lucidum* | smooth arrowwood, northern arrowwood |
| Amaranthaceae | *Amaranthus retroflexus* | red-rooted amaranth |
| Amaranthaceae | *Atriplex patula* | spearscale orach |
| Amaranthaceae | *Atriplex prostrata* | hastate-leaved orache |
| Amaranthaceae | *Bassia hirsuta* | hairy smotherweed |
| Amaranthaceae | *Chenopodium album* | lamb's-quarters, white goosefoot |
| Amaranthaceae | *Chenopodium rubrum* | red goosefoot, red coast-blite |
| Amaranthaceae | *Salicornia ambigua* | perennial glasswort |
| Amaranthaceae | *Salicornia depressa* | annual glasswort |
| Amaranthaceae | *Salsola kali* | prickly saltwort |
| Amaranthaceae | *Suaeda calceoliformis* | American sea-blite |
| Amaranthaceae | *Suaeda linearis* | annual sea-blite |
| Amaranthaceae | *Suaeda maritima* | herbaceous sea-blite, white sea-blite |
| Anacardiaceae | *Rhus copallinum* var. *latifolia* | winged sumac |
| Anacardiaceae | *Rhus glabra* | smooth sumac |
| Anacardiaceae | *Toxicodendron radicans* | poison ivy |
| Apiaceae | *Daucus carota* | Queen Anne's lace, wild carrot |
| Apiaceae | *Heracleum maximum* | American cow-parsnip |
| Apocynaceae | *Asclepias incarnata* | swamp milkweed |
| Apocynaceae | *Asclepias syriaca* | common milkweed |
| Apocynaceae | *Cynanchum louiseae* | black swallowwort |
| Aquifoliaceae | *Ilex verticillata* | common winterberry |
| Asparagaceae | *Asparagus officinalis* | asparagus |

| Asteraceae | *Achillea millefolium* ssp. *lanulosa* | American yarrow |
|---|---|---|
| Asteraceae | *Ambrosia artemisiifolia* | common ragweed |
| Asteraceae | *Artemisia stelleriana* | dusty miller, beach-wormwood |
| Asteraceae | *Baccharis halimilfolia* | groundsel-tree, eastern false willow |
| Asteraceae | *Chicorium intybus* | chicory |
| Asteraceae | *Cirsium vulgare* | common thistle, bull thistle |
| Asteraceae | *Erigeron canadensis* | Canada fleabane, horseweed, mare's tail |
| Asteraceae | *Erigeron annuus* | annual daisy fleabane |
| Asteraceae | *Eupatorium hyssopifolium* | hyssop-leaved thoroughwort |
| Asteraceae | *Eutrochium dubium* | coastal plain Joy-Pye weed |
| Asteraceae | *Eutrochium purpureum* | purple Joy-Pye weed |
| Asteraceae | *Iva frutescens* | maritime marsh-elder, hightide bush |
| Asteraceae | *Lactuca serriola* | prickly lettuce |
| Asteraceae | *Leucanthemum vulgare* | ox-eye daisy, white daisy, Marguerite, whiteweed |
| Asteraceae | *Nipponanthemum nipponicum* | Nippon-daisy, Montauk daisy |
| Asteraceae | *Pluchea odorata* var. *succulenta* | sweet-scented camphorweed, saltmarsh fleabane |
| Asteraceae | *Pseudognaphallium obtusifolium* | blunt-leaved rabbit-tobacco |
| Asteraceae | *Solidago nemoralis* | gray goldenrod |
| Asteraceae | *Solidago rugosa* | common wrinkle-leaved goldenrod |
| Asteraceae | *Solidago sempervirens* | seaside goldenrod |
| Asteraceae | *Sonchus arvensis* | field sow-thistle |
| Asteraceae | *Symphyotrichum ericoides* var. *ericoides* | heath American-aster |
| Asteraceae | *Symphyotrichum novae-angliae* | New England American-aster |
| Asteraceae | *Symphyotrichum pilosum* | awl American-aster |
| Asteraceae | *Symphyotrichum racemosum* | small white American-aster |
| Asteraceae | *Symphyotrichum tenuifolium* var. *tenuifolium* | perennial saltmarsh American-aster |
| Asteraceae | *Taraxacum officinale* | common dandelion |
| Asteraceae | *Xanthium strumarium* | rough cocklebur |
| Brassicaceae | *Arabidopsis thaliana* | mouse-ear thale-cress |
| Brassicaceae | *Barbarea vulgaris* | garden yellow-rocket |
| Brassicaceae | *Cakile edentula* | American sea-rocket |
| Brassicaceae | *Cardamine parviflora* | small flowered bitter-cress |
| Brassicaceae | *Lepidium campestre* | field pepperweed |
| Brassicaceae | *Lepidium virginicum* var. *virginicum* | poor-man's pepperweed |
| Brassicaceae | *Raphanus raphanistrum* spp. *raphanistrum* | wild radish, jointed charlock |
| Cactaceae | *Opuntia humifusa* | eastern prickly-pear cactus |

| | | |
|---|---|---|
| Caprifoliaceae | *Lonicera japonica* | Japanese honeysuckle |
| Caprifoliaceae | *Lonicera morrowii* | Morrow's honeysuckle |
| Caryophyllaceae | *Dianthus armeria* | deptford pink |
| Caryophyllaceae | *Honckenya peploides* ssp. *robusta* | seaside-sandwort |
| Caryophyllaceae | *Saponaria officinalis* | common soapwort, bouncing bet |
| Caryophyllaceae | *Stellaria media* | common stitchwort, common chickweed |
| Celastraceae | *Celastrus orbiculatus* | Asiatic bittersweet |
| Cistaceae | *Hudsonia tomentosa* | sand false heather, beach heather |
| Cistaceae | *Lechea maritima* | beach pinweed |
| Convolvulaceae | *Calystegia sepium* | hedge false bindweed, wild morning-glory |
| Cornaceae | *Nyssa sylvatica* | black gum, black tupelo, tupelo |
| Cupressaceae | *Juniperus virginiana* | eastern red cedar |
| Cyperaceae | *Carex* sp. | sedge |
| Cyperaceae | *Cyperus esculentus* var. *leptpstachyus* | nut flatsedge |
| Cyperaceae | *Cyperus grayi* | gray's flatsedge |
| Cyperaceae | *Cyperus strigosus* | straw-colored flatsedge |
| Euphorbiaceae | *Euphorbia polygonifolia* | seaside sandmat |
| Fabaceae | *Lathyrus japonicus* ssp. *maritimus* | beach vetchling, beach pea |
| Fabaceae | *Trifolium pratense* | red clover |
| Hypericaceae | *Hypericum perforatum* spp. *perforatum* | common St. John's-wort |
| Juncaceae | *Juncus gerardii* | saltmarsh rush |
| Lamiaceae | *Teucrium canadense* | American germander |
| Lauraceae | *Sassafras albidum* | sassafras |
| Lythraceae | *Lythrum salicaria* | purple loosestrife |
| Malvaceae | *Hibiscus moscheutos* spp. *moscheutos* | swamp rose-mallow |
| Molluginaceae | *Mollugo verticillata* | carpetweed, green carpetweed |
| Myricaceae | *Morella caroliniensis* | small bayberry, bayberry |
| Myricaceae | *Myrica gale* | sweetgale |
| Onagraceae | *Oenothera biennis* | common evening-primrose |
| Papaveraceae | *Glaucium flavum* | yellow horn-poppy |
| Phytolaccaceae | *Phytolacca americana* | American pokeweed |
| Pinaceae | *Pinus resinosa* | red pine |
| Pinaceae | *Pinus strobus* | eastern white pine |
| Pinaceae | *Pinus thunbergii* | Japanese black pine |
| Plantaginaceae | *Plantago lanceolata* | English plantain |
| Plantaginaceae | *Plantago maritima* ssp. *juncoides* | seaside plantain |
| Plumbaginaceae | *Limonium carolinianum* | sea-lavender |

| Poaceae | *Ammophila breviligulata* | American beach grass |
|---|---|---|
| Poaceae | *Cinna arundinacea* | sweet wood-reed |
| Poaceae | *Distichlis spicata* | saltgrass |
| Poaceae | *Elymus repens* | creeping wild rye |
| Poaceae | *Elymus virginicus* var. *halophilus* | common eastern wild rye |
| Poaceae | *Eragrostis spectabilis* | purple lovegrass |
| Poaceae | *Holcus lanatus* | common velvet grass |
| Poaceae | *Panicum* sp. | panicgrass |
| Poaceae | *Panicum virgatum* | switch panicgrass, switchgrass |
| Poaceae | *Phleum pratense* | common Timothy |
| Poaceae | *Phragmites australis* | common reed |
| Poaceae | *Spartina alterniflora* | saltmarsh cordgrass, smooth cordgrass |
| Poaceae | *Spartina patens* | saltmeadow cordgrass, saltmarsh hay |
| Poaceae | *Spartina pectinata* | prairie cordgrass |
| Polygonaceae | *Fallopia japonica* var. *japonica* | Japanese knotweed |
| Polygonaceae | *Fallopia scandens* | climbing bindweed, false buckwheat |
| Polygonaceae | *Polygonum articulatum* | coastal jointed knotweed |
| Polygonaceae | *Polygonum glaucum* | seaside knotweed |
| Polygonaceae | *Rumex acetosella* | common sheep sorrel, sheep dock |
| Polygonaceae | *Rumex crispus* var. *crispus* | curly dock |
| Rhamnaceae | *Frangula alnus* | glossy false buckthorn |
| Rosaceae | *Amelanchier canadensis* | Canadian serviceberry, eastern shadbush |
| Rosaceae | *Aronia arbutifolia* | red chokeberry |
| Rosaceae | *Prunus maritima* | beach plum |
| Rosaceae | *Rosa multiflora* | multiflora rose, rambler rose |
| Rosaceae | *Rosa rugosa* | beach rose |
| Rosaceae | *Rosa virginiana* | Virginia rose |
| Rosaceae | *Rubus* sp. | blackberry |
| Rubiaceae | *Galium aparine* | scratch bedstraw |
| Scrophulariaceae | *Verbascum blattaria* | moth mullein |
| Scrophulariaceae | *Verbascum thapsus* | common mullein |
| Simaroubaceae | *Ailanthus altissima* | tree-of-heaven |
| Solanaceae | *Datura stramonium* | Jimson-weed, thorn-apple |
| Solanaceae | *Solanum dulcamara* | climbing nightshade |
| Typhaceae | *Typha angustifolia* | narrow-leaved cattail |
| Vitaceae | *Ampelopsis glandulosa*, var. *brevipedunculata* | amur peppervine, porcelain-berry |
| Vitaceae | *Parthenocissus quinquefolia* | Virginia creeper, woodbine |

*Shrub islands in a matrix of beach grass.*
*Photo Credit: URI Environmental Data Center*

CHAPTER 25

# Assessing Shrub Island Dynamics on Napatree Point: 2019 Update

*Jessica Cressman-Greene,*[1,3] *Keith Killingbeck,*[2] *Peter August*[3] *& Janice Sassi*[1]

[1] Napatree Point Conservation Area, The Watch Hill Conservancy

[2] Department of Biological Sciences, University of Rhode Island

[3] Department of Natural Resources Science, University of Rhode Island

## INTRODUCTION

Cressman-Greene *et al.* (2016) reported on expansion of shrub islands in the Napatree Point Conservation Area over a two-year period. They found that islands of small bayberry (*Morella caroliniensis*) were expanding in extent more (mean ± 1 SD; 0.36 ± 0.34 m, 1.2 ± 1.1 feet) than islands of beach rose (*Rosa rugosa*; 0.11 ± 0.12 m, 0.4 ± 0.4 feet) or shrub islands that were a mix of rose and bayberry (0.16 ± 0.14 m, 0.5 ± 0.5 feet). They conducted another survey of shrub patches in the summer of 2017, and those results are reported in Cressman-Greene *et al.* (2018). In the summer of 2019, we did another survey of shrub patches, and those results are reported here.

## METHODS

In 2014, long-term monitoring plots were established by randomly selecting twenty bayberry, twenty rose, and fifteen mixed-species shrub islands. Individual shrub islands were defined as solitary patches of shrubs containing at least one of the two target species and separated by at least 0.3 m (0.9 feet) from another patch of shrubs. The matrix within which shrub islands occurred consisted of American beach grass (*Ammophila breviligulata*). In order to determine island expansion rates, 1.3 cm, (0.5 inches) diameter steel reinforcing rod (rebar) stakes were used to mark the boundaries of each island. One 0.6 m (2 feet) long rebar stake was placed at the farthest extent of each shrub island in each of the four cardinal directions. In the 2017 assessment conducted on 5 September, we measured the distance from each rebar stake to the nearest edge of the shrub island. This provided an estimate of the amount of expansion or contraction of the shrub island that had occurred since 2014. In addition, we measured the maximum height of the vegetation in the shrub island. The same measurements were taken in the 2019 sample conducted on 3

November. Statistical comparisons were done using analysis of variance. Statistical analyses were done using R Commander software in the R computing environment. Plant nomenclature follows Haines (2011).

## RESULTS AND DISCUSSION

In 2017, we found fifty-four of the fifty-five shrub islands. In one case, what had been a mixed-species shrub island was pure rose. In five cases, shrub islands were coalescing with adjacent shrub islands. Shrub islands increased in size in the twelve months between the 2016 and 2017 surveys (Table 1), typically 10–20 cm (4–8 inches). There was very little change in shrub height. There was a trend for most growth occurring in an easterly direction among all patch types, but not significantly so.

| Plot Type | East (± SD) | North (± SD) | South (± SD) | West (± SD) | Height (± SD) | Avg Growth (± SD) |
|---|---|---|---|---|---|---|
| Bayberry (n=9) | 17.5 (± 24.8) | 5.8 (± 35.5) | 10.9 (± 26.0) | 7.0 (± 18.8) | 5.1 (± 24.2) | 10.2 (± 15.3) |
| Mixed (n=19) | 31.5 (± 75.6) | 18.0 (± 16.7) | 18.5 (± 49.0) | 12.7 (± 18.8) | - 4.8 (± 29.2) | 19.5 (± 22.5) |
| Rose (n=18) | 23.4 (± 35.7) | 14.3 (± 23.8) | 20.7 (± 40.8) | 10.0 (± 22.1) | - 1.5 (± 16.6) | 17.2 (± 19.3) |

*Table 1. Twelve-month (2016 to 2017) growth rates in shrub islands on Napatree measured as expansion (positive values) or contraction (negative values) of the 2017 shrub island edge compared to the location of the shrub island edge in 2016. All measurements are in centimeters.*

| Plot Type | East (± SD) | North (± SD) | South (± SD) | West (± SD) | Height (± SD) | Avg Growth (± SD) |
|---|---|---|---|---|---|---|
| Bayberry (n=11) | 123.3 (± 173.0) | 52.2 (± 41.4) | 62.2 (± 58.2) | 87.3 (± 77.0) | 31.5 (± 23.9) | 84.6 (± 76.69) |
| Mixed (n=21) | 51.0(± 82.0) | 26.6 (± 31.2) | 38.3 (± 41.6) | 32.1 (± 26.9) | 2.1 (± 37.2) | 36.5 (± 25.70) |
| Rose (n=19) | 38.4 (± 35.5) | 20.3 (± 26.1) | 28.3 (± 42.2) | 25.5 (± 32.2) | 14.8 (± 18.1) | 28.1 (± 24.39) |

*Table 2. Three-year (2014 to 2017) growth rates in shrub islands on Napatree measured as expansion (positive values) or contraction (negative values) of the 2017 shrub island edge compared to the location of the shrub island edge in 2014. All measurements are in centimeters.*

| Plot Type | Height Difference Since 2014 (± SD) | Avg Growth(± SD) | Growth Rate (cm/year) |
|---|---|---|---|
| Bayberry (n=11) | 16.8 (± 24.4) | 81.5 (± 43.1) | 16.3 |
| Mixed (n=29) | 21.7 (± 16.7) | 73.6 (± 58.1) | 14.7 |
| Rose (n=13) | 9.8 (± 24.1) | 50.9 (± 37.3) | 10.2 |

*Table 3. Five-year (2014 to 2019) growth rates in shrub islands on Napatree measured as expansion (positive values) or contraction (negative values) of the 2019 shrub island edges compared to the location of the corresponding shrub island edges in 2014. All measurements are in centimeters.*

In 2019, we were able to measure expansion/contraction at fifty-three of the fifty-eight shrub patches. There was considerable expansion in shrub islands between 2014 and 2019 (Table 3). Bayberry patches have expanded more than rose or mixed-shrub islands, but not significantly so ($F=1.26$, $P=0.29$). Bayberry and mixed patches showed comparable increases in height but there were no significant differences among the three shrub patch types ($F=1.5$, $P=0.22$).

## CONCLUSIONS

All shrub patch types are getting larger. Since 2014, patches (species pooled) have expanded 60.7 cm (27.5 inches), or an annual rate of patch expansion of 13.9 cm per year (5.5 inches per year). In the 2019 sample, we observed a number of shrub patches that were distinct and isolated from neighbors in 2014 are now coalescing with adjacent shrub patches. We anticipate that shrub island expansion will continue until the next large storm event that exposes shrubs to intense salt spray. We would expect some level of die-back at that time. In the meantime, bayberry, a desirable native species that provides food and cover for wildlife, continues to increase in area in the dune habitat of the Napatree Point Conservation Area.

## ACKNOWLEDGMENTS

We are grateful for the assistance provided by Hope Leeson. The cover photo was provided by Christina Montello (Montello *et al.*, 2017).

## REFERENCES

Cressman-Greene, J., K. Killingbeck, P. August, and J. Sassi. 2016. "Assessing Shrubland Dynamics on Napatree Point, Watch Hill, RI: 2016." In *The State of Napatree Report: 2016*, edited by J. Sassi. The Watch Hill Conservancy. https://thewatchhillconservancy.org/napatree/napatree-resources/.

Haines, A. 2011. *New England Wild Flower Society's* Flora Novae Angliae: *A Manual for the Identification of Native and Naturalized Higher Vascular Plants of New England.* New Haven, CT: Yale University Press.

Montello, C., G. Bonynge, C. Damon, P. August, and J. Sassi. 2017. "Mapping Napatree Point with Small Unmanned Aerial Systems." In *The State of Napatree Report: 2017*, edited by J. Sassi. The Watch Hill Conservancy. https://thewatchhillconservancy.org/napatree/napatree-resources/.

*Habitat restoration volunteers at work.*
*Photo Credit: Hope Leeson*

CHAPTER 26

# Native Vegetation Restoration and Invasive Plant Control in the Napatree Point Conservation Area: 2017

*Hope Leeson,*[1] *Peter August*[2] *& Janice Sassi*[3]
[1] Rhode Island Natural History Survey
[2] Department of Natural Resources Science, University of Rhode Island
[3] Napatree Point Conservation Area, The Watch Hill Conservancy

## INTRODUCTION

Management of the numerous social trails created within the dune system has formed a focal point for the revegetation projects implemented in the Napatree Point Conservation Area (NTPCA). Social trails represent potential areas of degradation during storm events. Efforts have been made to limit the number of trails traversing the dune and to actively revegetate closed trails. In addition, our restoration focus since 2014 has been to increase vegetation diversity to provide a variety of food resources for pollinator and bird species by planting species commonly found on dune systems elsewhere in Rhode Island. Additional stewardship activities have included the control of invasive plant species, with a focus on the less-infested eastern end of the dune system.

Our work in 2017 included Phase III of the revegetation program within closed social trails as well as continued monitoring of the growth and survival of plants installed in Phases I (2014) and II (2015) (see map in Appendix 1). The 2017 planting effort took place in previously planted areas, in part, to compensate for losses due to human disturbance and to implement adaptive management strategies by increasing populations of species that our monitoring results have shown to have greater survival rates on Napatree Point. Monitoring of the growth and health of all planted populations continues to reflect impacts by herbivores and to illustrate the integration of all biological and ecological systems at play in the dune. Invasive species management in 2017 was limited to the continued control of three species: Asiatic bittersweet, tree-of-heaven, and yellow horn-poppy.

## METHODS

### Planting

The Phase III planting event occurred on 8 September and involved a combination of NTPCA and Rhode Island Natural History Survey staff and volunteers (Figure 1*A*). Ten people were involved throughout the day to plant a total of 1,374 plants and to carry water. Planting occurred in four closed social trails and within two back dune areas located on the east end of the NTPCA. The native species mix was comprised of three shrubs, two grasses, and one herbaceous perennial species. Our focus for Phase III was on augmenting previously planted social trails that, either because of harsh environmental conditions or human disturbance, had not begun to fill in as expected. Additionally, species not preferred by herbivores, such as bayberry (*Morella caroliniensis*) were added into the East End back dune area. Species that do experience browse were covered with woody twigs in an attempt to keep browsing at a minimum (Figure 1*B*).

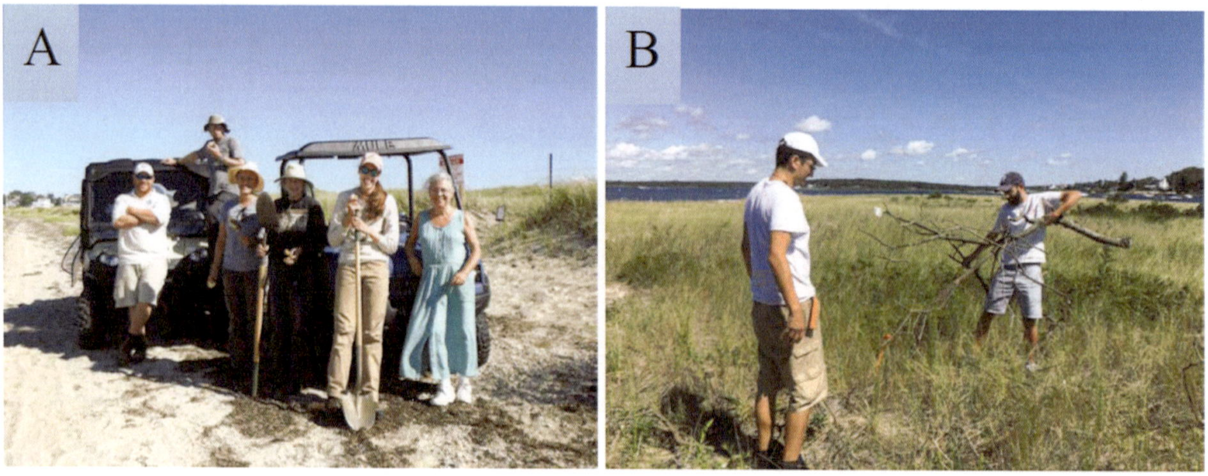


*Figure 1. A, Some Napatree Point planting crew members at the end of the day. B, Volunteers placing woody debris around beach plum (Prunus maritima) seedlings.*

*Photo Credit: The Watch Hill Conservancy*

As with the Phase I and II plantings, no soil amendments were added to the planting holes, and all shrubs were planted directly into the dune substrate. To retain water while roots become established, rice hull ash, a natural by-product of the rice production industry in Louisiana, was added to improve the water-holding capacity of the soil. It is expected that the cooler temperatures and increased rainfall typically experienced in fall will aid in the establishment of the woody planted material. All shrubs were watered after planting.

Plugs of American beachgrass (*Ammophila breviligulata*), saltmarsh hay (*Spartina patens*), and seaside goldenrod (*Solidago sempervirens*) were planted directly into the dune substrate without providing rice hull ash or watering after planting. The planting date occurred a day before a predicted rainfall, which we thought would provide sufficient rain for settling in the plugs. All plants had been well watered the morning of the planting.

| Species Planted in 2017 | | | |
|---|---|---|---|
| | Name | Location | Number Planted |
| Grasses | *Ammophila beviligulata* | Trails 1, 2, 3 | 650 |
| | *Spartina patens* | Trail 3 | 38 |
| Herbaceous Perennial | *Solidago sempervirens* | Trails 1, 2, 3 | 417 |
| Shrubs | *Morella caroliniensis* | East End | 37 |
| | *Prunus maritima* | East End | 33 |
| | *Rosa virginiana* | East End | 32 |

*Table 1. Species planted in Phase III of the NTPCA plant restoration project.*

## Monitoring

In 2015, a subset of woody plants within each planted species populations were tagged and numbered to enable relocation of individual plants from year to year for the purposes of tracking their progress and monitoring survivorship. In 2017, a total of fifty-four tagged plants, representing ten distinct species, were monitored. For each of the tagged plants, it was noted whether it was alive or dead, with a note as to overall health of living individuals. Plant height was measured to the nearest centimeter at the tallest living bud. Impacts to leaves and stems by herbivores was recorded, along with the height at which stem browse occurred (Figure 2A). The density of surrounding vegetation was recorded, along with plant type (shrub, tree, grass, or perennial) estimated. Of additional interest was the proximity within 1 m (3.3 feet), of either bayberry (*Morella caroliniensis*) or beach rose (*Rosa rugosa*). Root and underground stem growth was assessed by quantifying the number of stems (ramets) growing off of the principal root system.

In addition to the individual replicate data points, we collected data on the health and vigor of eighty-eight representative plants from all species populations located throughout the planted areas. For each population of a planted species, the total number of individual plants found to be alive and dead were counted. From the population, a single representative was randomly selected, and data were collected consistent with that described above for the individually tagged plants.

As a part of our 2017 monitoring, individual shrubs were selected as proxies for each species grouping planted in September 2017. Plants were tagged and given a unique number (Figure 2B). Plant height was measured to the nearest centimeter at the tallest living bud. Any impact to leaves and stems by herbivores since the planting date was recorded, along with the height at which stem browse occurred. The density of surrounding vegetation was recorded, along with plant type (shrub, tree, grass, or perennial). Of additional interest was the proximity (within 1 m, 3.3 feet) of either bayberry (*Morella caroliniensis*) or beach rose (*Rosa rugosa*). Root and underground stem growth were assessed by quantifying the number of stems (ramets) growing off of the principal root system.

*Figure 2. A, Beach plum showing evidence of stem browse by white-tailed deer. B, Tagging a representative bayberry (Morella caroliniensis) along an East End social trail for future replicate monitoring.*

*Photo Credit: Hope Leeson.*

Along social trails where perennial grass and herbaceous species were planted as a part of the Phase III planting, the total number of plants installed was recorded, with photographs taken to document the density of plantings within the social trails. The photographs will serve as a visual reference for future monitoring (Figure 3). Since herbaceous stems die back in fall, the plants placed into these trails are inappropriate for our tagging system. Additionally, over time it will be increasingly difficult to discern planted from naturally occurring individuals, so a more meaningful metric will be the overall stem density within the trail as an indication that the trail is no longer in use.

*Figure 3. Northern portion of the East End social trail, filling in naturally with American beachgrass (Ammophila breviligulata), the trail received additional plantings of bayberry (Morella caroliniensis) and seaside goldenrod (Solidago sempervirens).*

*Photo Credit: Hope Leeson.*

## Invasive Species

Invasive species management is an effort that involves consistent, multiple years of treatment and monitoring to successfully eradicate the species' population. Since 2014, management of invasive species within the dune system in the NTPCA involved a combination of mechanical and chemical controls. In all cases, mechanical control has been the first method applied. This has served in many cases to eliminate populations, and for others has reduced the overall biomass of populations requiring chemical herbicide to control growth. In 2017, all invasive species populations that were treated in 2014 and 2015 were monitored to determine the extent of survival and need for continued management. Treatment of persisting populations of invasive species was concentrated on tree-of-heaven (*Ailanthus altissima*) located in the eastern-most portion of the dune (Figure 4A) and Asiatic bittersweet (*Celastrus orbiculatus*). Yellow horn-poppy (*Glaucium flavum*) (Figure 4B) continues to be found above the high tide wrack line along portions of the beach at the western end of the point. Hand-pulling before the long seed pods are ripe is the best method to prevent establishment. For the tree-of-heaven, all above ground stems were cut and left to dry on the dune. A glyphosate-based herbicide was then applied directly to the cut stem surface. Stems of Asiatic bittersweet found along the southern end of planting Trail 3 were hand-pulled and removed from the dune system.

*Figure 4. A, Tree-of-heaven in the eastern end of the dune. B, Yellow horn-poppy (Glaucium flavum).*
*Photo Credit: Hope Leeson.*

## RESULTS AND DISCUSSION

### Monitoring and Planting

At the inception of the dune restoration project, species selected for planting at Napatree were based on three criteria. We chose plant species which: (a) are already present on Napatree Point, (b) exist elsewhere in Rhode Island on dune and coastal lagoon systems, and (c) are adapted to disturbance regimes and readily resprout when aboveground vegetation is compromised. Since 2014, the survival and vigor of the species planted has been the focal point of our annual monitoring efforts. The measure of plant vigor has been plant height and the number of ramets produced. As a form of vegetative reproduction, ramets produced by an individual plant provide a second measure of plant success. Notations as to evidence of sexual reproduction (production of flowers and seeds) for herbaceous and grass species has also provided a meaningful metric of plant health and vigor. The documentation of propagules of species not previously present in the eastern portion of the dune will be the ultimate measure of success for the establishment of those species.

Two herbaceous species—hyssop-leaved boneset (*Eupatorium hyssopifolium*) (Figure 5*A*) and purple love-grass (*Eragrostis spectabilis*) (Figure 5*B*)—that were not previously present in the Napatree dune system have shown high survivorship (in 2017, 75 percent and 40 percent, respectively), are increasing in stem density, and have consistently produced flowers and seeds. Both species are adapted to growing in sandy and poor soils, provide seed for seed-eating birds, and in the case of the boneset, nectar and pollen for insects. Over time, we hope to document additional propagules of both species in the dune, which would reflect establishment success for these two species at Napatree Point.

*Figure 5. A, Multiple stems of mature Hyssop-leaved boneset in fruit, with seed dispersing, in the East Dune. B, Purple love-grass, measuring 20 cm (7.9 inches) across, in fruit. Photo Credit: Hope Leeson.*

Measurements gathered on the proximity to planted individuals of two pre-existing shrub species, bayberry (*Morella caroliniensis*) and beach rose (*Rosa rugosa*), as well as the density of all shrub, herbaceous and grass species stems (largely seaside goldenrod and American beachgrass) has been included in the monitoring effort to determine if adjacent species negatively (by competing for light, water, and nutrients) or positively (by serving as nurse plants and increasing the availability of water, nutrients, and protection from herbivores) impact planted individuals.

Prior to 2017, the data collected did not differentiate between the vigor of plants in close proximity to pre-existing shrubs compared with those placed at a greater distance, or between those placed where grass and perennial vegetation was sparse versus those placed into more densely vegetated areas. As trails have begun to naturally fill in around the planted individuals, the data we collected in 2017 reflected these differing competition scenarios. The data in Figure 6 depict a two-year trend in growth measured from replicate shrubs in 2016 and 2017. Mean height differences (in centimeters) are reported for each species for which we collected data from four or more individuals in both years. Where stem densities across all growth habits (grass, perennial, and shrub) were greater than ten stems within 1 m of the planted individual, shrub height reflects positive growth. Where stem densities were less than ten stems within a 1 m area, height reflects negative growth. Beyond the metric of herbivore browse, it is not possible at this point to say whether the increased stem density is performing in a positive sense for the shrubs, in any way other than providing cover from herbivores.

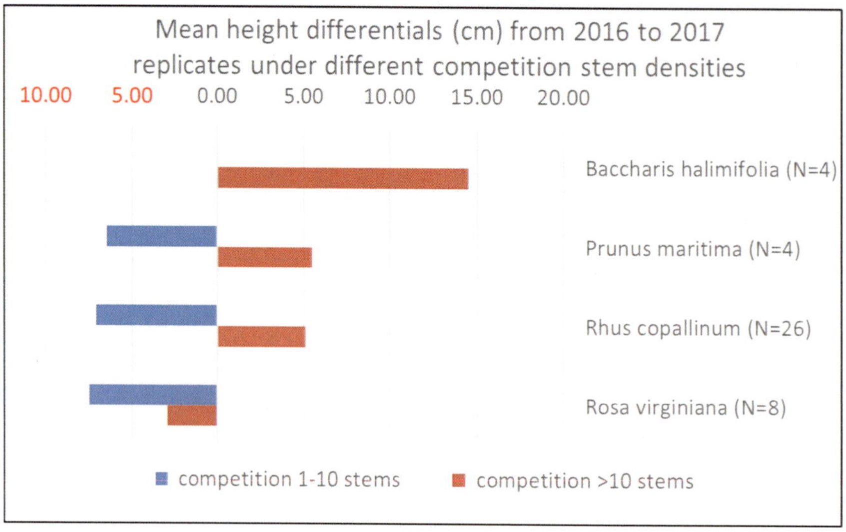


*Figure 6. Relationship between positive and negative growth with competition stem density.*

Notably, Virginia rose (*Rosa virginiana*) experienced negative growth under both scenarios, although to a lesser degree where vegetation was more dense. Herbivore preference for Virginia rose has been observed in all three years of data collection (Figure 7). To date, and despite the preference, survivorship for this species has remained relatively high. Of the seventeen individuals monitored in 2016, thirteen were relocated in 2017. However, 2017 did mark the first year when, anecdotally, it appeared there was an increased preference for the species, which was borne out by our monitoring data and revealed a higher proportion of individuals in the 51 percent to 95 percent consumed category.

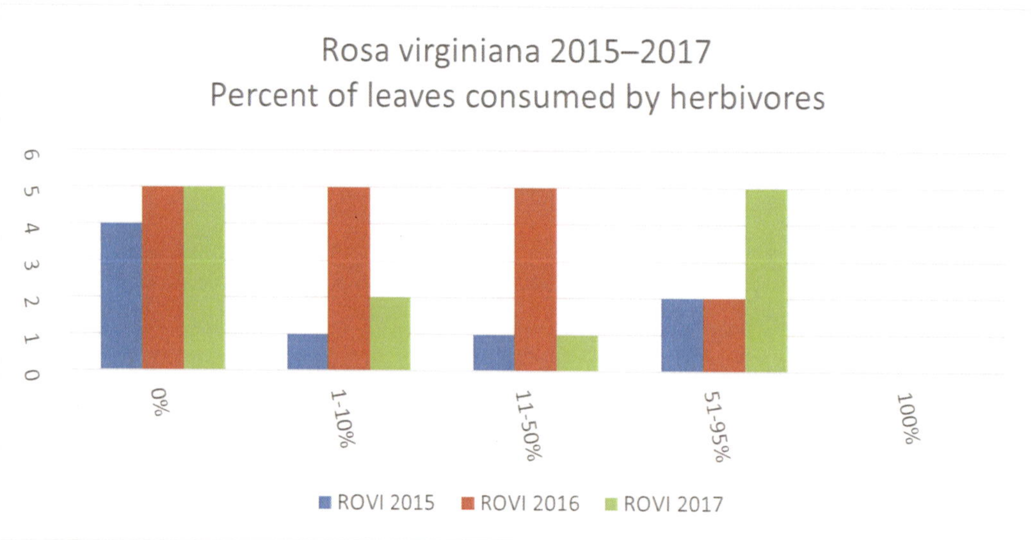


*Figure 7. Herbivore pressure 2015–2017 for Virginia rose (ROVI - Rosa virginiana).*

Browsing on the stems and leaves of the shrub species planted in the NTPCA continues to be a concern for the establishment and survival of the species. Evidence of scat and cut patterns indicate that browsing herbivores include meadow voles, rabbits, and white-tailed deer. For species that are located in open habitat and those that have been shown to represent preferential forage species for mammals, it may be worth considering some sort of temporary caging to allow plants to become established and grow above the height of the animals feeding on them. Figures 8 and 9 show levels of herbivory to leaves and stem browse for all shrub species. It should be noted that although bayberry (MOVI) and groundsel tree (BAHA) showed evidence of browse to 1–10 percent of their leaves, browsing was due to insect larvae rather than mammalian herbivores.

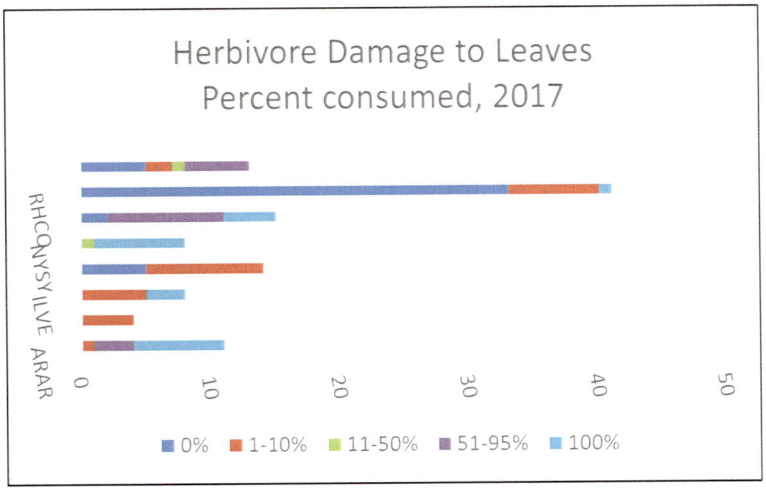


*Figure 8. Herbivore pressure on leaves of all shrub species. Species abbreviations are: ROVI - Rosa virginiana, RHCO - Rhus copallinum, PRMA - Prunus maritima, NYSY - Nyssa sylvatica, MOCA - Morella caroliniensis, ILVE - Ilex verticillata, BAHA - Baccharis halimifolia, and ARAR - Aronia arbutifolia.*

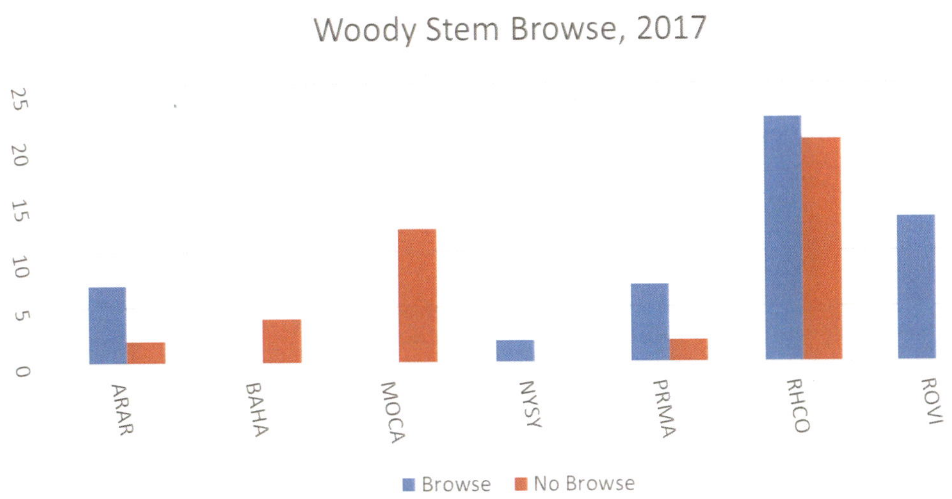


*Figure 9. Herbivore pressure on stems of all leaves. Units are number of plants browsed or not browsed. Plant abbreviations as in Figure 8.*

Overall growth in height is an indicator of plant health and can also be a measure of the degree of herbivore browse pressure over time. Figure 10 shows the average height changes measured from our set of replicate individuals since 2015. Initially, all woody species planted at Napatree Point, except beach plum (*Prunus maritima*), showed an increase in height over the course of the 2015 growing season. However, due to the impacts of browsing in 2016 and 2017, only groundsel tree (*Baccharis halimifolia*) grew taller on average. Beach plum, winged sumac, and black gum all showed insignificant changes in height since 2016, which could indicate a leveling off of herbivore pressure.

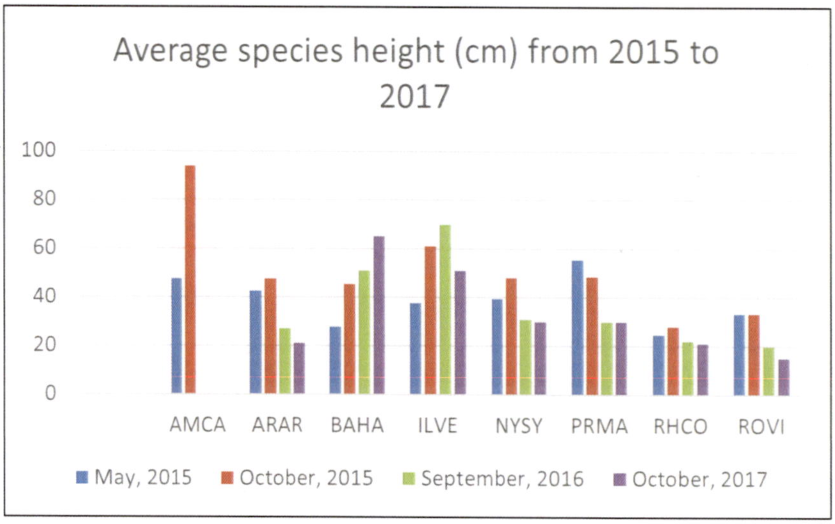


*Figure 10. Growth height (centimeters) over time in the NTPCA. Species abbreviations are: PRMA - Prunus maritima, ROVI - Rosa virginiana, RHCO - Rhus copallinum, NYSY - Nyssa sylvatica, ILVE - Ilex verticillata, BAHA - Baccharis halimifolia, ARAR - Aronia arbutifolia, and AMCA - Amelanchier canadensis.*

In addition to herbivore browse, growth in height can be limited due to environmental stresses, such as drought or droughty soils. Although winged sumac (*Rhus copallinum*) performed relatively well in 2017 based on survivorship data, it is apparent that overall growth in height among the replicate plants sampled has been limited when compared with their growth in 2015 (Figure 10). It is interesting to note that the data collected from all individuals monitored show that on average, where winged sumac did not show significant positive growth, individual growth instead was focused on the production of ramets. Conversely, individuals showing positive growth did not produce ramets. For example, the smallest winged sumac measured 9 cm (3.5 inches) in height and produced six ramets, while the tallest measured 50 cm (19.6 inches) and produced no ramets (Figure 11). The number of ramets produced ranged from zero to as many as seven, with the average being three ramets per plant. In three instances, a single ramet was observed extending 0.25 m (0.8 feet) away from the parent plant. Vegetative reproduction is a common feature of plants, such as winged sumac, that are adapted to stochastic disturbance events. It is their ability to occupy extensive areas via underground stems and the ability to regrow aboveground stems from the underground portion of the plant that facilitates their survival. So although our past two years of data collected from winged sumac height differentials, as compared with 2015 data, is not impressive, the majority of plants monitored did show positive growth of ramets, which reflects an allocation of resources to the production of the underground portion of the plant.

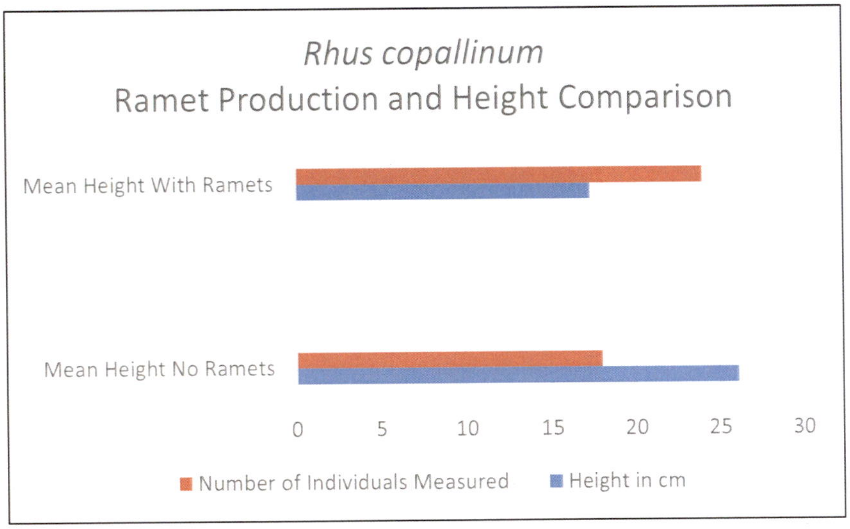


*Figure 11. Relationship between ramet production and plant height.*

Losses of plants over time is inevitable, as stressful environmental factors select for more tolerant individuals, and as we have seen this year with Virginia rose, herbivore preferences change. Since our initial planting efforts in 2014, plants have been lost due to a combination of factors, including changes in salinity and flooding on the salt marsh and at the edges of the lagoon, drought, trampling and removal of plants by humans, and herbivore consumption. The 2016 evaluations of our monitoring revealed that the species exhibiting the least amount of impact were those which were already present at Napatree. In 2017, we implemented a response to this realization through our selection of Phase III species. Species that exhibited a high percentage of mortality, such as shad (*Amelanchier canadensis*), chokeberry (*Aronia arbutifolia*), and black gum (*Nyssa sylvatica*) were not replanted. Table 2 presents overall mortality of species in comparison to the number of individuals planted during all phases of the revegetation project.

| Species | For All Plants | | 2015–2017 |
|---|---|---|---|
| | Alive | Dead | Number Planted |
| *Aronia arbutifolia* | 9 | 6 | 30 |
| *Amelanchier canadensis* | 0 | 8 | 8 |
| *Baccharis halimifolia* | 4 | 0 | 51 |
| *Eragrostis spectabilis* | 29 | 0 | 80 |
| *Eupatorium hyssopifolium* | 30 | 0 | 44 |
| *Ilex verticillata* | 5 | 0 | 8 |
| *Morella caroliniensis* | 37 | 0 | 40 |
| *Nyssa sylvatica* | 1 | 1 | 8 |
| *Prunus maritima* | 52 | 5 | 75 |
| *Rhus copallinum* | 150 | 0 | 200 |
| *Rosa virginiana* | 46 | 0 | 47 |

*Table 2. Mortality of all tree and shrub species planted in the NTPCA.*

Unanticipated changes in environmental factors can bring about unexpected impacts to individual species. In 2017, we experienced a large loss of groundsel tree (*Baccharis halimifolia*). In 2014, a total of fifty-one individual plants were placed along the southwestern edge of the lagoon. Placement was to the water-side of the existing shrub line of beach rose (*Rosa rugosa*) and sand dune. In 2015, all plants were observed to have shown positive growth (Figure 10). In 2016, the species continued to grow in height, and the majority of plants began flowering and producing seed (Figure 12*A*). At that point, it was becoming difficult to discern untagged planted individuals from naturally occurring individuals growing in the same location. Survivorship was recorded at 80 percent, based on data collected from tagged replicates. In 2017, we were able to relocate fewer than 10 percent of the originally planted individuals for monitoring. The loss reflects changes that have apparently occurred in the tidal and storm regime at the southwestern end of the lagoon. In 2017, increased heights of tidal inflow and storm washover added thick layers of eelgrass wrack exactly in the location where many groundsel were planted.

A second unexpected environmental consequence occurred in relation to a portion of the Phase III plantings. In mid-September, Rhode Island experienced a long duration of heavy north winds from Hurricane Jose. As a part of the Phase III planting, plugs of American beachgrass (*Ammophila breviligulata*) had been planted at the top of the dune crest in an attempt to stabilize and vegetate exposed sand. Strong wind funneling through the channel created by the social trail we were hoping to revegetate (located across the high point of the dune in the east end), scoured away 10–12 cm ( 3.9–4.7 inches) of sand. Grasses planted at the dune crest were exposed to the drying wind and sun (Figure 12*B*). However, most survived because of the length of the roots on individual plants, encouraged at the greenhouse by the deep tubes in which the grass seedlings had been grown, and we were able to rebury the exposed plugs.

*Figure 12. A, Planted groundsel tree (Baccharis halimifolia) in fruit along the lagoon in 2016. The same location is now covered with eelgrass wrack. B, American beachgrass (Ammophila breviligulata) plugs exposed by heavy winds during Hurricane Jose in 2017.*

*Photo Credit: Hope Leeson*

## Measuring Project Goals

In addition to the mortality, growth, and reproduction of the plants placed into the Napatree dune system, there exist two additional measures of success for the revegetation project: (a) the question of whether closed social trails are becoming less visible over time due to the introduction and growth of supplemental vegetation and (b) whether the increased diversity of species is providing pollinator and bird species with additional food resources.

To evaluate the first metric, we have compiled photographs taken at the outset of the project along with those taken annually while monitoring. Figures 13, 14, and 15 illustrate significant changes in planted areas as a result of a combination of plantings, reduced human disturbance, and the natural ability of the dune to revegetate itself.

*Figure 13. A, Plants ready for installation along the First Trail in 2014. B, The trail in 2017, filling in with hyssop-leaved boneset and naturally seeded sedges and American beachgrass.*

*Photo Credit: The Watch Hill Conservancy*

*Figure 14. A, The First Trail in 2015 after one year of growth. B, The First Trail in 2017, filling in with winged sumac and naturally expanding colonies of American beachgrass.*

*Photo Credit: The Watch Hill Conservancy*

*Figure 15. A, Trail 2 beginning to fill in in 2016, after being planted in 2015. B, Trail 2 in 2017, filling in with winged sumac, seaside goldenrod, and naturally expanding colonies of American beachgrass.*

*Photo Credit: The Watch Hill Conservancy*

There remain some difficult areas, such as the dune crest at the southern end of the east end social trail, where wind and deep sands make revegetation a challenge, and along portions of Trails 1, 2, and 3 where beachgoers appear to continue to disturb the sand and impact the revegetation process. Supplemental plants were placed along these trails in 2017 to fill in the trail, and signage and roping have been fortified to discourage foot traffic. Human impacts may remain problematic (it is in this area that in 2016 and 2017 we observed that plants were pulled out of the ground and strewn aside) and may be diminished by more outreach and education on the importance of dune integrity.

The second metric, that of additional food resources being provided by an increase in the diversity of species, has a longer time schedule for measurement, and will depend on the survival and reproductive capabilities of the novel, and augmented numbers of pre-existing, shrub and herbaceous species placed into the Napatree dune system. Species which seem likely to survive at this point, and which should be observed for evidence of contribution to the ecosystem, are hyssop-leaved boneset, seaside goldenrod (Figure 16), purple love-grass, winged sumac, Virginia rose, and beach plum. In 2017, University of Rhode Island graduate student Aya Rothwell spent time observing bee species in the NTPCA and documenting the plant species on which they were collecting nectar and pollen. Her data should shed some light in this regard.

## Invasive Species Management

Several invasive species continue to be present in the eastern portion of the dune system. With the exception of tree-of-heaven, overall numbers are diminishing. Three of the eight species targeted for treatment have been controlled and showed no regrowth in 2017 (Table 3). This includes a population of Japanese knotweed (*Fallopia japonica*), which although a single stem sprouted in June of 2016 and died back by July, no stems were observed in 2017. The single plant of Morrow's honeysuckle (*Lonicera morrowii*) treated in 2014 has not resprouted, nor has a population of porcelain berry (*Ampelopsis brevipedunculata*). The single large glossy buckthorn (*Frangula alnus*) continues to resprout from the root system and will require treatment in 2018 to keep its growth under control. Additionally, tree-of-heaven populations will need focused management effort in 2018, as they appear to be rebounding within the locations where individual stem and root systems were pulled in 2015. For both glossy buckthorn and tree-of-heaven, it would be desirable to apply a combination of mechanical and chemical treatments. The single population of black swallowwort (*Cynanchum louiseae*) has been managed by hand-pulling the stems and roots, although this activity should continue annually to prevent it from becoming more widespread. Seedpods, which are dispersed by the wind, should be picked off before they mature to prevent further spread of the species in the east dune. Populations of Asiatic bittersweet (*Celastrus orbiculatus*) will also need to be monitored and managed either through hand-pulling or by cutting stems and applying a glyphosate-based herbicide. Yellow horn-poppy (*Glaucium flavum*) is an annual species with buoyant and salt tolerant seeds. The species will likely continue to wash ashore from other locations in the area and germinate. The beachfront should be monitored annually in early summer when the plant is in bloom and hand-pulled to prevent seed dispersal.

*Figure 16. A, Documenting pollinators along the First Trail in the east end. B, Bee and wasp species on seaside goldenrod.*

*Photo Credit: The Watch Hill Conservancy*

*C, Monarch butterfly feeding on the nectar of seaside goldenrod.*

*Photo Credit: Hope Leeson*

| Species Name | Common Name | # of Populations 2014 | # of Populations 2015 | # of Populations 2016 | # of Populations 2017 |
|---|---|---|---|---|---|
| *Ailanthus altissima* | tree-of-heaven | 5 | 2 | 2 | 3 |
| *Celastrus orbiculatus* | Asiatic bittersweet | 8 | 5 | 3 | 3 |
| *Cynanchum louiseae* | black swallowwort | 1 | 1 | 1 | 1 |
| *Glaucium flavum* | yellow horn-poppy | 3 | 1 | 1 | 1 |
| *Lonicera morrowii* | Morrow's honeysuckle | 1 | 0 | 0 | 0 |
| *Fallopia japonica* | Japanese knotweed | 1 | 0 | 0 | 0 |
| *Frangula alnus* | glossy buckthorn | 4 | 2 | 1 | 1 |
| *Ampelopsis brevipedunculata* | porcelain berry | 0 | 1 | 0 | 0 |

*Table 3. Invasive plant species managed within the eastern dune and western beach communities in the NTPCA, 2014–2017.*

## CONCLUSIONS

Once again, we see the demonstrable importance of monitoring to inform and provide feedback on our restoration and stewardship efforts. The set of data gathered revealed that browsing by herbivores remains an important consideration in terms of plant survival and affirmed the resiliency and adaptability of several species. The increase in ramets produced by winged sumac is very encouraging, as is the vitality and reproduction of hyssop-leaved boneset and purple love-grass.

In addition to documenting the results of changes made to a system, monitoring and data collection is a way of facilitating concentrated observation and building a greater understanding of an ecological system. Of course, along with increased awareness of the system's intricate workings come more questions. For example, is the positive growth evidenced by individual plants growing in areas of surrounding stem density merely the result of physical camouflage, or could the shading capabilities of an increased density of stems retain more soil moisture and lower soil temperatures, thereby providing a mutual benefit to all plants? And what factors influence the apparent allocation of energy by winged sumac individuals toward either ramet production or vertical height?

Additionally, the losses of groundsel tree suggest a need for greater understanding of the effects of sea level rise and its impact on soil salinity and hydrology. Of particular interest is its effects on the composition of the shrub-scrub habitat located to the west and southwest of the lagoon. How will the vegetation respond to salinity changes over time? How will the transition progress? As we have seen, the system is constantly changing.

The experiences gained through this work continue to be of interest to organizations, such as the Natural Resources Conservation Service, United States Fish and Wildlife Service, Rhode Island Department of Environmental Management Division of Fish and Wildlife, University of Rhode Island Coastal Institute, Rhode Island Natural History Survey, and the Rhode Island Wild Plant Society. Over time, it is hoped that we will be able to share further knowledge gained through the restoration of social trails and observation of change on the Napatree Point dune system with all of these organizations.

## ACKNOWLEDGMENTS

Fieldwork for monitoring vegetation management at Napatree Point in 2017 was supported through a combination of funds and assistance provided by the Coastal Resources Management Council Habitat Restoration Trust Fund, University of Rhode Island Coastal Institute, Watch Hill Fire District, and The Watch Hill Conservancy. Staff of the Napatree Point Conservation Area provided valuable field assistance and logistical support. We are sincerely grateful for the support that has been generously provided by Grant Simmons and Judith Swift.

***Appendix 1.*** *Map of Napatree Point Conservation Area Restoration Sites*

*Crew members from the Youth Conservation League.*
*Photo Credit: Melissa Cote*

CHAPTER 27

# Invasive Plant Control in the Napatree Point Conservation Area: 2019

*Melissa Cote,*[1] *Hope Leeson,*[2] *Janice Sassi*[1] *& Kevin Rogers*[1]
[1] Napatree Point Conservation Area, The Watch Hill Conservancy
[2] Rhode Island Natural History Survey

## INTRODUCTION

Invasive plants are non-native species that readily reproduce, have the potential for rapid growth, and once established, can be harmful to native species in an ecosystem. Invasive plants have the potential to outcompete native plants for space and resources, thus altering the ecosystem, which can negatively impact other native organisms. The resulting condition at Napatree Point Conservation Area (NTPCA) is one of continued routine management. Our current focus for management of invasive species on NTPCA is within the maritime herbaceous dune system, which is located between the entrance of NTPCA and the lagoon (August *et al.*, 2016). A total of thirteen invasive plant species have been found in NTPCA's dune system since 2014 and one species, yellow horn-poppy, on the rocky beach of the western point (Table 1). Since 2014, we've used a combination of mechanical and chemical treatments to prevent invasive plants from becoming widespread and altering the dune system.

| Common Name | Scientific Name | Plant Type |
|---|---|---|
| Asiatic bittersweet | *Celastrus orbiculatus* | woody perennial vine |
| black swallowwort | *Cynanchum louiseae* | herbaceous perennial vine |
| climbing nightshade* | *Solanum dulcamara* | herbaceous perennial vine |
| common reed* | *Phragmites australis* | grass |
| crab apple* | *Malus* sp. | tree |
| glossy false buckthorn | *Frangula alnus* | shrub |
| Japanese knotweed | *Fallopia japonica* | herbaceous perennial |
| Montauk daisy* | *Nipponanthemum nipponicum* | herbaceous perennial |
| Morrow's honeysuckle | *Lonicera morrowii* | shrub |
| multiflora rose | *Rosa multiflora* | shrub |
| porcelain berry | *Ampelopsis brevipedunculata* | woody perennial vine |
| sycamore maple* | *Acer pseudoplatanus* | tree |
| tree-of-heaven | *Ailanthus altissima* | tree |
| yellow horn-poppy | *Glaucium flavum* | herbaceous perennial |

*Table 1. List of invasive plant species found in the NTPCA dune system since 2014.*
**Denotes a newly found invasive plant species from the 2018 survey.*

## METHODS

### Management 2014 to 2017

Invasive species management is a multiyear effort consisting of continuous treatment and monitoring. Management of invasive plants within NTPCA's dune system began in 2014 with a focus on the less-infested eastern end of the barrier beach (Leeson *et al.*, 2014). Treatment in 2014 and 2015 consisted of mechanical and chemical methods, with mechanical treatment being the first method of control (Leeson *et al.*, 2015). Monitoring in 2017 of all previously treated invasive species populations showed diminishing numbers with the exception of tree-of-heaven (*Ailanthus altissima*) (Figure 1A) and Asiatic bittersweet (*Celastrus orbiculatus*) for which further treatment was administered. For tree-of-heaven, all aboveground stems were cut and left to dry on the dune. A glyphosate-based herbicide was then applied directly to the cut stem surface. Stems of Asiatic bittersweet were hand-pulled and removed from the dune system. Yellow horn-poppy (*Glaucium flavum*) (Figure 1B) found just above the high tide wrack line on the western point of NTPCA was also a noted concern. Hand-pulling before the long seed pods are ripe is the best method to prevent establishment of yellow horn-poppy (Leeson *et al.*, 2017).

A

B

*Figure 1. A, Tree-of-heaven (Ailanthus altissima) in the eastern end of the dune. B, Yellow horn-poppy (Glaucium flavum).*

*Photo Credit: Hope Leeson*

## Management 2018 and 2019

On 26 October 2018, we conducted an invasive plant species survey of the NTPCA dune system. The dune was walked from east to west from the beach entrance to the lagoon, which more than doubled the previous survey area (Appendix 1). The area of the 2018 survey is shown in Figure 2. Locations of each population of invasive plant were recorded with a Garmin etrex20 and input into ArcGIS Desktop and mapped (Figure 2). We noted that the population of tree-of-heaven on the eastern end of the dune appeared to be spreading.

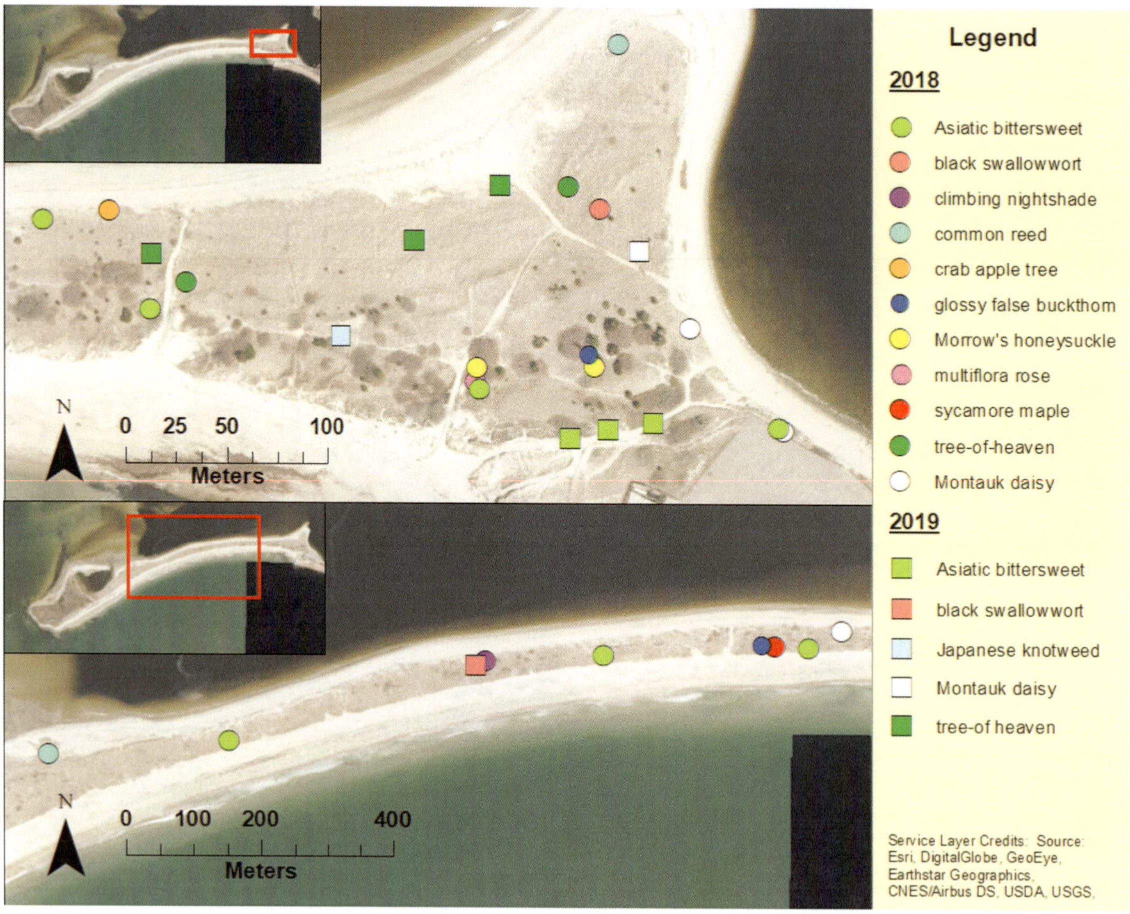


*Figure 2. Survey area for 2018 and locations of invasive plant species found.*

In April 2019, stems of the tree-of-heaven population were cut and left to dry on the dune in an attempt to contain the population until herbicide could be applied. The Rhode Island Youth Conservation League (YCL) joined us in July 2019 to help with mechanical control of the remaining invasive plants found on the 2018 survey. We observed that the tree-of-heaven population cut in April had regrown. The stems were cut again and left on the dune to dry. Smaller trees (~ <1-inch diameter) were dug up; their roots and rhizomes were followed and disposed of. Mechanical treatments administered to the other invasive plants are detailed in Table 2.

| Common Name | Scientific Name | Manual Treatment |
|---|---|---|
| Asiatic bittersweet | *Celastrus orbiculatus* | • Large vines were cut at the base and left on the dune to dry<br>• Small vines were dug up and the roots disposed of |
| black swallowwort | *Cynanchum louiseae* | • Seed pods removed and disposed of<br>• Dug up, especially barbed roots, and disposed of |
| climbing nightshade | *Solanum dulcamara* | • Hand-pulled and disposed of due to many ripe berries on the vine |
| common reed | *Phragmites australis* | • Hand-pulled and dug up as much of the roots and rhizomes as possible and disposed of |
| crab apple | *Malus* sp. | • Removed with a weed wrench and disposed of |
| glossy buckthorn | *Frangula alnus* | • Berries were removed and disposed of<br>• Roots were removed with a weed wrench and disposed of |
| Montauk daisy | *Nipponanthemum nipponicum* | • Dug up (shallow root ball) and disposed of |
| Morrow's honeysuckle | *Lonicera morrowii* | • Berries were removed and disposed of<br>• Roots were removed with a weed wrench and disposed of |
| multiflora rose | *Rosa multiflora* | • Removed with a weed wrench and disposed of |
| sycamore maple | *Acer pseudoplatanus* | • Dug up (was small) and disposed of |
| tree-of-heaven | *Ailanthus altissima* | • Trees with stems >2.5 cm (1 inch) diameter were cut and left to dry on the dune<br>• Trees with stems <2.5 cm (1 inch) diameter were dug up; roots were disposed of |

*Table 2. Mechanical treatments administered to each invasive plant population in July 2019.*

## RESULTS AND DISCUSSION

Invasive species continue to be present in the NTPCA dune system. With the exception of porcelain berry, the populations of invasive species in NTPCA increased from 2017 to 2019 (Figure 3). These results are likely influenced by the increase in the survey region and due to the additional help from the YCL in 2019. With more people searching the dune system there was greater potential for locating invasive plants that had possibly been missed in past surveys. Additionally, no mechanical or chemical treatments were administered in 2018, which would have allowed populations more time to rebound and possibly spread after treatment in 2017.

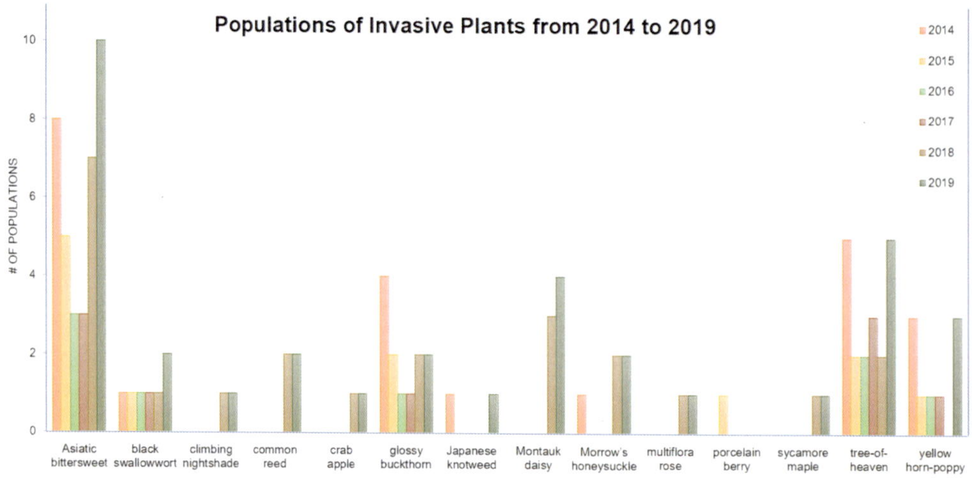


*Figure 3. The number of populations of each invasive plant species in the NTPCA dune system from 2014 to 2019.*

Asiatic bittersweet (*Celastrus orbiculatus*) and tree-of-heaven (*Ailanthus altissima*) continue to be the most widespread and have the largest number of populations (Figures 2 and 3). The tree-of-heaven populations have consistently regrown following mechanical cutting. Populations also appear to be expanding. A concentrated effort to manage Asiatic bittersweet and tree-of-heaven is necessary to hinder their population growth and expansion. Yearly mechanical treatment through hand-pulling smaller plants and cutting larger stems is necessary to control these populations. Additional chemical treatment via application of a glyphosate-based herbicide to cut stems is desirable and has been scheduled for spring 2020.

The single population of black swallowwort (*Cynanchum louiseae*) in the eastern dune continues to resprout and has been managed by hand-pulling the stems and roots on a yearly basis. That effort should continue. A second population of black swallowwort was found in the western portion of the dune in 2018 (Figure 2) and should also be managed by annual hand-pulling. Additionally, for both black swallowwort populations the wind-dispersed seed pods should be removed before they mature to prevent the species from becoming more widespread in the dune system. The single glossy buckthorn (*Frangula alnus*) located in the eastern dune continues to resprout, and a second tree was found farther west in the dune system in 2018 (Figure 2). Both trees were removed with a weed wrench in 2019. However, if they show signs of regrowth, chemical treatment should be administered. The treatment of the single Japanese knotweed (*Fallopia japonica*) population in 2015 was successful until October 2019 where a single stem was found at the location of the original population. The new stem needs close monitoring and chemical treatment if it continues to grow. Two plants of Morrow's honeysuckle (*Lonicera morrowii*) were found in 2018 (Figure 2), and both were removed with a weed wrench in 2019. These locations will be monitored for regrowth. The single population of porcelain berry (*Ampelopsis brevipedunculata*) has not resprouted since it was hand-pulled in 2014; however, its original location should still be monitored during surveys. Yellow horn-poppy (*Glaucium flavum*) was not found in our survey area, but it has been found on the rocky shore of the western point. It is an annual species with buoyant and salt tolerant seeds. The species will likely continue to wash ashore from other locations in the area and germinate. The beachfront should be monitored annually in early summer when the plant is in bloom and hand-pulled to prevent seed dispersal.

The identification of several new invasive plant species during the 2018 survey are likely due to the expansion of the survey region. Exceptions to this are Montauk daisy (*Nipponanthemum nipponicum*), multiflora rose (*Rosa multiflora*), and one population of common reed (*Phragmites australis*), all of which had been found in the original survey area and were determined to be new locations as of 2018. The climbing nightshade (*Solanum dulcamara*) should be monitored, as it can regrow from any roots left behind. Any new plants should be hand-pulled to prevent it from spreading. Although there is an established population of common reed on the western point of NTPCA, we are trying to prevent any populations from becoming established east of the lagoon. Currently, there are two populations that we are managing through hand-pulling and digging up the roots and rhizomes (Figure 2). These populations require annual management to prevent them from spreading. The single crab apple tree (*Malus* sp.) could have been leftover from the houses that originally occupied NTPCA prior to the hurricane of 1938. We may have missed it in the past since it was only a few feet tall and not a flourishing tree. After removing it with a weed wrench in July 2019, it is unlikely that the crab apple tree will regrow. Montauk daisy is an increasing concern as the number of plants on the dune has increased over time (Figures 2 and 3). These plants are easily removed by hand-digging as they have a shallow root ball—a practice that should continue when new plants or resprouts are found. The single sycamore tree found was a sapling that was hand-pulled and hopefully will not regrow.

## CONCLUSIONS

Managing invasive plant species in NTPCA is important to protect the ecological integrity of the dune system and its native plant species. Management activities, including monitoring and treatment, should be done on an annual basis to control the current invasive plant populations, prevent them from becoming widespread, and to locate any new invasive plant species. Specifically, mechanical treatment of tree-of-heaven and Asiatic bittersweet should be done annually to prevent the populations from expanding. We plan on administering chemical treatment to both tree-of-heaven and Asiatic bittersweet populations by applying herbicide directly to cut stems in the spring of 2020. This activity will be done by a licensed herbicide applicator and in a manner that will prevent harm to surrounding native vegetation. Herbicide will also be applied in the same manner to the glossy buckthorn or Japanese knotweed populations if there is evidence of regrowth.

In the future, it would be interesting to quantify invasive plants on NTPCA in other metrics besides the number of populations. For example, if we quantified biomass during mechanical removal, it would provide us with another metric and a better understanding of the extent of our invasive plant populations. Additionally, since many of the tree-of-heaven populations contain more than one stem, we could calculate stem density before and after treatment to determine if the treatments are effective.

## ACKNOWLEDGMENTS

Special thanks to the Youth Conservation League for their help with field work. Funding for this project was awarded by the Rhode Island Coastal Resource Management Council from the Coastal and Estuarine Habitat Restoration Trust Fund and the University of Rhode Island Coastal Institute.

## REFERENCES

August, P., J. Sassi, K. Rogers, and J. Cressman Greene. 2016. "The Geography of the Napatree Point Conservation Area (2016)." In *The State of Napatree Report: 2016*, edited by J. Sassi. The Watch Hill Conservancy. https://thewatchhillconservancy.org/napatree/napatree-resources/.

Leeson, H., P. August, and J. Sassi. 2017. "Native Vegetation Restoration and Invasive Plant Control in the Napatree Point Conservation Area: 2017." In *The State of Napatree Report: 2017*, edited by J. Sassi. The Watch Hill Conservancy. https://thewatchhillconservancy.org/napatree/napatree-resources/.

Leeson, H., S. Ruhren, D. Gregg, J. Sassi, P. August, J. Killian, and J. Cressman. 2014. "Native Vegetation Restoration and Invasive Plant Control in the Napatree Point Conservation Area: 2014." In *The State of Napatree* Report: *2014*, edited by J. Sassi. The Watch Hill Conservancy. https://thewatchhillconservancy.org/napatree/napatree-resources/.

***Appendix 1.*** *The 2015 survey area and locations of invasive plant species found.*

*Eelgrass mapping team on RV Simmons.*
*Photo Credit: The Watch Hill Conservancy*

CHAPTER 28

# Eelgrass or Macroalgae: An Assessment of Submerged Aquatic Vegetation Mapping Technologies*

*Peter August,[1] Michael Bradley,[1] Christopher Damon,[1] Bryan Oakley,[2] Grant Simmons,[3] Melissa Cote,[3] Nicole Rohr,[4] Caitlin Chaffee[5] & Lindsay Green-Gavrielidis[6]*

[1] Department of Natural Resources Science, University of Rhode Island

[2] Department of Environmental Earth Science, Eastern Connecticut State University

[3] Napatree Point Conservation Area, The Watch Hill Conservancy

[4] Coastal Institute, University of Rhode Island

[5] Rhode Island Coastal Resources Management Council

[6] College of the Environment and Life Sciences, University of Rhode Island

## INTRODUCTION

Rhode Island's largest contiguous patch (33.3 ha, 82.4 acres) of eelgrass (*Zostera marina*) occurs between the Napatree Point Conservation Area and Sandy Point in Little Narragansett Bay. Eelgrass is a critical habitat for fish and shellfish (Bradley *et al.*, 2013). Because of the ecological importance of eelgrass, habitat restoration and monitoring has been an important activity in the state for two decades. Montello *et al.* (2017) experimented with sUAS (small unmanned aircraft system), or drone, imagery to determine if low altitude, high-resolution aerial photography would be a practical way to map eelgrass. They found that the sUAS imagery showed exceptional detail and texture in eelgrass beds. It appeared to be possible to distinguish eelgrass from macroalgae in the patch she experimented with.

Large areas of macroalgae can occur within or adjacent to eelgrass beds, and the two can be difficult to distinguish, especially when the source imagery does not have high spatial resolution and detail. In 2017, Bradley mapped eelgrass off the Connecticut coastline from Watch Hill to New London. The imagery collected by aircraft for that project was superb, and Bradley found a decrease in overall area of eelgrass in the region. He surmised that in many locations, eelgrass had not declined, but instead, earlier mapping

* Originally published 2018

delineations had mistakenly included large patches of macroalgae. The similar visual signatures between eelgrass and macroalgae can result in inaccurate estimates of the extent of this important habitat.

The purpose of this study was to obtain detailed imagery of a 4 ha (10-acre) patch of eelgrass and macroalgae that sits off the northwest corner of Sandy Point. We used a variety of imaging technologies to discern where eelgrass/macroalgae boundaries were, and when combined with the ground truth data, we hoped to identify unique spectral signatures in drone photography to distinguish these two submerged habitats.

*Figure 1. Location of the eelgrass patch studied in this project.*

## METHODS

### Study Site

The 4 ha (10-acre) patch of eelgrass we studied is located off the northwestern corner of Sandy Point (Figure 1). Examination of the 2017 aerial photography obtained for the Long Island Sound eelgrass mapping project (sponsored by the United States Fish and Wildlife Service, USFWS) showed this site to be a complex mosaic of macroalgae and eelgrass. The water depth at this location was less than 3.5 m (11.5 feet).

### Imaging

sUAS—A DJI Phantom 4 Pro sUAS with a 20-megapixel camera was used for this project. A dual strategy was employed for image collection: Flight 1 would capture a series of still images in an aerial survey grid pattern (Figure 2) that would be mosaicked together in a traditional manner, while Flight 2 would record several minutes of video footage to assess the efficacy of sUAS video for the identification and delineation of submerged seagrasses. For both flights the effects of glint, or sunlight reflecting from the water's surface, were minimized to provide maximum light penetration by limiting flights to an operating window when the sun's angle of incidence was between 20 and 38 degrees above the horizon.

Flight planning and control for the aerial grid survey was accomplished using the Map Pilot App for iOS. This third-party software managed all aspects of the survey, including flight speed and camera triggering. Parameters within the software were configured to achieve 75 percent overlap of adjacent images in the forward direction and 70 percent side-lap between consecutive flight lines. For video capture, the sUAS was manually piloted using the standard DJI Go App for iOS. All flights were flown at an altitude of 120 m (394 feet) with a forward speed of approximately 9.6 m/s (31.5 feet/s). Post-flight image processing was done using Pix4Dmapper software.

*Figure 2. Flight plan for sUAS imaging of the study site. The pilot's position is shown by the purple dot.*

Underwater Video—A SeaViewer HD Pro Package underwater video system with GPS and record capability (www.seaviewer.com) was towed along transect lines through the eelgrass patch using a 5.5 m (18-foot) Zodiac motorboat (Figure 3). Transects were chosen to pass through areas of eelgrass, macroalgae, and sand. Video images were stamped with a geographic coordinate. The geographic coordinates along the track the camera was towed were recorded and entered into a GIS. The imagery was used to identify the submerged habitat types along the transects. These data were entered into a GIS so known habitats could be evaluated against the spectral signature of imagery. Macroalgae in the video images were identified by Lindsay Green-Gavrielidis (Appendix 1).

*Figure 3. Transects used for the underwater video recording. Colored symbols indicate the submerged habitat type at the location.*

Diver-based imagery—Divers with GoPro cameras (GoPro Hero 3, GoPro Hero 5) swam the eelgrass patch and recorded images of the macroalgae and eelgrass that were present.

2017 USFWS Aerial Photography—In support of the Long Island Sound eelgrass mapping project in 2017, the USFWS commissioned new imagery that was obtained on 21 June 2017. The aircraft flew at an altitude of 5,181 m (17,000 feet). The resulting imagery had a pixel size of 25 cm (9.8 inches).

## RESULTS

The underwater video camera did an excellent job of recording submerged habitats along the transects. The resolution of the video was exceptional, and habitat types were easily discernable (Appendix 1). The resulting map of verified bottom habitats superimposed on the aerial photography allowed us to identify the differences in the spectral signatures between eelgrass and macroalgae (Figure 3). The macroalgae were uniform in color and very dark, while the eelgrass was mottled in texture and lighter in color than the macroalgae.

The GoPro imagery allowed careful examination of different bottom habitats (Appendix 2). The video recordings from the GoPro, while not the high resolution of the underwater video system described above, allowed divers to swim transects that encompassed different bottom types. Unfortunately, we were unable to tag GoPro images with a geographic coordinate so we could identify where an image was taken on the map. We hope to remedy this during the next monitoring session by towing a GPS receiver along the water's surface and by using the time stamp in the GPS and the photo metadata to geolocate each image.

The sUAS still imagery did not produce reliable results. As found earlier (Montello *et al.*, 2017), creating an image mosaic over water is difficult, as water lacks distinct points that can be used to tie together overlapping images. Image mosaicking was acceptable close to shore where land features could be used to tie images together, but as the distance from shore increased, mosaicking became difficult. The sUAS video recording was clear and of high resolution, allowing the eelgrass and macroalgae patches to be easily distinguished (Figure 4). The video flight took less than thirty minutes. We feel this will be a very promising method to pursue for eelgrass mapping. Software exists (esri Full Motion Video) that allows the digitization of information off from video frames (e.g., eelgrass patches), resulting in a planimetrically correct polygon that can be used in a GIS. The sUAS video of the eelgrass patch can be viewed at: https://youtu.be/DFF2ZmdkYvQ.

*Figure 4. Screenshot from video recorded by the sUAS.*

## CONCLUSIONS

This was the first attempt to gain insight on how best to monitor relatively small, local eelgrass patches in a cost-effective manner that capitalizes on recent image acquisition technologies. These experiments provided valuable information as to methodologies we want to further pursue in the future. We confirmed that sUAS-derived still photography is not a promising methodology to continue because of the challenges related to mosaicking images. sUAS-derived video, on the other hand, was an excellent image collection methodology. Once we develop a workflow to digitize from the video to create a planimetrically accurate polygon, this will be a rapid and accurate method to map eelgrass. The underwater video system provided reliable ground-truthing on the distribution of eelgrass and macroalgae. Georeferencing of these images allowed us to compare known submerged vegetation patches with spectral signatures on the imagery. One drawback of the underwater video system is relatively high cost (~ $7,000) when compared with handheld instruments such as GoPros. The GoPro camera systems produced very good still and video images of the eelgrass and macroalgae habitats; when we develop a method to georegister the imagery, this will be a rapid and inexpensive means to ground-truth presumed patches of submerged aquatic vegetation to use in place of the more expensive underwater video system.

## ACKNOWLEDGMENTS

This project was supported by the University of Rhode Island Coastal Institute, the USDA Renewable Resources Extension Act, The Watch Hill Conservancy, and the URI College of the Environment and Life Science Extension Program in Natural Resource Conservation and Management.

## REFERENCES

Bradley, M., M. Ekberg, and P. August. 2013. "Eelgrass in Little Narragansett Bay." In *The State of Napatree Report: 2013*, edited by J. Sassi. The Watch Hill Conservancy. https://thewatchhillconservancy.org/napatree/napatree-resources/.

Montello C., G. Bonynge, C. Damon, P. August, and J. Sassi. 2017. "Mapping Napatree Point with Small Unmanned Aerial Systems." In *The State of Napatree Report: 2017*, edited by J. Sassi. The Watch Hill Conservancy. https://thewatchhillconservancy.org/napatree/napatree-resources/.

**Appendix 1.** *Representative photos from the underwater video with algae and eelgrass species noted on the images. Identifications provided by Dr. Lindsay Green-Gavrielidis.*

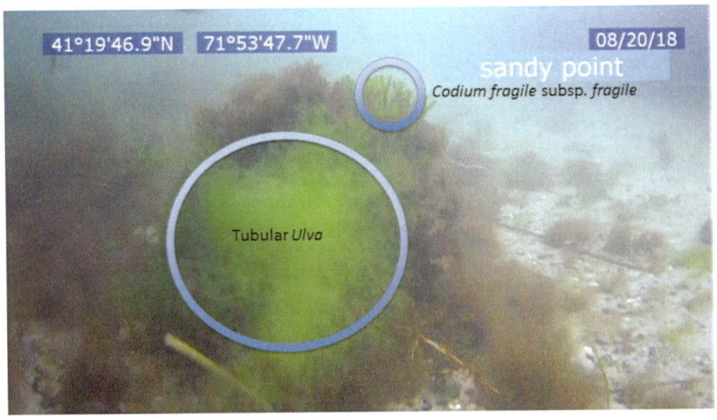

**Appendix 2.** *Representative photos of macroalgae and eelgrass habitats obtained from the GoPro cameras.*

*Diver with GoPro camera filming submerged aquatic vegetation.*

*Macroalgae in foreground, eelgrass behind.*

*Patch of eelgrass.*

*Eelgrass off the Napatree Point Conservation Area.*
*Photo Credit: The Watch Hill Conservancy*

CHAPTER 29

# Assessing Seagrass Field-Mapping Techniques in the Napatree/ Sandy Point Eelgrass Meadow*

*Peter August,*[1,3] *Michael Bradley,*[1] *Bryan Oakley,*[2] *Charles LaBash,*[1] *Beck LaBash,*[4]
*Christopher Damon,*[1] *Grant G. Simmons,*[3] *Braden Fleming*[1,3] *& Janice Sassi*[3]
[1] Department of Natural Resources Science, University of Rhode Island
[2] Department of Environmental Earth Science, Eastern Connecticut State University
[3] Napatree Point Conservation Area, The Watch Hill Conservancy
[4] Cohasset Center for Student Coastal Research

## INTRODUCTION

Eelgrass (*Zostera marina*) is a rooted, flowering perennial that occurs in shallow waters in the coastal zone. Eelgrass obtains nutrients (especially nitrogen) from the subaqueous soils that it grows in and removes nutrients from the water column. Through photosynthesis, eelgrass produces large quantities of oxygen. One square meter of eelgrass can produce 10 liters of oxygen in a day. Eelgrass meadows perform many important ecosystem services (Burdick *et al.*, 2020; Reynolds, 2018; Bradley *et al.*, 2013). It cleans our waters by absorbing nutrients and provides habitat to a host of organisms. Juvenile fish of many species that are important to our commercial and recreational fisheries safely hide from predators in the dense eelgrass vegetation. Invertebrates live on the grass blades and are food for larger animals. The seagrass beds off Napatree are important for our wildlife too. Terns feed on the small fish, Osprey feed on the larger fish, and the Brant Geese that are so abundant off Napatree in the winter are vegetarians and graze on the eelgrass itself. In terms of ecosystem services, seagrass beds are the third most productive ecosystem on the planet (after estuaries, like Little Narragansett Bay, and wetlands).

A large area of eelgrass (37.6 ha, 93 acres) occurs in the shallow waters between Sandy Point and Napatree Point (August *et al.*, 2020). Indeed, the single largest contiguous patch of eelgrass in the state of Rhode Island (31.5 ha, 78 acres) dominates this seagrass meadow (Figure 1). Because of the importance of eelgrass beds to the marine and terrestrial biodiversity of the region, we have been monitoring this habitat

* Originally published in 2020

for a number of years (August *et al.*, 2018). In addition, a consortium of environmental organizations (CRMC, Save the Bay, URI, USFWS, and Narragansett Bay Estuary Program) map and monitor eelgrass in all of Rhode Island's coastal waters at five-to-six-year intervals. This provides evidence that seagrass habitats are increasing or decreasing in extent throughout the state. A complete remapping of Rhode Island's seagrass beds will be undertaken in 2021.

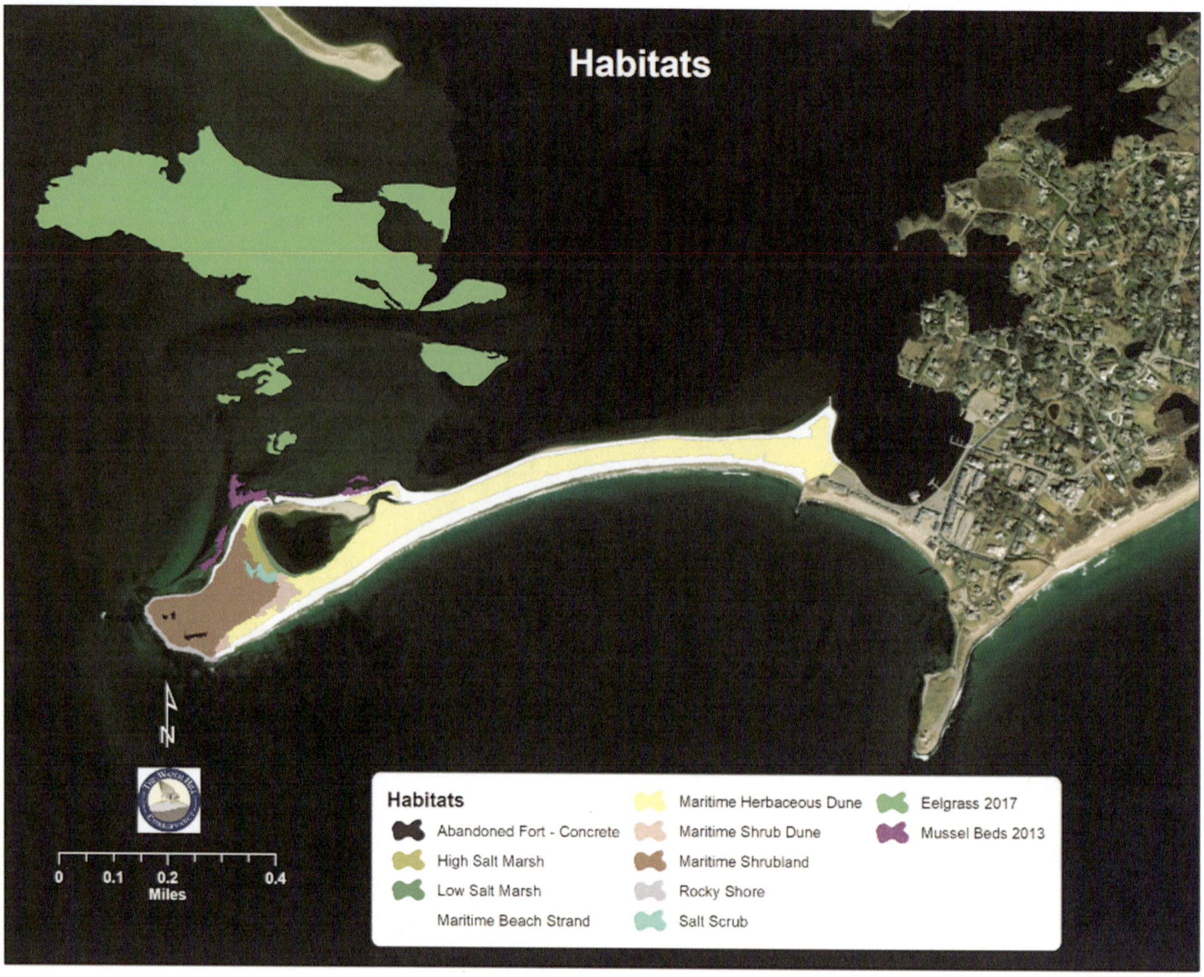


*Figure 1. Eelgrass meadows between Sandy Point and the Napatree Point Conservation Area.*

Mapping and monitoring seagrass habitats requires high-resolution orthophotography obtained in spring or early summer under calm wind conditions and low levels of glint on the ocean surface. These requirements are quite exacting, and the time window to obtain high-quality imagery can be very short. Once the imagery is obtained, expert aerial photo interpreters examine the orthophotography for the unique visual signature of seagrass beds. Polygons are digitized to mark the extent of eelgrass patches in a geographic information system (GIS). Co-author Michael Bradley has been doing the air photo interpretation of seagrass habitats in Rhode Island for the past twenty years.

Once the seagrass beds have been delineated, most of the patches are chosen for in-the-field validation and verification. This involves navigating to an eelgrass patch in a boat, and with divers or an underwater video camera, examining the seafloor to determine if seagrass actually occurs there and if the GIS polygon accurately depicts the extent of the habitat. Macroalgae has a very similar visual signature as eelgrass and can be hard to distinguish on the aerial photography (August *et al.*, 2018). The field validation part of the project is a critical step but can require considerable time and effort. Field verification using divers is slow and exhausting. Underwater video recording systems, such as the SeaViewer Pro device used for seagrass mapping in Rhode Island, are expensive, complex, and require towing behind a boat.

The purpose of this study was to determine the efficacy of using an inexpensive underwater drone to video record the eelgrass around a point location. The results of the drone surveys are compared to video recordings obtained by divers at the same locations. If the underwater drone is proven to be a viable method of field-verifying eelgrass beds, this technology might be considered for use in the 2021 statewide seagrass mapping project.

## METHODS

We used a Trident Underwater drone system to video record the extent of eelgrass around a sampling location (Figure 2). The system is relatively inexpensive (<$2,000), can operate for three hours on a single charge, record 1080p (30 fps) video, and tethers to a small controller system on the boat by a 25 m (82-foot) cable (Figure 2). Three thruster engines on the Trident are used to move the drone through the water column. The operator controls the device using a handheld Android-based JXDs192k game controller, which communicates to the controller system on the tether case through a Wi-Fi link. Video recorded by the Trident can be stored on the controller or the Trident itself. In dark filming conditions, the drone has six LED lights to illuminate the scene.

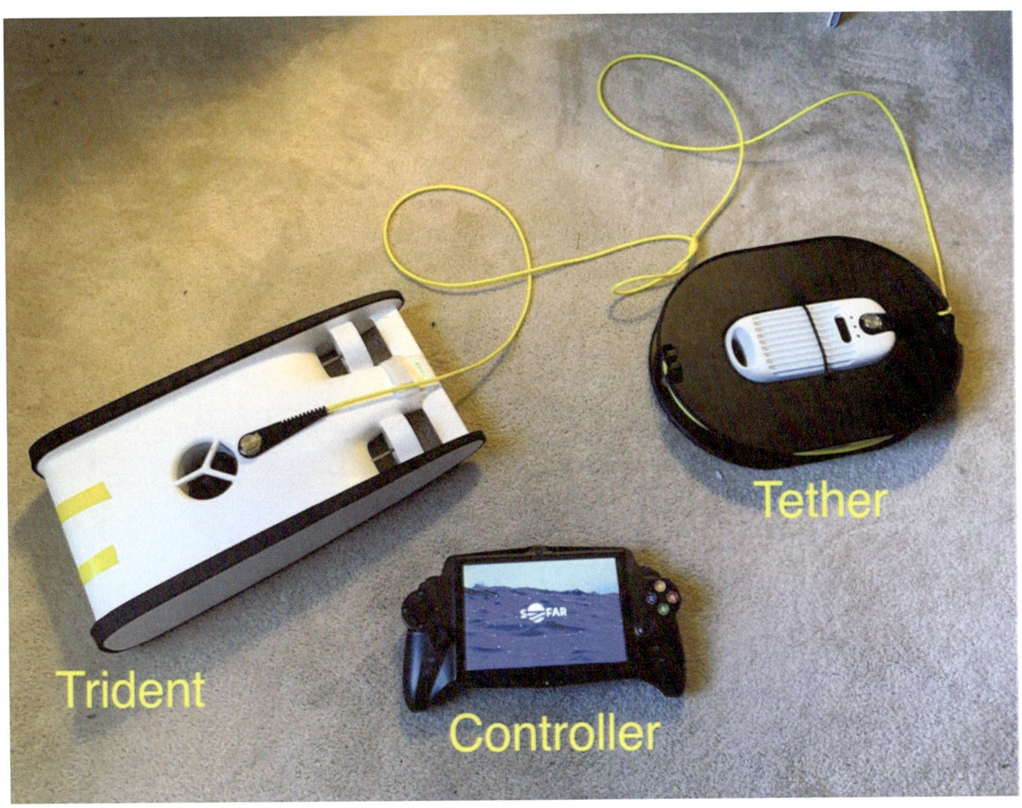


*Figure 2. Trident underwater drone system. The eelgrass meadow off Napatree was mapped in 2017.*

Additional work on the site in 2018 and 2019 showed very little change since the 2017 mapping. We chose eleven locations to examine the extent of eelgrass. Sites were chosen to maximize variation in eelgrass cover. Four of the locations were in dense patches of eelgrass, four locations had little to no eelgrass, and three locations were in moderate cover eelgrass. The eleven sites were uploaded into an ArcGIS.com web mapping application, and this was used to navigate to the locations in the field.

At each validation site, the Trident was run along the seafloor for 10 m (32 feet ) on the starboard, port, and stern sides of the boat (18-foot Grady-White center console). Approximately two to three minutes of video were recorded during each of the three transects. Concurrently, a diver (snorkel) with a handheld GoPro camera (both Hero 3 and Hero 5 models were used) swam along each of the three transects and recorded photographs of the bottom at five-second intervals using the time-lapse mode of the GoPro cameras. The positions of the GoPro images were logged using a Bad Elf Differential GPS towed behind the diver as part of the dive float. Two individuals (Bradley and August) estimated the percent cover of eelgrass in each of the Trident videos. The mean percent cover of the transects at each site was used as the overall eelgrass cover score for a location. Similarly, each GoPro image obtained by the diver was examined, and an overall percent of eelgrass cover score was assigned for each validation site. A Pearson coefficient of correlation (r) was computed to assess the relationship between the two methods of measuring eelgrass cover. An r value of 1 indicates both measures yield identical estimates. An r value of 0 indicates no relationship between the two measures.

As part of another study, the Napatree/Sandy Point eelgrass meadow was imaged using side-scan sonar and a small unmanned aerial system (sUAS drone aircraft) (Figure 3). A DJI Phantom 4 Pro sUAS with a 20-megapixel camera in video mode was used for this project. The side-scan imagery was obtained using an EdgeTech 4125 dual-frequency side-scan sonar system. Eelgrass will be delineated from both sources in 2021. These were going to be student projects during the summer of 2020 but had to be postponed because of the Covid-19 pandemic. These projects will continue in the 2020-2021 academic year.

*Figure 3. Additional imagery of the Napatree/Sandy Point eelgrass patch. A, Single frame from sUAS video. Dark patches are eelgrass, light patches are sand. Project boat (5.5 m, 18 feet long for scale) is in the center. B, Extent of side-scan imagery obtained in the summer of 2020.*

## RESULTS AND DISCUSSION

For each of the eleven sampling sites, we obtained an average of ninety-three GoPro photographs and 7.5 minutes of Trident video. It took ten minutes per station to collect the Trident video data. Estimates of eelgrass cover made by the diver-based GoPro images and the Trident video recording were highly correlated (r = 0.95, Figure 4). Both methods yielded very similar estimates of the extent of seagrass at a site. The results shown here were based on the observations of Michael Bradley, a very experienced eelgrass ecologist. The results obtained by Peter August were essentially identical (correlation between Trident and GoPro, r = 0.95).

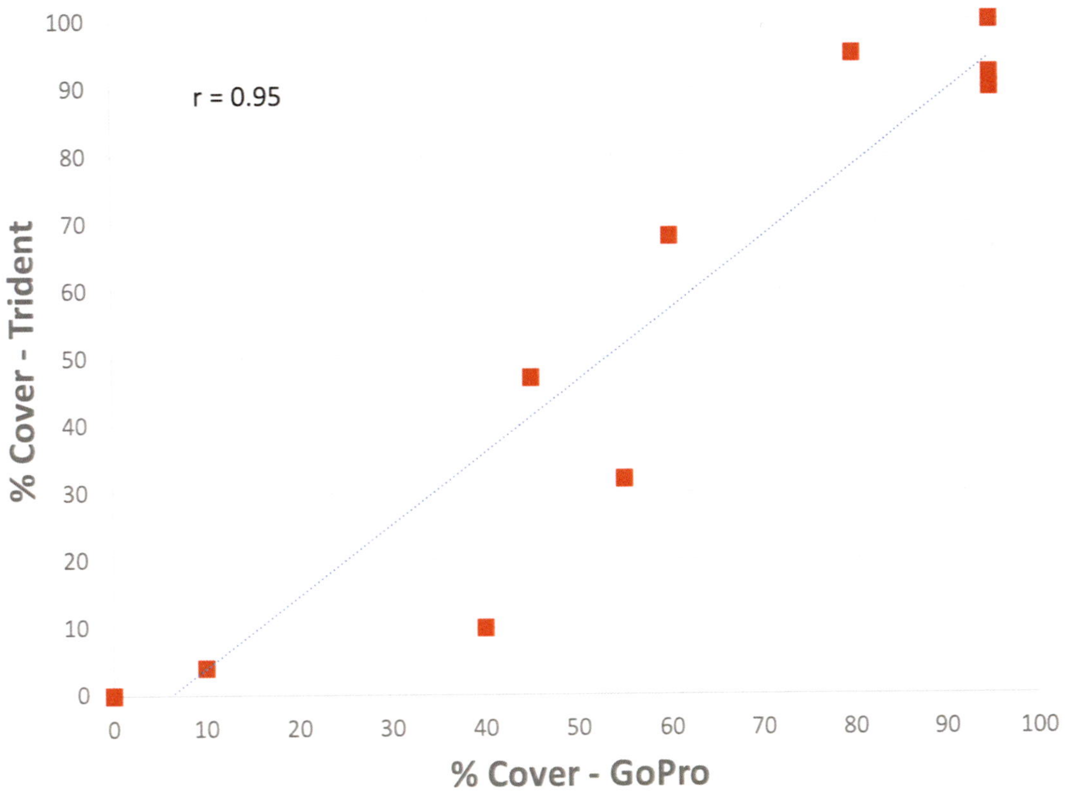


*Figure 4. Scatterplot of the relationship between estimates of eelgrass cover based on diver GoPro photographs and Trident video imagery. The blue dashed line is the best fit regression line. The strong positive correlation between the two assessment methods has important practical implications. The timetable for 2021 Rhode Island eelgrass mapping mission will require field validation of the polygons delineating seagrass beds in the fall months. Diver surveys will be difficult when water temperatures drop. The Trident-based assessment will be safer and faster.*

The quality of the Trident video images was excellent. Each frame was stamped with the date, time, depth, heading, and water temperature (Figure 5).

*Figure 5. A frame from the Trident video imagery.*

We were able to make a number of observations on the operation and use of the Trident drone.

1. Getting skilled at controlling the Trident in the water requires practice. Anyone who uses a computer game control device will be adept, but if the operator is not a gamer, it will take some time (4-5 hours) to master the controls.
2. The Trident thrusters can be overwhelmed by strong tidal velocities. Operating at slack tides can be desirable. We found that in Little Narragansett Bay water clarity was best during flood (incoming) tides. At ebb tides (outgoing) water clarity declined due to tannins and sediments from the Pawcatuck River.
3. The Trident can dive up to 100 m (330 feet). Most of our eelgrass sites were in very shallow water (<2 m, 6.5 feet). One of our challenges was accessing sample sites. The Grady-White has a draft of 0.6 m (2 feet). The mean tidal range in little Narragansett Bay is approximately 0.7 m (2.5 feet). Working near high tide was beneficial when sampling points were in shallow water.
4. The thruster propellers would frequently become entangled in eelgrass blades and stop turning. This was easy to discern since the operator would lose control of the Trident. The drone could easily be retrieved with the tether and the propellers cleared of eelgrass.
5. The controller screen shows a real-time video image of what the Trident is "seeing." The image, however, is quite dim and in full sun, hard to make out detail. This was not a major problem since the only important information the operator needed to know was if the device was on the bottom. It was easy to control the direction of the Trident by focusing on the bright yellow tether cable.
6. A 100 m (328-foot) tether cable might be a good investment if operating the Trident in deep water.
7. If the validation protocol requires a linear transect of video imagery rather than a point sample, a towable underwater video system, such as the SeaViewer system, might be a better solution. Point versus transect validation protocols is a large, complex issue and beyond the scope of this study.

## CONCLUSIONS

The Napatree/Sandy Point eelgrass complex is important for many reasons. It is a very high-value marine habitat for fish, shellfish, and nutrient removal from the water column. The eelgrass meadow is easily accessible, and because of the shallow depth of most of the site, boat traffic is light. It is an excellent location to experiment with various mapping and monitoring systems (Montello *et al.*, 2017).

The results of this study indicate that a Trident underwater drone system provides video imagery that can be used to assess the extent of seagrass cover at a site. Cover estimates were almost identical to those obtained from a diver. It appears to be a viable method to assess the accuracy of eelgrass maps derived from digital orthophotography.

## ACKNOWLEDGMENTS

This project was supported by the Sea Grant program of Rhode Island, the University of Rhode Island Coastal Institute, The Watch Hill Conservancy, and the A. M. Roberts Foundation.

## REFFERENCES

August P., M. Bradley, C. Damon, B. Oakley, G. Simmons, M. Cote, N. Rohr, C. Chaffee, and L. Lindsay Green-Gavrielidis. 2018. "Eelgrass or Macroalgae: An Assessment of Submerged Aquatic Vegetation Mapping Technologies." In *The State of Napatree Report: 2018*, edited by J. Sassi. The Watch Hill Conservancy. https://thewatchhillconservancy.org/napatree/napatree-resources/.

August P., J. Sassi, K. Rogers, and J. Cressman Greene. 2020. "The Geography of the Napatree Point Conservation Area (2020 Update)." In *The State of Napatree Report: 2020*, edited by J. Sassi. The Watch Hill Conservancy. https://thewatchhillconservancy.org/napatree/napatree-resources/.

Bradley, M., M. Ekberg, and P. August. 2013. "Eelgrass in Little Narragansett Bay." In *The State of Napatree Report: 2013*, edited by J. Sassi. The Watch Hill Conservancy. https://thewatchhillconservancy.org/napatree/napatree-resources/.

Burdick, D., K. Edwardson, T. Gregory, K. Matso, T. Mattera, M. Paly, C. Peter, F. Short, and D. Torio. 2020. "A Case for Restoration and Recovery of *Zostera marina* L. in the Great Bay Estuary." *PREP Reports and Publications*, 441. https://scholars.unh.edu/prep/441.

Montello, C., G. Bonynge, C. Damon, P. August, and J. Sassi. 2017. "Mapping Napatree Point with Small Unmanned Aerial Systems." In *The State of Napatree Report: 2017*, edited by J. Sassi. The Watch Hill Conservancy. https://thewatchhillconservancy.org/napatree/napatree-resources/.

Reynolds, P. 2018. "Seagrass and Seagrass Beds." Smithsonian Institution Ocean (website). https://ocean.si.edu/ocean-life/plants-algae/seagrass-and-seagrass-beds.

*Photo Credit: Daniel Cole, The Watch Hill Conservancy*

SECTION SIX

# NEWS

*Youth Conservation League team showing off extensive root system of an invasive plant—Asiatic bittersweet (Celastrus orbiculatus).*
*Photo Credit: The Watch Hill Conservancy*

CHAPTER 30

# Notable News and Sightings of Fauna and Flora at Napatree Point

*Janice Sassi,*[1] *Daniel Cole*[1] & *Peter August*[1,2]

[1] The Watch Hill Conservancy

[2] Department of Natural Resources Science, University of Rhode Island

## INTRODUCTION

Each annual *State of Napatree Report* contains brief items of interesting news or natural history observations that occurred during the year. Here is a collection of snippets recorded over the past ten years. All photos by The Watch Hill Conservancy.

## 2013

### PRICKLY PEAR CACTUS

Prickly pear cactus (*Opuntia sp.*) is found on the east coast from Florida to the sand dunes of Cape Cod. A Rhode Island endangered species, prickly pear cactus has never been documented on Napatree Point.

Numerous spiny, flat paddles (cladodes) of prickly pear washed ashore on Napatree during Hurricane Sandy, which occurred 29 October 2012. These pieces may have originated from a population of plants on Long Island.

By early July 2013, several young prickly pear cactus plants were growing on both the ocean and bayside of Napatree. The locations of these plants were recorded using a GPS, and they will be checked in 2014 for regrowth.

## 2014

### CHAPLIN B. BARNES AWARDED DISTINGUISHED SENIOR FELLOW OF URI COASTAL INSTITUTE

Chaplin B. Barnes was celebrated at a reception at the Ocean House on 6 March following a special showing of *Watch Hill: Portrait of a New England Seaside Village*, a Rhode Island PBS documentary based on his book *Watch Hill Through Time*.

Professor Peter V. August presented Chap Barnes with a lengthy citation on behalf of Governor Lincoln D. Chafee, recognizing Chap's many accomplishments and honoring him for ". . . all that he has done to protect an historically important, ecologically significant, and beautiful corner of the Ocean State."

Professor Judith Swift, director of the URI Coastal Institute, presented Chap with the Institute's highest award: the first Distinguished Senior Fellow of the University of Rhode Island Coastal Institute. Professor Swift remarked, "The work that Chap has done to protect and preserve the natural and cultural resources of Watch Hill embodies the vision and values of the Coastal Institute."

Although Chap was forced to step down as executive director of The Watch Hill Conservancy due to illness, he continued to work hard for the Conservancy, and especially his beloved Napatree, until his death in July.

The Land Trust Alliance chose the Napatree Point Conservation Area as one of eleven conservation areas in the nation to serve as case studies of effective stewardship to achieve resilience to climate change. Dr. Peter V. August presented the work we are doing on Napatree at the 2014 National Land Trust Annual Conference in a well-attended session entitled "Learn, Plan, Adapt, and Inspire: How Land Trusts Are Finding Successes and Opportunities to Support Climate Adaptation and Coastal Resilience."

## 2015

### NAPATREE IS DESIGNATED A GLOBALLY IMPORTANT BIRD AREA BY THE NATIONAL AUDUBON SOCIETY

The Audubon Society has designated the Napatree Point Conservation Area a Globally Important Bird Area. In its ornithological description of the site, the Audubon notes that "Napatree Point and Sandy Point have high ornithological significance in every season." The Audubon Society reports that the "Important Bird Areas Program was initiated by BirdLife International in Europe in the 1980s. Since then, over 8,000 sites in 178 countries have been identified as Important Bird Areas, with many national and regional IBA inventories published in nineteen languages. Hundreds of these sites and millions of acres have received better protection as a result of the Important Bird Areas Program." The importance and ecological value of the Napatree Point Conservation Area has now reached global significance!

## NANOTAG TELEMETRY TOWER ON NAPATREE YIELDS EXCITING RESULTS

The radio telemetry tower on the western tip of Napatree remained operational in 2015 and will continue monitoring the passage of birds tagged with radio transmitters in 2016. The study, headed by UMass Amherst PhD candidate Pamela Loring and funded by the US Bureau of Ocean Energy Management, is yielding incredible results. The Napatree tower is one of seventeen towers located along the southern New England shoreline from Provincetown on Cape Cod to Fire Island on Long Island. The towers detect passing birds that have been tagged with very small electronic transmitters called nanotags. In 2015, the Napatree tower recorded 187,776 events of marked birds passing the site. The majority of these (174,000 detections) were Piping Plovers that were tagged by Napatree Science Advisor Dr. Peter Paton. The remaining were Common Terns and Roseate Terns.

## 2016

## NAPATREE PART OF THE US FISH AND WILDLIFE SERVICE JOHN H. CHAFEE COASTAL BARRIER RESOURCES SYSTEM (CBRS)

As a "System Unit" in the CBRS network, Napatree is afforded an important level of protection. With System Unit status, federal funds cannot be used to develop or damage the environmental quality of Napatree. Federal funds can, however, be used to support certain conservation and recreation projects in the NTPCA.

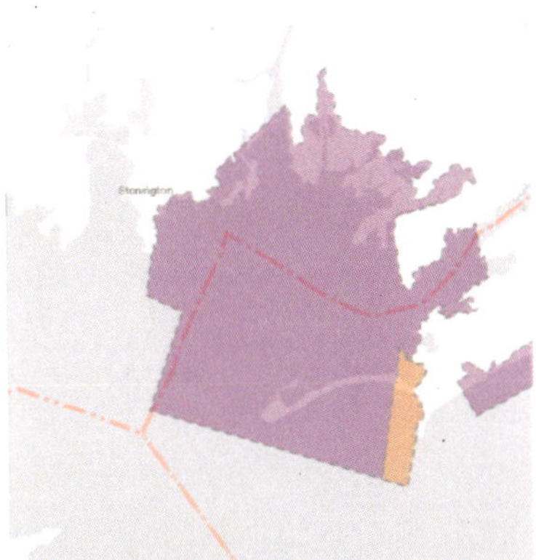

## 2017

## SAND-FRIENDLY FENCING ON NAPATREE

One of our management challenges is clearly educating visitors about the best paths to get around Napatree and making sure they do not accidently wander into ecologically fragile sites or areas where they will be exposed to large numbers of ticks or dense poison ivy. Snow fencing is not desirable because it traps sand and changes the natural geological processes that keep Napatree a dynamic ecosystem. With funding from the URI Coastal Institute and the Rhode Island Coastal Resources Management Council, we were able to add 100 m (330 feet) of split rail fencing to guide visitors from the entrance to the bayside of Napatree. Besides being warm and welcoming, split rail fencing requires no maintenance and does not trap sand.

## 2018

### LAND TRUST DAYS MIGRATION WALK ON NAPATREE

Almost fifty people joined us for our annual RI Land Trust Days event, a "migration walk on Napatree Point" led by veteran science photojournalist and environmental educator Bruce Fellman. Bruce is a semi-retired magazine editor for Yale University and the author of the syndicated natural history column "A Naturalist's Journal."

## 2019

### RECORD SEAL ACTIVITY ON NAPATREE IN 2019

Across Rhode Island, seal sightings on beaches were extremely frequent. Numerous gray seals and harp seals (see photo of a beautiful young harp seal) came ashore on Napatree. Most appeared healthy and were simply napping on the beach. It is not clear what precipitated this surge of seal activity. Never approach a seal on the beach. They are probably resting and should not be disturbed. Always keep your dog leashed on Napatree. A dog-seal encounter could result in serious injuries to the dog.

### YOUTH CONSERVATION LEAGUE REMOVES INVASIVE PLANTS ON NAPATREE

With a generous grant from the RI CRMC Habitat Restoration Fund and URI Coastal Institute, we were able to deploy the Youth Conservation League (YCL) on Napatree for three days in the summer. Under the guidance of Science Advisor Hope Leeson, the YCL team removed invasive plants on Napatree. The YCL is a summer program that provides training to high school students on the science and practice of conservation land stewardship.

## 2020

### NUISANCE TIDES: OUR NEW NORMAL

One manifestation of climate change is an increase in nuisance tides—tides that exceed our normal high tide levels. Over the past five years, we have averaged thirty-eight days per year where the Napatree entrance is totally inundated by nuisance tides.

## 2021

### EELGRASS RECOVERY OFF NAPATREE

Eelgrass is a critical habitat for juvenile fish and other marine organisms. Prior to Superstorm Sandy, 81 ha (200 acres) of eelgrass occurred between Napatree and Sandy Point; 14 ha (35 acres) were between the Napatree bayside shore and The Cut. This was one of the largest eelgrass patches in Rhode Island. After Superstorm Sandy, only 3.6 ha (9 acres) of the 14 ha (35 acres) remained. Surveying in 2021 suggests the Napatree eelgrass patch is now recovering. We will monitor this in 2022.

## 2022

### NAPATREE LAGOON CLOSES AT LOW TIDE

On 30 December 2022, the inlet of the Napatree Lagoon went dry at low tide. This inlet had been flowing freely since Superstorm Sandy in 2012, but the extra-tropical storm Elliott on 22 December is thought to have deposited additional sand at the mouth of the inlet, resulting in the change in flow. This unexpected event serves as a reminder that the only constant with the Napatree Lagoon is that it is always changing.

# About the Authors

DR. PETER AUGUST is Professor Emeritus in the Department of Natural Resources Science at the University of Rhode Island. He taught and conducted research in landscape ecology and the application of geographic information system technologies to land conservation and stewardship for almost 40 years. He has chaired the Napatree Science Advisors since 2011.

JANICE SASSI managed the Napatree Point Conservation Area from 2009 to 2021. With a staff of 8-10 seasonal educators and naturalists, she oversaw educational activities on Napatree, stewardship programs, and informing visitors of the natural history of the site. Prior to her appointment as Napatree Manager, Janice worked in the public sector.

DANIEL COLE is the current manager of Napatree and has served as a park ranger, educator, and site coordinator of a number of natural areas in Florida for over a decade, most recently the 11,500-acre Nature Conservancy Disney Wilderness Preserve in Kissimmee Florida. A native New Englander, Daniel has academic degrees in biology and secondary education.

# About BRG Scientific

BRG SCIENTIFIC was founded to advance scholarly works of the highest intellectual quality. We place priority on books showcasing the topics of ecology, natural history, evolution, animal behavior, and comparative biology written by vetted colleagues. Our team is driven by excellence of scholarship, not by volume of sales. We strive to reach technical audiences as well as curious, non-technical readers wishing to expand the horizons of their knowledge. We recognize that rigorous, well-researched science can also be pleasing to the eye and soul and that beautiful and informative graphic illustrations complement and enhance the written narrative.

***—Current BRG Scientific Releases—***

**CARVING NATURE AT ITS JOINTS**
Theodore I. Grand & Carol E. Underwood

**A CAMERA TRAPPER'S COMPANION**
Chris Wemmer

**SUBSTANCE OVER STYLE: A FIELD GUIDE TO LEADERSHIP IN HIGHER EDUCATION**
Donald DeHayes

***—New Releases by BRG Scientific Coming Soon—***

**MAJESTY, POWER & RESILIENCE: A PHOTOGRAPHIC ESSAY OF THE AMERICAN WEST**
Vince Warren

**SPIRIT OF NATURE IN NORTHERN NEW MEXICO**
Brian Miller

BRG
SCIENTIFIC

www.ingramcontent.com/pod-product-compliance
Lightning Source LLC
LaVergne TN
LVRC090358230926
848486LV00046BA/396

* 9 7 9 8 9 8 9 0 3 6 9 0 5 *